W9-AEX-323

Animal Studies

Animal Studies

An Introduction

Paul Waldau

OXFORD
UNIVERSITY PRESS

Oxford University Press is a department of the University of Oxford.
It furthers the University's objective of excellence in research, scholarship,
and education by publishing worldwide.

Oxford   New York
Auckland   Cape Town   Dar es Salaam   Hong Kong   Karachi
Kuala Lumpur   Madrid   Melbourne   Mexico City   Nairobi
New Delhi   Shanghai   Taipei   Toronto

With offices in
Argentina   Austria   Brazil   Chile   Czech Republic   France   Greece
Guatemala   Hungary   Italy   Japan   Poland   Portugal   Singapore
South Korea   Switzerland   Thailand   Turkey   Ukraine   Vietnam

Oxford is a registered trademark of Oxford University Press
in the UK and certain other countries.

Published in the United States of America by
Oxford University Press
198 Madison Avenue, New York, NY 10016

© Oxford University Press 2013

All rights reserved. No part of this publication may be reproduced, stored in a
retrieval system, or transmitted, in any form or by any means, without the prior
permission in writing of Oxford University Press, or as expressly permitted by law,
by license, or under terms agreed with the appropriate reproduction rights organization.
Inquiries concerning reproduction outside the scope of the above should be sent to the
Rights Department, Oxford University Press, at the address above.

You must not circulate this work in any other form
and you must impose this same condition on any acquirer.

CIP data is on file at the Library of Congress
ISBN: 978–0–19–982701–5 (hardcover)
ISBN: 978–0–19–982703–9 (paperback)

For Judith,
who so often provides the wings

Great ideas, whether insights illuminating one of the communities amid which we live, introducing specific living beings in our larger community, or helping us recognize the frontiers of our hearts and minds, often slip into the world as gently as doves. If we quietly, lovingly observe our neighbors, listening to them as fully as we can, we hear a faint flutter of wings amid the daily uproar of greed, the attempts to engineer our consent, the building of personal empires—on these wings fly gentle stirrings of life and hope.

Such life and hope offer us much, although some will take this to mean only that through the marketing of such ideas, we can make a profit.

But believe differently—ideas, insights, hopes, and profound stirrings of life-for-life are, rather, more personal. They are awakened, revived, nourished by billions of us, solitary individuals or small groups whose caring about others, made real through our deeds and generosities, every day crosses frontiers of caring, and thereby pushes back the stark, crude implications of our species' ugly history of harm to others.

When one patiently seeks and, yes, finds such aspirations and visions so widespread, there emerges the ever-threatened truth that each and every human, on the foundation of her or his own sufferings and joys, builds for all living beings. When we notice such possibilities, we can soar and thereby join our larger community even as we arrive at our fullest selves.

Contents

Introduction

A book introducing Animal Studies should increase everyone's abilities to achieve three aims. Most obviously, such a book should foreground the nonhumans with whom we share this planet—these are sometimes friendly, more often elusive and mysterious beings. Their realities as individuals and members of nonhuman communities have had a deep allure for many humans and have motivated different forms of Animal Studies no matter how one defines it. Meeting this aim of foregrounding the beings outside our own species is challenging for profoundly important reasons to be explained throughout this book.

As a second aim, an introduction to Animal Studies necessarily must engage the many different dimensions of humans' interactions with animals of all kinds (that is, both nonhuman and human). This second aim shares some features with the humanities and social sciences, which have traditionally and overwhelmingly focused on human abilities and human-to-human interactions. Animal Studies inevitably expands this focus by keeping other living beings in the foreground (the first aim) even as human-to-human interactions that involve other-than-human animals are also brought into the foreground. The human-to-human interactions to be studied in this way include not only past and present traditions but future possibilities as well.

This second aim of introducing human-level issues to Animal Studies may initially seem easy to achieve because it includes a focus on ourselves, but achieving this aim requires one to meet formidable challenges that rival those of foregrounding other-than-human animals. A principal problem in meeting this second aim stems directly from our wonderfully capacious but-ever-so-complex human language capabilities. The existence of multiple languages and dialects has, from time immemorial, complicated humans' sharing and transmission of views of other living beings; moreover, peculiarly modern forms of this problem today continue to create challenges for all forms of Animal Studies. The result has been the existence of stratum after stratum of differences among the humans who discuss other living beings.

Animal Studies can, accordingly, seem bafflingly layered. Not only are there layers to be identified and worked through because scholars studying animals used different languages (such as Latin versus Sanskrit) in the past; there are also layers produced by evolving languages (such as ancient Greek versus modern Greek, or early English versus today's globalized English). There are more layers yet, for even if one chooses to work within only a single modern language, one will still encounter multiple discourse traditions (such as science versus literature versus law, and on and on). Finally, in each of the languages or discourses one encounters in seeking to communicate about other living beings, one will assuredly find abstractions coined and advanced by the uninformed; generalizations, practices, and stories that carry familial and cultural freight; and myriad claims and overtones anchored in innumerable theoretical ruminations that humans have constructed to describe what the perceptive William James in the late nineteenth century called the "buzzing, blooming confusion" that a human experiences in looking about the world.[1]

It might seem that addressing the layered world of humans' views regarding other living beings, so unduly complicated by these features of language, could not possibly be more challenging. Yet one more major challenge remains, for humans' attempt to understand the living beings beyond our species is carried out, as is the entire project of Animal Studies, in a world where virtually every nook and cranny is full of unbelievably diverse beings, only the tiniest fraction of which can be seen by the unaided human eye.

Buried within these layers of complexity, however, is good news. Engaging other living beings is possible because our species has substantial, even if sometimes unrealized, skills of self-reflection and communication. This good news leads directly to a third aim that any introduction to Animal Studies needs to meet—prompting each student to explore humans' possibilities with other animals in personally relevant ways. Individual students need permission to explore their own reactions and abilities regarding the nonhuman living beings they meet. Further, they benefit greatly from exploring how their own familial background impacts what they already have experienced of the more-than-human world. Students also need to explore the histories of their social and cultural heritages.

Such personal explorations can deepen each human's ability to engage the larger issues of what our species has been doing generally at the human-nonhuman intersection. And here is more good news—meeting the third aim is possible because while Animal Studies can indeed be a bewildering world, what saves the day (and night) is that the world each of us shares with other living beings is by any measure an astonishing world. This in particular makes Animal Studies a joy. Indeed, work pursued in Animal Studies can enable both scholars and students to recognize their own abilities to explore and develop our species' intersection with other living beings precisely because it underscores the basic fact that humans have choices in the way they interact with other living beings. An introduction to Animal Studies, then, needs to make clear the inevitable personal, ethical, and communal dimensions of Animal Studies, for it is the collection of individual human responses that determines how our communities act today and thereby shapes our species' future possibilities with other-than-human animals.

The aims of this introductory text—introducing other animals' realities, unpacking the complexities of the human side of Animal Studies, and calling out the personal dimensions of our responses to other living beings—coincide with the driving forces that make Animal Studies an inherently challenging exploration requiring the full range of individual and communal human skills. In 1928, Henry Beston movingly described why this exploration is needed:

> We need another and a wiser and perhaps a more mystical concept of animals. Remote from universal nature and living by complicated artifice, man in civilization surveys the creature through the glass of his knowledge and sees thereby a feather magnified and the whole image in distortion. We patronize them for their incompleteness, for their tragic fate for having taken form so far below ourselves. And therein do we err. For the animal shall not be measured by man. In a world older and more complete than ours, they move finished and complete, gifted with the extension of the senses we have lost or never attained, living by voices we shall never hear. They are not brethren, they are not underlings: they are other nations, caught with ourselves in the net of life and time, fellow prisoners of the splendour and travail of the earth.[2]

This book argues that in the coming decades and centuries, our species has prospects of realizing such a vision only if each of these aims is met—in other words, we must seek out and factor in other animals' realities as we consider our own views, practices, and possibilities. Each of us must also take personal responsibility for our responses to other living beings as we notice them and take seriously our present and future possibilities with them. It is, in fact, the actual, local world that each person inhabits that sparked both Beston's hope and other surpassing visions of humans' community with the rest of life.

Any form of Animal Studies that aspires to such breadth can be attempted only through investigations that interrelate a variety of approaches. Such work must be both open ended and humble, for developments in individual fields, disciplines, and arts potentially enrich each other. Work done in many cooperating precincts, then, gives Animal Studies the best prospects of developing suitably multifaceted approaches that can meet the many-layered challenges which the field faces.

Today, the emergence of extraordinarily democratic and cost-effective communication capabilities opens up possibilities of information exchange that have produced the animal-related developments referred to in chapter 1 as "ferment." These diverse worldwide developments not only allow but actually prompt students of Animal Studies to see the ubiquity and diversity of other animals, to learn of research developments, and to encounter the astonishing range of humans' cultural attitudes toward other living beings. Such developments thereby renew humans' long-standing interest in other living beings even as they nurture the growth of awareness of animal issues in high-profile circles such as law, education, public policy debates, popular artistic expression, and much more. Thus, as explained in the following chapters, one easily finds diverse discussions and materials from a great variety of contexts where researchers, active citizens, students, and leaders of organizations notice the presence of nonhuman animals as important factors to be considered directly or indirectly.

All of these developments have stimulated awareness in many influential circles of previously unnoticed connection possibilities. For example, humans' intelligence-driven grasp of the universe, which is what has prompted so much research into other animals' realities and our own cultures' diverse thinking about other-than-human animals, combines in a special way with humans' capacity for self-reflection. When one examines the views of other-than-human animals now widely held in principal institutions of industrialized societies—law, education establishments, businesses, government, and public policy think tanks—one often finds that mere caricatures dominate. Established views may at first seem superficially reasonable but, upon reflection, they are revealed to be uninformed guesses, dismissive generalizations, or biased accounts that have failed to take any and all nonhuman animals seriously. The result has been the prevalence in many circles of self-inflicted ignorance about other-than-human animals.

A willingness to inquire, especially when one is presented with responsibly developed and verifiable information, is mandated by the special skills referred to in this book as critical thinking (discussed in chapter 2). When such skills are employed, they make only too apparent that the radical dismissals of nonhuman animals that prevail in so many influential institutions have been underdetermined by actual facts easily discerned by those who choose to look carefully.

This is one reason that enabling each student's exploration of Animal Studies is part of good education. As Animal Studies unpacks and describes the past, present, and future dimensions of humans' intersection with other living beings, it necessarily prompts each of us to employ key forms of thinking that help us become more fully aware of our encounters with both nonhumans and humans.

There is, then, a certain timelessness and timeliness to Animal Studies that invest it with great potential. Further, the confluence of certain factors undergirding contemporary ferment on animal issues—the affluence of certain countries, the availability of science, the breadth of globalized communication, the deepening of critical thinking, the availability of traditions of academic freedom—are producing Animal Studies programs of unparalleled power and range. Thus, with the kinds of humility and cross-disciplinary cooperation needed to pursue the three aims listed above, the present era can be the most auspicious time ever for exploring the realities and mysteries of other animals and human animals.

A corollary of these possibilities is that the future of Animal Studies lies with individual humans who will, with imagination and attention, push Animal Studies to forms of understanding that do not today prevail in many circles. Such forms of understanding may have previously prevailed, in one guise or another, in unfamiliar cultures or even in subcultures of the industrialized world. But most citizens and educational institutions in the industrialized sectors of society have had to reimagine such visions—so have many impoverished people whose daily lives require focus on survival rather than the important challenges laid out in this book. But Animal Studies has a remarkable heritage—in a surprising range of cultures, people have achieved deep understanding of both the local nonhuman animals and ways of coexisting with them. It is true, of course, that in the societies today widely held to be the most "advanced," such awareness has often been forgotten or simply repudiated.

Yet again, however, there is good news. Contemporary developments in Animal Studies reveal that many people today desire to learn about nonhuman animals—some seek to recover lost perspectives; others work to ignite creative thinking and artistic sensibilities regarding other living beings; and many work through one or more of the impressive sciences that our species has nurtured. The upshot is that today a great variety of people who think other animals are important in and of themselves now share their unique vision of how best to study other living beings.

One possibility, then, is that people alive today will develop convincing insights about our relationships with other animals. Another is that those to be born in the coming decades will shift paradigms by standing on the shoulders of those who pursue Animal Studies today, offering undreamed-of insights and options that move human understanding of other living beings ever further beyond what we now think and feel and guess.

Thus Animal Studies has much potential for a different kind of education—it suggests the humbling possibility that, from the vantage point of the future, present-day practices may well seem those of uneducated, uncaring, self-absorbed consumers. In this, Animal Studies has a kind of negative potential to reveal that many of the people we today call educated are the most serious vandals of the earth. Far more positively, however, Animal Studies makes obvious why studying the nonhumans with whom we share this planet is valuable to humans in a great variety of ways.

Chapter 1 provides a definition of Animal Studies that focuses on the ways human individuals and cultures are now interacting with other-than-human animals, have in the past interacted with living beings beyond our own species, and in the future might interact with them. Topics that fall easily and fully under this definition are found in so many different areas of human life, however, that it is helpful to think of Animal Studies as an umbrella term that goes beyond the common notion of a single, discrete discipline. Even a little reflection will reveal, then, that Animal Studies will have great breadth. Humans live amid an astonishing array of lives that are so diverse they defy description. Any one society will have developed its view of "animals" in relationship to only some of these nonhumans. Some individuals and societies have grown up amid complex, cognitively capable nonhumans, such as elephants, chimpanzees, and coastal dolphins. But other individuals and societies came to maturity in parts of the earth dominated by altogether different and far simpler living beings. So the views found in any one culture (such as one's birth culture) are by no means likely to inform one about the full range of life beyond the species line.

Chapter 1 elaborates on these themes as it answers the question, what is Animal Studies? This answer discusses four basic tasks that Animal Studies attempts to achieve. The first of these tasks will seem human-centered to some, for it requires telling a full history of humans' interactions with other living beings. The second task is other animal–centered, providing fundamental questions about how we generate meaningful perspectives on other living beings' individual and communal lives.

The third task is animal-centered in the broadest sense, that is, centered on both non-humans and humans, for its focus is exploration of future possibilities of a shared, more-than-human world. The fourth task for Animal Studies returns to an issue centered solely on human animals, though in a humble form—how can we recognize the nature of, and accept, the obvious limits as to what humans might know about other living beings?

Chapter 1 then opens three doors—who and what "animals" are, why Animal Studies is important, and how meeting other animals creates fundamentally personal connections. Chapters 2 and 3 then go through these open doors to explore fundamental challenges raised in the central human endeavors of history, culture, education, science, and politics. Chapter 4 uses the work of chapters 2 and 3 to introduce three areas of inquiry about other-than-human animals that today are the cutting edges of contemporary Animal Studies. As pointed out in chapter 4, however, these areas are themselves developing so quickly that they also prompt questions about the limits and future of Animal Studies.

This sets the stage for chapters 5 through 9 as they explore the human-nonhuman intersection in additional areas. Chapter 5 looks at how nonhuman animal issues appear throughout the creative arts. Chapter 6 explores philosophical reflection on our engagement with other lives. Chapter 7 turns to important comparative endeavors that look at legal systems, religious traditions, and our many human cultures. Chapter 8 engages the multifaceted problems and limits grounded in humans' rich social natures. Chapter 9 looks at the fields of geography, anthropology, and archaeology.

Chapters 10 and 11 turn to two very different challenges. The first is telling the whole story, that is, getting beyond histories that are merely human-centered and therefore dysfunctional for us as we attempt to thrive in a multispecies universe. Chapter 11 argues that Animal Studies needs to explore connections between, on the one hand, the marginalization of certain humans and, on the other hand, interwoven forms of violence and oppression that impact both these humans and nonhuman animals.

Chapter 12 examines questions of leadership and vision. Our species' self-image reflects pride in the level of rich individuality so evident in each human person even as we tout the unity of the human species. The chapter asks how individuals, private and public institutions, societies in general, and our species as a whole might come home to our own animality and the inevitability of our encounter with other animals.

Chapter 13 concludes this book by posing questions about the future of Animal Studies even as we admit that it is still to be chosen. Does the fact that, amid our working out of the details, we can now see the outline of Animal Studies in the near future help us in any way in guessing at the longer-term futures that are coming?

Animal Studies

Opening Doors

Animal Studies engages the many ways that human individuals and cultures are now interacting with and exploring other-than-human animals, in the past have engaged the living beings beyond our own species, and in the future might develop ways of living in a world shared with other animals. Seeing these pasts, presents, and futures requires a great deal of us—we need the utmost in human humility about our abilities and limits, just as we need complete candor about our complicated heritages of compassion and oppression. We also need our most careful forms of thinking and the best of our soaring imagination because at one and the same time, we are in some respects like all other animals, like only some other animals, like no other animal.

The important human skills of rigorous and critical thinking have by no means dominated past thinking about the living beings beyond our own species. Indeed, our record of human-on-human oppression tells us such skills are sometimes absent for prolonged periods of time in our institutions, law, education, public policy, religion, and so much more. Animal Studies, therefore, faces constant challenges and risks as it attempts such work.

One of these challenges is, as chapters 2 and 10 suggest, telling the whole story of the almost countless ways that different human individuals and cultures have interacted in the past with neighboring other-than-human animals in local, shared habitats. To date, the full story has not yet been even closely approximated. Another challenge faced by Animal Studies is providing adequate information regarding the many ways human cultures, societies, nations, and local communities today are interacting with lives beyond the species line.

Perhaps the greatest challenge facing Animal Studies, however, is seeing our possible futures in such interactions—discerning what is possible is an important exercise that will help us see much about our own spirit. Such work is fraught with difficulty, for while some humans have long proposed that, for a variety of reasons, our species needs to interact in new, more protective ways with living beings beyond our own species, others have long reacted against any suggestion of problems in the past and therefore resist calls for change. Such resistance is often anchored in the long-prevailing—and thus now socially and psychologically comfortable—dismissals that are the heartbeat of so many claims to human superiority.

These and many other challenges create both difficulties and opportunities for Animal Studies as it pursues its encompassing task of looking at past, present, and future dimensions of the human-nonhuman intersection. One of the greatest opportunities is outlining the pervasive human-centeredness that now dominates much thinking in certain circles. There are, as explained below, powerful but dysfunctional forms of human-centeredness that comprise an attitude often described as human exceptionalism—the prevalence of this attitude has made it hard for many people to admit not only past problems but also obvious limitations that we, as humans, have in grasping the features of some other animals' lives. Such limits exist for a variety of reasons, one of which is that Animal Studies involves the human study of living beings who are sometimes only partially like us in awarenesses, intelligences, perceptions, personalities, societies, allegiances, emotions, and so much else. Often, the other-than-human living beings engaged by Animal Studies possess such abilities, but in astonishingly different ways than we do, or they possess altogether different abilities that are fundamentally alien to us. Some of these living beings are fearsome in the extreme, while others are gentle but shy and even fearful of us—perhaps because their way of life is disrupted or harmed by our mere presence. Others may seem repulsively ugly to us even as, we might humbly surmise, our own beauty goes unnoticed by them.

The good news is, of course, that humans have a capacious spirit. We can attempt to study other beings even when we recognize that some of their features—perhaps most of them—are only partially available to us. Even as we face limits, we can try again and again, individually and collectively, to learn as much as possible about them. It is even part of our genius to use recognition of our own limits in ways that are helpful to us.

Four General Issues

Basic to Animal Studies are a number of issues that, on their face, are simple to state but that will require this entire book to uncover in depth and breadth.

1. The question "What is Animal Studies?" is only partially answered by the definition with which this chapter opens. A full answer requires that one explore all three of the following questions.
2. Who and what are "animals"?
3. Why is Animal Studies important?
4. What explains the personal connection so evident when meeting animals?

These issues are most productively seen and addressed in combination, and it is this multifaceted, multilevel inquiry that drives Animal Studies.

Four Basic Tasks

To illuminate and explore these four general issues, Animal Studies takes on four fundamental tasks. The first task has already been mentioned—telling the entire story about our past with other living beings. This task attempts what amounts to a shared history, moving across many human cultures and many different kinds of nonhuman animals.

A more complicated second task is to develop perspectives on other living beings' individual and communal lives. Going beyond our own history requires much imagination and the deepest of commitments to seek out other animals' realities—this issue is raised throughout this book, for the actual realities of other animals are so diverse that they are seen through and reflected in many different human endeavors and, most importantly, in humans' daily lives. Thus while this topic is given its most thorough development in chapter 3 when our human sciences are discussed, the exploration of other animals' realities is a task so fundamental that it belongs just as fully to many other, nonscience human endeavors that explore both our daily lives and our grandest generalizations.

These first two tasks, telling the entire story and developing perspectives on other animals' realities, work together in several ways. By pursuing them together, we can recognize how harsh many chapters of our own history of dealing with nonhuman animals have been. We also notice that humans have often impacted other-than-human creatures that can, in astonishing ways, share many of the traits we value most in ourselves as living beings.

In combination, these first and second tasks create a third basic task of exploring future possibilities. The possible futures are, of course, diverse, ranging from living in a shared, more-than-human world to living in human-centered ways begun by some of our forebears.

The third task in turn makes clear that a fourth basic task must also be accomplished—we need to be frank about the nature and extent of the inevitable limits of what humans might know about other living beings, and then work as diligently as we can within these limits. Such constraints on our knowing are sometimes clear, but sometimes vague. We can openly appreciate that some of the limits on our present knowledge may yield to a future human's creative gifts, or the efforts of a group's imaginative work, either of which could open our minds to undreamed-of possibilities of human awareness of certain other-than-human lives.

Deeper into the First Task

Exploring these four tasks individually makes it clear that while each is distinct from the others, work on each task prompts one to see the others better. For example, accomplishing the initial task—telling the entire story—will require multiple skills. Some of the story has been told but much "history" has been unduly stilted because it is one-dimensionally biased (chapter 2). Not only is the story overwhelmingly biased toward humans but, as contemporary historians recognize, there is also a recurring, debilitating tendency to favor merely one group or class of humans to the exclusion of other humans.

So the whole story is only now being contemplated. A respectable outline of this complicated tale will take years, for much of the past was barely noticed, let alone recorded and preserved. An elaboration of the more salient features will likely take many decades, perhaps even centuries. Putting together the entire larger-than-human story will require many character traits and skills, including a willingness to be honest, the political wisdom to ensure academic freedom in learning centers, and the imagination to look past our now centuries-long tradition of human-centeredness in education. It will require a robust exploration of many cultures and religious traditions, both interfaith and secular-religious dialogues, and much more.

Fathoming the Second Task

The challenges of developing perspectives on other living beings' individual and communal lives are, as discussed throughout this book, as formidable as they are important. This is so because without attempting to know other animals' lives to the utmost of our abilities (admittedly limited), we cannot know all of the consequences of our actions—we will not know which harms we cause, what kinds of communities we destroy, how much pain we cause, and so on. In effect, without an informed perspective on other living beings, we cannot know the neighbors with whom we share ecological community.

Developing perspectives on other living beings may in fact be the most challenging of the four tasks. In the other tasks, our imaginations can and must soar—telling the whole story requires imaginative exploration of our past, as does exploring future possibilities. So does determining the limits regarding what humans might know about other living beings, but this kind of self-evaluation is peculiarly within our abilities.

Under the second task, however, *our* inquiry is about *their* realities. Attempting to know the actual biological, communal, individual, and even personal realities of other beings forces us beyond ourselves and the parts of the world most easily accessible to our natural abilities. No doubt, some will find any proposal that we seek out extrahuman realities to be counterintuitive because they agree with Alexander Pope that "the proper study of Mankind is Man."[1] There are, however, multiple grounds for humans to take stock of other animals' actual realities. These grounds go beyond recognizing that many people naturally wonder about what other living beings are like, and they also go beyond the commonsense proposition that we ought to acknowledge whatever realities we happen to observe. Some people wish to know if the lore they have inherited about other animals is in any way accurate. Some wish to track other animals to feel connected, and others hunt them for food or to prevent them from harming one's family or livelihood. Many take responsibility to learn something about the lives of the nonhuman animals near us so that they can then factor such information into an evaluation of whether our actions impact these living beings. Especially complex ethical questions are raised when humans choose to dominate other animals. The impact of such domination can be evaluated only if one is informed about the lives captive animals lead when not dominated by humans, which is why one of the twentieth century's most respected voices on zoo issues suggested in 1950 that the way to evaluate zoos is to measure the life that zoos afford their captive animals against the lives those same animals have in the wild.[2]

Perhaps most compellingly, many humans have recognized that humans are fulfilled by acknowledging connections to other animals. Such connections are enhanced greatly when one is informed about what other animals are really like. Challenging philosophical questions arise as we assess just how certain one can be about other animals' inner lives, and whether these are in any respect like the inner realities that humans experience. Because our species is capable of rich, frank self-reflection about such matters, attempting to know other-than-humans causes us to grow, to get outside ourselves, and surely at times to understand ourselves better. Such benefits also flow from the fact that pursuing other animals' actual realities requires a number of additional skills that fit naturally into chapter 2's discussion of critical thinking. Such skills create optimal circumstances for recognizing what we really know versus what is operating in our societies as mere wishful thinking, bias, prejudice, or self-inflicted ignorance.

Further, in a crucial sense, much else in Animal Studies depends directly on our willingness to develop and coordinate our abilities to respond to this second task. In this venture, we are greatly enabled by modern sciences (chapter 3), but also by insights from literature (chapter 5) and comparative studies of religions and cultures (chapters 7 and 9). Any attempt to ascertain other animals' realities will be further enhanced through recognition that some cultures have long worked to develop insights about other animals' actual lives. For example, one can only wonder how much earlier than 1984 Western scientists might have discovered the subsonic communications of elephants if the science establishment had been open minded about observations made by indigenous peoples regarding elephants' constant communication with each other (see chapter 13).

Many different considerations, then, support the conclusion that humans should, when they can, learn about other animals' realities.

Ranging Widely in the Third Task

Discerning future possibilities of living in a shared world might be considered by some to rival the second task as the most important of these four tasks. This task has an implicitly ethical dimension and, without recognition of this salient feature, some people may find no reason to tell the whole story, seek out other animals' realities, or recognize the limits and humilities with which the fourth task is concerned.

Trying honestly to call out future possibilities requires a mix of realism and conscience. In this respect, this third task calls forth the fundamental human abilities of imagination and caring. These two abilities in concert prompt the root question of all ethics, "Who are the others?" (see chapters 7 and 10). Caring about "others" in some modern societies is understood implicitly, sometimes explicitly, as involving only humans. But exploration of history and cultures tells us again and again that caring for others outside our own species has long been recognized as a particularly enabling form of making community. Importantly, caring only about nonhuman others no more invokes our full ethical abilities than does caring exclusively within the species line. Said another way, failing to care about humans is, in fact, a failure to care about animals of a rich and complex sort.

Caring both within and across the species line is, thus, the form of self-transcendence that prompts the richest, fullest, most human forms of making community. As such, caring so broadly offers the fullest prospects for realizing a key insight about human fulfillment that every wisdom tradition has noted—Viktor Frankl made a classic observation that "self-actualization is possible only as a side-effect of self-transcendence."[3] This is the same wisdom that animates the so-called Golden Rule as it appears in various forms like "love others as you love yourself." This wisdom is also the heartbeat of the encompassing claim that "we cannot be truly ourselves in any adequate manner without all our companion beings throughout the earth. The larger community constitutes our greater self."[4]

Subsequent chapters suggest two further features of caring about others. First, such caring can enhance critical thinking because it prompts open-mindedness (critical thinking, in turn, of course, helps foster important processes of self-evaluation about the range and reach of our abilities to care about others). Second, for similar reasons, caring about others can be valuable for science and other empirical explorations.

The Humilities of the Fourth Task

The first three tasks beg a fourth—identifying the nature of any and all limits as to what humans might know about other living beings. This last of the four central tasks of Animal Studies has features of a philosophical problem that has been identified in one culture after another—exploration of what it is that humans can and do know. Such problems have been examined most fully in a subfield of philosophy known as epistemology, but they also appeal to common sense—how is it that one can distinguish mere opinion and psychological certainty from true knowledge? A classic problem in this area is knowing, as opposed to guessing or feeling that one probably senses correctly, what another human experiences or thinks. Absolute certainty about such a matter is, upon reflection, very elusive. For example, one never really knows if others are telling the truth or lying about their own inner experiences. Even if we are convinced they are not lying, they might be delusional or otherwise misreporting their own internal state. We can, based on circumstances and our own observations, guess another's feelings or inner thoughts. But even a little reflection reveals that our guessing about their realities clearly falls short of absolute knowledge, for we might guess wrong. These and other reasons explain why wisdom traditions have counseled humans again and again to be cautious when claiming to really know many elusive things.

When it comes to claims about nonhumans, then, the epistemological challenges are magnified greatly for both obvious and subtle reasons. We do not have a shared language through which to communicate, and even if we did, we still would not know if what was being reported was accurate. This is why identifying the nature of any and all limits as to what humans might know about other living beings is a key task requiring honesty, constant work, and liberal doses of humility.

Prospects of accomplishing this fourth task will increase greatly when our species creates richer forms of history (chapter 10). Similarly, this task benefits from good-faith efforts to identify the realities of other animals and familiarity with different cultures' understanding of our possibilities with other living beings. In and through such efforts, we recognize our human abilities and limits. In essence and in practice, then, this fourth task is a humbling one for our species. It constantly forces us to face our finitude, to call out how we so often have claimed without any evidence that the world was designed for our clan, our race, our nation, our religion, or our species and its global society. Finally, just as the first, second, and third tasks require liberal doses of critical thinking along the continuum called out in chapter 2, so, too, this fourth task is greatly enhanced by mature reflection on one's own thinking and that of one's fellow species members.

The Exceptionalist Tradition

Animal Studies as an academic discipline proceeds against a human-centered backdrop. When one explores this backdrop, it becomes clear that there are a number of different human-centerednesses, some of which are mild while others are both virulent and dysfunctional. Particularly noteworthy is one "basic idea that forms the core of Western morals, and that is expressed, not only in philosophical writing, but in literature, religion, and the common moral consciousness."[5] A prominent American moral philosopher unpacks this basic idea into its component parts:

> This core idea has two parts, and involves a sharp contrast between human life and non-human life. The first part is that human life is regarded as sacred, or at least as having a special importance; and so it is said the central concern of our morality must be the protection and care of human beings. The second part says that non-human life does not have the same degree of moral protection. Indeed, on some traditional ways of thinking, non-human animals have no moral standing at all. Therefore, we may use them as we see fit.[6]

Human-centered orientations have taken many forms as various cultural traditions have attempted to explain why humans, as ethical beings, can favor their own kind more than other living beings.

Pointing out that some of these explanations are dysfunctional (because, ironically, they are counterproductive for humans in important ways) does not require a wholesale condemnation of each and every form of human-centeredness. A focus on our own species can clearly be constructive, healthy, and productive. Developing greater loyalty to one's own family or local community, which can be seen as one kind of human-centeredness, produces some very positive results. In fact, our family has the important place in our lives of first, most important, and thus primary home. Yet few argue that our duties to members of our own family exhaust our duties of compassion and respect such that we can ignore all beings outside our own family.

Another form of human-centeredness commonly talked about today is loyalty to all members of our own species. Clearly, such loyalty to other humans has become a very important ethical, religious, and political factor. Among the greatest achievements of humankind have been successful social movements that sought to abolish slavery, racism, and sexism and to establish both moral and legal rights for individual humans. In a very real sense, these extraordinary achievements had to be consciously chosen to override the matter-of-fact realities of humans not treating each other as important, let alone as equals (these social movements achieved, then, social constructions of the kind described in chapter 8). The long history of some humans dominating other humans, which of course continues in many ways today, provides evidence that individual humans do not naturally develop a powerful loyalty to all members of their species, but must choose this loyalty. What everyday life realities suggest strongly, then, is that humans consistently develop allegiance only to some humans, such as family members or their local community. This is our biological reality, while affirmations of the fact that we can, with effort, develop a species-wide loyalty in order to affirm human dignity is a choice and today an important achievement of the human species.

There are, however, other, far more aggressive and troubling forms of human-centeredness that have been confused with affirmations that underscore the value and dignity of each and every human. One of these is human exceptionalism, which drives dysfunctional human-centeredness of the kind spoken of so often in this book. Human exceptionalism is the claim that humans are, merely by virtue of their species member-ship, so qualitatively different from any and all other forms of life that humans rightfully enjoy privileges over all of the earth's other life forms. Such exceptionalist claims are well described by Rachels as "the basic idea" that "human life is regarded as sacred, or at least as having a special importance" such that "non-human life" not only does not deserve "the same degree of moral protection" as humans, but has "no moral standing at all" whenever human privilege is at stake.

Such claims about humans' rightful place and privileges are sometimes anchored in religious beliefs that humans were invested with superiority by a divinity who prizes humans more than other living beings. Exceptionalist claims also have been based on the conclu-sion that humans simply have enough power to impose their domination on others, thereby making humans' privileged place just and moral.

Human exceptionalism has today become widespread while taking a variety of forms. Many people have developed the self-serving rationalization that human domination is the order of nature, much as Aristotle once assumed that slavery and female inferiority were an integral part of the design of nature. A corollary of this view is that humans naturally possess an allegiance to all other humans, which fully justifies humans in dominating all animal spe-cies outside our own. Yet others recognize that this important but harmful form of human-centeredness is an unnatural development that humans now impose on the rest of earth's life community, but one which is congenial to human communities' thriving and expanding. These different claims today work together in ways that have created a loosely cohesive excep-tionalist tradition by which humans simply ignore the harms that our species does to other living beings and ecosystems more generally.

This book makes a variety of arguments that humans need to get beyond the species line because the exceptionalist tradition has spawned great harms—to other animals, of course, but also to both our relationship with the more-than-human world and our own freedom, creativity, and imagination. Human exceptionalism is, then, worse than problematic—it pro-duces deeply imbalanced views that have often been, and continue to be, unrealistic, danger-ous, and harsh. Beyond the obvious problems created for the most visible nonhumans and their communities, hidden harms are also done to less visible nonhuman communities and our own communities, selves, and children.

The exceptionalist tradition leads our modern, industrialized societies to fail in rela-tionships with the more-than-human world. It pushes our societies and citizens out of bal-ance with the larger world, promoting "progress" in ways that harm us, other living beings, and the earth as a whole. It supports out-of-kilter forms of education that equip people to be, as one educator puts it in chapter 2, "more effective vandals of the earth."

Such a "humans only" orientation to the world is not unfamiliar to anyone these days, for, as argued in chapters 3 and 4, present public policy and law are dominated by such exclu-sions. In some public policy and law circles, people assume it is pointless to raise a protest against so fashionable a creed, especially because claims of superiority give us a sense of importance and privilege. Historically, once a human group has developed a sense of its own importance and entitlement, it has always been difficult for its members to look at the world and then, on the basis of fairness and conscience, divest themselves of their privileges. The exceptionalist tradition threatens to make such self-serving narrow-mindedness a species-wide trait.

Because the more virulent human-centerednesses impoverish the world and us, there is much at stake in developing more balanced views, including strikingly important benefits on the human side of the ledger. This is why challenges to the exceptionalist tradition can draw strength from several domains—the most obvious is realism about other animals' abilities and the harms our worst forms of human-centeredness are doing to nonhumans. But animal protection can also draw strength from the benefits that humans as individuals and as a community derive from caring in such broad ways. Many people have recognized this in countless ways—one of the most evocative is Robinson Jeffers's short, pithy comment, "Man and nothing but man is a sorry mouthful." As noted throughout this book but particularly in chapter 8, dysfunctional forms of human-centeredness have impoverished us in ways that have led to a predictable reaction, namely, the worldwide social movement to recognize humans' possible connections with other-than-human animals and thereby to reinstate compassion as a leading human virtue.

Because it dominates so much of the industrialized world, critiques of the exceptionalist tradition trouble many people. Animal Studies is pushed, however, by virtue of its scientific and ethical commitments to engage problematic features of the human claim to have surmounted animality that is so typical of theologies, philosophies, secular materialism, and educational institutions of certain cultures.

Interdisciplinary Openings

If one ponders these four tasks and works through the astonishing breadth and depth they require of us, one can see why Animal Studies is something more than a single discipline. It is, instead, a collaboration of many different disciplines pursuing phenomena that comprise every human's daily milieu. Each of the four tasks requires an interdisciplinary approach, for each task is better seen, and then better pursued, in environments that pay attention to multiple approaches, multicultural awareness, and the benefits of communal effort.

At the heart of Animal Studies, then, are fundamental disciplinary humilities, for no single set of ideas or concepts, no isolated vocabulary scheme or traditional way of talking, and no single theory or traditional set of generalizations provides the tools needed to accomplish any one, let alone all four, of these tasks. These disciplinary humilities are distinct from the deeply personal humilities imposed on any one individual human who pursues Animal Studies. They are also distinct from the species-level humility needed to pursue Animal Studies in its fullest form.

Collectively, these interdisciplinary realities help those who pursue Animal Studies recognize that their work will often be partial and inevitably subject to revision after further exploration. But the compensating benefit of the interdisciplinary features of Animal Studies is that those who study animal issues will work within and across a vibrant, multidisciplinary group of approaches featuring diversity of many kinds, a theme elaborated in chapter 13.

How to Define a Growing Field?
Exploring the First Issue

This chapter's definition of Animal Studies is intended to open doors and minds. In our present era, which is dominated by the power and achievements of science, providing definitions is common because many impressive sciences have thrived by centering their work on definitions dominated by quantification, measurement, and statistics (often needed to be taken seriously in certain scientific circles). But a definition centered on such approaches is impossible in Animal Studies for multiple reasons, not the least of which is that many important features in life cannot be defined quantitatively. For example, we rarely attempt to define or measure love, beauty, friendship, and the like.

Relatedly, many people dislike, even distrust, definitions despite grasping how important they are. When philosophers or other analysts of human language discuss the nature of definitions, they often list many different kinds of definition. Such discussions invariably point out that while some definitions are good, others are poor and confusing. In many polarized discussions, for example, those who consciously or unconsciously seek a particular result can cast a loose definitional net in order to, as it were, reel in an argument. Definitions can, it turns out, distort as well as illuminate.

Since ideally a definition both clarifies and liberates us to seek certain realities, bear in mind this chapter's answer to the question, "What is Animal Studies?"

> Animal Studies engages the many ways that human individuals and cultures are now interacting with and exploring other-than-human animals, in the past have engaged the living beings beyond our own species, and in the future might develop ways of living in a world shared with other animals.

How humans now are intersecting with other-than-human animals is, only a little inquiry shows, very closely connected to how humans in the past thought about and treated other living beings. We are, to our core, cultural animals—in other words, all humans inherit the traditions of their birth culture, including ways of thinking and speaking and impacting other-than-human animals who happen to share their birth locale. Seeing these traditions as a psychologically invested heritage, rather than the absolute, unchangeable order of the natural world, is an important lesson for each human. Such realities become even more obvious when one reflects on our capacity to choose different futures as we engage our world's other living beings.

Besides prompting students to look at past, present, and future issues, this definition of Animal Studies makes it clear that Animal Studies will grow in ways that even acknowledged experts today cannot anticipate. Understanding why Animal Studies will grow and metamorphose again and again in the future is important. This rich and kaleidoscopic endeavor has recently evolved dramatically, and it will surely continue this trajectory in the coming decades and centuries. Because Animal Studies begins with and is anchored by both the diverse realities of other-than-human animals' lives and the great variety found in humans' engagement with other living beings, it must be able to explore widely and unceasingly as it engages both other-than-human animals and humans' great cultural variety. Such exploration is described in chapters 2 through 10 as they survey a wide range of contemporary human fields and their possibilities under the Animal Studies umbrella.

Animal Studies' Complementary Journeys

Answering the question, "What is Animal Studies?" then, must go well beyond definition because Animal Studies, at its core, involves multiple journeys. One of these journeys is in a very real sense toward other animals. The other journey is toward humans' self-recognition as morally capacious, imaginative animals capable of the tasks of Animal Studies.

It is true that, as noted throughout this book, humans have a mixed past in the matter of ethics. Any careful study of the last several thousand years of our history will show that many human cultures have been dominated by cruelty to living beings—those seriously harmed have included countless humans, to be sure, but earth's nonhuman communities have been harmed to an even greater extent. This past is in no way remote today, for it exists within us, evident in practices, speech, closed minds, and a continuing arrogance.

Of course such harms by no means exhaust human attitudes toward other-than-human animals—just as humans have so often richly engaged unfamiliar humans, so too are there innumerable instances of behavior toward nonhuman animals that display human potential quite fully. Such information makes it clear that dismissals of nonhuman animals, such as those that characterize today's industrialized societies, need not dominate our future. Competing with the most lurid aspects of both past and present human mistreatment of nonhuman animals is a bright future made possible by our capacity for compassion. Our future will, no doubt, continue to be impacted by our historical shortcomings in understanding and coexisting with the other-than-human animals with which we share the earth. But the future possibilities before us are far richer than most visions of our future have imagined.

Humans have undertaken both of these journeys—one toward other animals, one toward ourselves—in all sorts of ways. Just as human cultures feature great variety in their accounts of who humans are, so, too, our view of who nonhumans are reflects comparable diversity.

Importantly, it is this diversity in human approaches to nonhuman animals that makes particularly clear how rich human abilities are in terms of future forms of Animal Studies. In this diversity is much evidence of human capacities for learning, humility, community, and compassion, all of which are key tools for developing a robust form of Animal Studies. Because of these abilities, Animal Studies promises to be, when mature, a truly broad, multifaceted, constantly renewing inquiry.

On Slowing the Human Heartbeat in Contemporary Animal Studies

Given the prevalence and political dominance of the exceptionalist tradition, it will not surprise too many that Animal Studies today features forms that are starkly human-centered. This is in part because education is Animal Studies' natural ecological niche and has been dominated by inquiries whose common heartbeat is the exceptionalist tradition (see chapter 2).

Many contemporary versions of Animal Studies are, in fact, parasitic on categorization of living beings in overtly human-centered ways—for example, the concluding section of this chapter explores ways in which the category "companion animals" carries decidedly human-centered features. Similarly, chapter 7 points out that highly theoretical approaches to Animal Studies risk certain forms of human-centeredness.

Such human-focused emphases in today's forms of Animal Studies are not in any major sense the largest or most important feature of the field. They are, instead, the by-product of habit, compromise, political expedience, and marketing ploys to get students interested in courses and programs. In general, then, human-centered versions of Animal Studies reflect transitional approaches that consider other-animal-centered concerns even as they foreground traditional and admittedly human-centered forms of study. Chapter 2 reveals much about such approaches in today's education. There are risks in such forms of Animal Studies, however, for in environments already dominated by the exceptionalist tradition, a too-heavy emphasis on the human side of human-nonhuman interactions can morph back into merely another form of the humanities or social science.

Such a preoccupation with human-centered concerns should not be confused with the historical work described above as the second journey of Animal Studies. That work necessarily focuses on humans for the purpose of identifying how members of our own species have related to, connected with, and at times harmed other living beings on our shared planet. This work is a crucial element of Animal Studies, for it leads to the altogether healthy process of humans, as a dominant species, coming home to our heritage and thereby learning about both our limitations in knowing other living beings' actual realities and possibilities for coexisting in mixed-species communities with some nonhumans. When this journey remains realistic and humble, it is a productive one that confers much health on and balance to human outlooks.

Critical thinking plays a role in calling out the significant risks that lie in versions of Animal Studies that start and end with human-centered features. They also underscore that striking a better balance among nonhuman animals' realities, human animals' realities, and the intersection of the two is not only possible but also affords those who pursue this field meaningful opportunities to grow through engaging other living beings and their own framing of issues. This growth in turn creates possibilities for reducing the myopias and dysfunctions of the exceptionalist tradition and other human-centerednesses. Such possibilities can prompt many active citizens, whether they be individuals acting alone, members of organizations, or educators, to develop important ways of focusing on other-than-human animals that contribute to today's forms of Animal Studies.

Although our species is surely capable of developing balanced approaches within Animal Studies where human-centeredness does not remain the leading element, nonetheless programs and individual courses will in the foreseeable future likely remain tinged with milder forms of human-centeredness. This is so for multiple reasons—our Western cultural heritage is weak on animal issues, and contemporary education, business, public policy, and the practice of scientific research are rooted in human-centered habits and traditions.

So, What's in a Name?

Work of the kind introduced in this book under the title Animal Studies goes under a variety of other names today, including human-animal studies, animal humanities, animality studies, the human-animal bond, companion animal studies, anthrozoology, posthumanism, critical animal studies, species critique, biopolitics, and more.[7] While the diversity of names signals that the field is so new that it has not reached any consensus on either specific topics or its outer limits and borders of inquiry, in general ways all of these approaches share certain features. All reflect the inevitability of interactions between humans and some nonhumans, just as each of these approaches in one way or another signals the impossibility of exploring all aspects of all nonhuman animals. The four tasks outlined in this chapter suggest, however, that work going forward under any of these names will be unproductive or irrelevant if it fails to in some way engage other animals' realities in a relatively informed rather than ignorant manner. Similarly, work on animal issues needs to take account of the fact that humans characteristically have a range of options for treating other-than-human animals.

While some of the names listed above accommodate a broad, shared notion of "animality," others favor humans, possibly reinforcing the exceptionalist tradition. For example, when studies of other animals use the name "human-animal studies," there are significant risks tied directly to the fact that the lead element in the field's name is "human." In a human-centered environment, this name may have advantages, such as marketing appeal. But putting the word "human" first in any study of humans' intersection with other living beings creates the impression that the endeavor's first and most important inquiry is human animals. Such an approach also reinforces the distancing of humans from other animals because it invokes the artificial dualism "humans and animals" discussed later in this chapter.

That such risks and other uncertainties exist can be seen in various definitions of "human-animal studies." Notice, for example, the human-centered features of the following definition taken from a 2008 collection titled *Social Creatures: A Human and Animal Studies Reader*: "So what is Human-Animal Studies? The focus of HAS is the study of human-animal interaction. Ultimately, HAS asks: what can we learn about ourselves from our relationships with other animals? What does the way we think about and treat other animals teach us about who we are?"[8] The editor restricts "human-animal studies" in important ways by asserting, "HAS is not biology or animal behavior.... Neither is the emphasis on other animals' social relationships with human animals."[9] Perhaps most revealing is the fact that in these passages the meaning of the word "animal" changes back and forth. In the phrase "human-animal studies," "animal" works as a reference to nonhuman animals only. This is also true of how the word works in the phrase "animal behavior" (in other words, the behavior at issue is only nonhuman animals' behavior, not that of human animals). Even the subtitle *A Human and Animal Studies Reader* picks up on the tradition of separating human animals from all other animals.

Yet the text also occasionally includes the terms "other animals" and "human animals." In these latter instances, the word "animal" invokes the commonsense notion and scientific certainty that humans are vertebrates, mammals, primates, and great apes, all of which are eminently animal categories. Such inconsistencies risk reinforcing denials of humans' animality and thereby create uncertainties. By contributing to such problems, the title "human-animal studies" continues the long tradition of denying what everyone knows—humans are animals—with the corresponding risk that many who hear the term "human-animal studies" will use this field as just another vehicle to focus primarily on humans.

Whether a field with a name that foregrounds humans must inevitably proceed in a biased way is up for debate, but any name that remains tinged by human-centered concerns opens the door to the principal focus being animals within the species line, especially as these humans connect merely with those other living beings that are pleasing to humans. In effect, then, the name "human-animal studies" keeps humans in the foreground, making it hard to bring home the salient fact that in Animal Studies humans are but one of the animals studied.

The same risks attend alternatives like "animal humanities," "the human-animal bond," and "companion animal studies." Another alternative that has broad possibilities is anthrozoology—this newly coined word follows the Western educational tradition of invoking Greek words to name disciplines. The three Greek root words that form anthrozoology mean, respectively, humans (*anthropos*), living beings (*zoon*, which is broad enough to include humans), and study or science (*logos*). Based on these roots, the term anthrozoology can be read generously to mean "study by humans of all living beings."[10] In practice, though, one finds different styles among those willing to be called anthrozoologists. A tame form of the field includes nonhuman animals that have been chosen because they interact with humans, or are similar to humans, or because human imagination has historically been fascinated with these nonhumans. A less tepid form includes more nonhuman animals, but some nonhuman animals are excluded because humans are not interested in them—here the linchpin holding the field together remains human-centered. Finally, a more robust form of anthrozoology considers other animals' realities as a factor or, at times, even uses them to set the primary agenda of the field—here other animals are considered even if their realities are not like humans' realities in any way, or there is no issue of human interaction with such living beings. In effect, this is the equivalent of the field as described in this book.

Getting Beyond Human-Centerednesses

Whatever name one chooses for this kind of work, and if education touching on nonhuman animal issues is to reach the level of a robust and mature field, the motivation for study surely cannot be humans primarily—such a concentration is already the preoccupation of the humanities as a megafield. This particular observation has important implications—Animal Studies is not a subfield of the humanities, nor is it merely another human-oriented social science. To be sure, as already underscored, it is important within a robust form of Animal Studies to concentrate at times heavily, even exclusively, on human interests. But making such a focus the dominant factor turns Animal Studies into a mere subdivision of the much larger projects we know as the humanities and the social sciences. Animal Studies has enriching affinities with, but different goals than, these important educational domains (these issues and the relationship of Animal Studies to the modern university's megafields—science and the humanities—are addressed in chapter 13).

Just as there are, then, reasons to avoid choosing a name along the lines of "human-animal studies," there are advantages to other names. Animal Studies is scientifically correct if the living beings concentrated upon are all animals. But if *only* nonhuman animals are to be the focus, then the name "Animal Studies" will potentially mislead in the ways that "human-animal studies" does. This is the reason that this introduction again and again makes the point that Animal Studies necessarily includes the study of important human issues.

A broad notion of Animal Studies, then, avoids perpetuating human-centeredness because it sends an underlying message that humans are only one of the animal species to be studied. By avoiding subliminal messages that Animal Studies starts with, and is primarily focused on, humans, the name "Animal Studies" avoids perpetuating the very mentality that has radically subordinated all nonhuman animals to humans.

The endeavor of Animal Studies has, in fact, multiple focal points, interactive components, and core emphases that require not only each of the complementary journeys already described (toward other-than-humans as well as toward humans) but an effort to weave together what one learns on both journeys. This weaving together is implicit in the term "studies," as pointed out most explicitly in chapter 2's discussion of critical thinking and chapter 13's discussion of Animal Studies as an encompassing educational pursuit.

A Tandem of Interested Minds and Disinterested Motives

Much of this chapter is driving toward the conclusion that open-mindedness plays a key role in Animal Studies. Yet open-mindedness alone, which some humans possess as a matter of character and personality, will not suffice. Though it is what philosophers call a necessary condition to knowing other living beings, it is not a sufficient condition. In layman's terms, one needs a combination—caring enough to look, patient observation, imagination, and communal sharing on top of the humility of open-mindedness—to inquire adequately. With this combination, "what a thing is the interested mind with the disinterested motive."[11]

Beyond open minds, then, Animal Studies needs a willingness to explore the world's great diversity, freedom from biased motives and human-centeredness, and a context in which commitments to academic excellence are matched by commitments to academic freedom. Only with such flexibility in its commitment to explore realities can Animal Studies develop the humility and integrity by which it can sustain itself as it encounters the difficulties mentioned throughout this book.

A corollary is that Animal Studies has ample room for creativity and imagination as we take careful, honest looks at the world around us. As part of the journey toward both other animals and our own human animal abilities, it invites into its center a willingness to foster creative attempts to inquire about other animals and then engage them seriously.

Benefits of Animal Studies

A corollary of the broad definition of Animal Studies as immersed in open, critical thinking is that it provides considerable space for study of human animals. Though this inquiry is, as already noted, but one topic alongside other equally important topics such as nonhuman animals' realities, focusing on humans in this manner is fully relevant for many reasons other than the obvious reason that humans are animals in a more-than-human world. It is also important because illuminating humans' abilities to harm or coexist with others, no matter what their species membership, brings humans into much greater awareness of the extraordinary human capacity for caring.

Humans have benefited greatly by noticing other animals and taking them seriously—chapter 2 discusses examples of how seminal religious figures have deemed inevitable interactions between humans and other living beings to be of the utmost ethical importance. The benefits of such interactions are often unnoticed in modern times because of a prevailing but facile conclusion that humans gain "the most" through a preoccupation with their own interests.

The benefits of Animal Studies are, in fact, quite diverse. They include the development of educational forms that prompt rich human thinking and imagination. They open hearts and minds to forms of compassion that strengthen character, enrich the human mind and creative impulses, and enhance key reflective capabilities like critical thinking. Animal Studies also puts students in challenging contexts (human-centered or not) and thereby creates one opportunity after another for self-actualization through self-transcendence and connection to a larger, more-than-human community.

Arrayed against the most developed forms of compassion and character are myriad forms of human selfishness. Similarly, the benefits of enhanced critical thinking skills and self-actualization through self-transcendence stand opposite self-indulgence and other self-aggrandizement. Finally, connection to our larger community contrasts well with ignorance-driven forms of the exceptionalist tradition that prompt so many of our species to commit what might be called the fallacy of misplaced community—in essence, the notion that the human species alone should be our focus.

Thus even if today's versions of Animal Studies are merely a first step that harbors some forms of human-centeredness, there is a more robust, benefit-laden sense of Animal Studies that is waiting offstage in the wings, as it were.

Exploring the Second Issue: Who and What Are "Animals"?

Although virtually everyone is aware that the word "animal" in the best-known modern human languages has dual meanings in tension with one another, many circles bury this tension in counterproductive ways. Ironically, some science-focused enterprises promote what amounts to antiscientific language practices along the lines of "humans and animals" to eliminate the likelihood of ethical challenges (see chapter 3). In everyday situations, too, even though we sometimes talk of humans as animals, far more often we talk in ways that separate humans from all other animals. Phrases equivalent to the English "humans and animals" are staples in the most widely spoken languages. This practice has its peculiarities, since many human language traditions are congenial to naming humans as primates, mammals, or other generalized animal categories drawn from scientific terminology.

But even though many can easily and often speak of other living beings as "animals," most humans today balk when it comes to employing a phrase like "human animals." Choosing this scientifically correct option or alternatives such as "other living beings" or "other-than-human animals" is, in some circles, viewed as antagonistic, even politically incorrect. This is so because the science-based way of speaking contends with the fashion of separating humans from the larger community of life. Further, most humans today continue to train their children and grandchildren to speak of the traditional categories of "humans and animals." There is, in effect, a kind of schizophrenia that at once embraces and repudiates the obvious truth that we are animals.

Talking as if humans are not animals remains possible because many people appear to deem the claim that we are, in fact, animals merely trivially true—the reasoning appears to be that since humans are considered so different from any other animals, it is reasonable and therefore right to ignore humans' obvious animality as confirmed by common sense, so many cultures, and our science traditions. One can ask, however, why the obvious important differences between humans and other beings are allowed to obscure, even eclipse, the even more obvious important similarities.

Those who travel in the center and at the margins of Animal Studies must repeatedly negotiate such questions and their awkwardnesses. Further, in the academic world and even in some circles of the scientific establishment (see chapter 3), those who speak of "nonhuman animals" or some equivalent take political risks with their careers. By choosing scientific terminology, they remind others in their circle of the pervasive, disingenuous denial that sits at the heart of phrases like "humans and animals."

The fact that there are risks for those who choose to speak of "nonhuman animals" and "human animals" rather than "humans and animals" helps foreground a series of crucial issues at the very threshold of Animal Studies. First, such risks exist in a wide range of contexts— politics, religious institutions, corporate boards, trade associations, educational contexts, scientific research settings, the local and national gatherings of professions, and even general social circles.

Second, such risks mean that those who pursue Animal Studies inevitably are confronted with loaded choices. Will the student of Animal Studies employ or ignore scientific terminology? How much focus needs to be given to the competing meanings of "animal" and in which contexts? How frank should any instructor's consideration be of the simple question, "Are humans animals?"

Because one widely accepted answer is, "Of course humans are animals," these questions are hard to ignore. This is particularly true when one advocates that critical thinking skills be taken seriously. Because so many important human realms continue to promote ways of thinking and speaking that obscure and override our own animality—such habits remain business as usual in politics, many religion-focused realms, commercial enterprises, professional circles, schools, and an astonishing number of science-committed circles—choosing to confirm or ignore humans' animality in core human activities like everyday speech habits will likely continue to be an unavoidable dilemma for decades to come.

A Critical Thinking Issue: On Talking of Humans as Animals

The "humans and animals" language habit is more than a misapprehension of vocabulary. Further, it is not merely nonscientific but actively antiscientific, because it ignores humans' clear membership in the animal world. But a far more serious tragedy is that, by passing along this habit, we teach our children to commit the fallacy of misplaced community. This happens because this phrase and its kin implicitly distance our species from other-than-human cousins—thereby, our children's worldview is anchored in a misleading, harmful dualism that we have inherited and now perpetuate.

The domination of the "humans and animals" mentality includes more than its prevalence and political correctness in scientific circles, education, public policy programs, and mainline religious institutions. One hears this vocabulary employed by animal protection advocates, even those who work hard at being radical. A revealing example comes from a book full of challenges to establishment harms to nonhuman animals—the author begins a strident critique of capitalism-based harms to nonhuman animals with a chapter that carries the challenging title "Taking Equality Seriously." The author, whose lifestyle has been chosen in part according to "a desire to live [his] life critically as a social anarchist," begins the chapter with the human/nonhuman division: "As a species, our relationship with animals is admittedly odd. We have 24-hour cable television channels devoted to shows about animals, and at least in the global North, the institution of companion animal ownership is deeply embedded in our cultural traditions."[12] Although this author works hard to be frank about the astonishing harms done to "animals," employing the word "animals" to mean "all of them" creates a "we/they" dynamic that draws its power from the very dualism that the author challenges in so many other ways.

The answer, then, to the question of who talks this way is everyone, since our ordinary, everyday ways of speaking continue to feature so many pieces and remnants of this antiscientific point of view. Said plainly, it is, at best, difficult to avoid such bad verbal habits. This can be explained in part by the fact that we are apes (our scientific classification puts us in the group known as "the great apes"), a term we use in its verb form to mean copy or imitate. Note in the following example how even exciting scientific discoveries are framed in language that foregrounds humans in ways that cause all of the earth's other animals to disappear.

The January 31, 2011, edition of the *New York Times* ran a front-page article with this opening line: "In a building at NASA's Ames Research Center here [Moffett Field, California], computers are sifting and resifting the light from 156,000 stars, seeking to find in the flickering of distant suns the first hints that humanity is not alone in the universe."[13] Alone? Why would anyone say such a thing in a world so obviously populated with, literally, millions and millions of species beside our own? No one is unaware that our earth is a shared world. One might, then, be tempted to conclude that the journalistic approach here is merely rhetorical excess. Such linguistic habits, however, reveal that major sectors of our social and scientific discussion remain squarely within the exceptionalist tradition.

This article's language is but one of thousands upon thousands of subtle exclusions that each of us is exposed to over the course of years. Such repetition is subtle in the way it anchors and reinforces both milder forms of human-centeredness and the exceptionalist tradition that eclipses other-than-human animals entirely. Our peculiar habits of thinking and speaking bewitch even sophisticated scientists—the motto of the program known as SETI (Search for Extraterrestrial Intelligence), which was set up in the 1960s and 1970s, is "Are we alone?" This motto was chosen by scientists who know only too well that all humans live with and amid countless mammals, birds, insects, and, especially, many trillions of microorganisms that we cannot see.

The tendency to employ language choices that make other living beings disappear exists in the face of some astonishing facts about our relationship with those other-than-humans we can in fact see. For example, in some industrialized countries the number of households that include nonhuman companion animals is larger than the number of households with children.[14] "Companion animals" are clearly central for us today—indeed, our use of "companion" for this category of living beings suggests clearly that we make a choice to bring them into our homes and places of business and even to name them "family" members. So in no meaningful way are humans "alone in the universe" even if SETI spends the rest of its days monitoring extraterrestrial silence.

Note, too, how the very way we talk about these treasured "family" members also reveals a denial of humans' animality. The prevailing way of speaking is to talk of them as companion animals, but never to call humans "companion animals" despite the obvious fact that we, too, are animals who are companioning our owned dogs and cats. This inconsistency draws its validity from, even as it anchors ever more firmly, the habit of dividing up the world into "humans and animals."

In many different ways, then, Animal Studies faces the psychologically important reality that the vast majority of humans in the industrialized world have been trained to use "animals" to mean "nonhuman animals." The training has been accomplished by moral authorities who nurture and dominate our individual lives (parents, educators, elders in various communities to which we belong, and prominent voices in government-based circles). It stands to reason, then, that it may be hard to see how peculiarly narrow are the ways we think and reason about our own self-importance in one field after another. If we feature the dualism "humans and animals" in our philosophies, animal protection movement activism, theologies, businesses, educational establishment, scientific practices, and public policies, it stands to reason that many of us will struggle to see how fully dominated we are by the exceptionalist tradition.

Animal Studies' commitment to critical thinking requires that such verbal habits be called out because they have major consequences. They isolate us unrealistically, create false divisions, and effectively obscure that we are not "alone" because we so obviously live in a more-than-human world. Such effects impoverish our imagination and the ability of our children to learn about the world they inhabit.

Any commitment to critical thinking requires, of course, much more than honest engagement with the impacts of our verbal habits. It mandates examination of human-centered practices of any kind, such as concentration on those nonhumans commonly put into the role of humans' pets rather than on a wider range of nonhumans, that may directly or indirectly foster habits of mind and action that reinforce the exceptionalist tradition. Further, by examining the mildest forms of human-centeredness that one finds in contemporary Animal Studies, one can assess whether they benefit nonhumans or, instead, prompt some students of Animal Studies to slide imperceptibly toward, and eventually into, harsher forms of human-centeredness. Animal Studies has been developed, and will surely continue, in those societies where a majority of the citizens think of verbal habits like "humans and animals" as merely reflecting a feature of the world, not a long-standing tradition that obscures humans' own animality. A key issue, then, is assessing what causes different people to subscribe to exceptionalist forms of ethics that favor human privilege and luxury that cause problems for nonhuman lives.

Treating the "human-animal divide" as part of nature can be challenged not only as misleading and harmful, but as grounded in ideology rather than truth.[15] The divide at times operates in the manner of a journalistic factoid, that is, a bit of unverified or inaccurate information that is presented as factual and then repeated often enough that it becomes widely accepted as obvious and a matter of common sense. This factoid, of course, has consequences. It causes many to ignore that we are animals who can, if we accept our animality, reach out in myriad ways to other living beings.

It is the case, then, that whenever Animal Studies goes forward amid the irrational claim that the "animals" referred to in the very term "Animal Studies" are all and only nonhuman animals, the field plays to the kinds of fallacies and self-inflicted ignorance that lead to serious dysfunctions and, worse, irreparable harms. But there is good news for those who despair of the awkwardness of finding new ways of speaking—both the history of challenging verbal exclusions and violence (such as the habits of speaking that kept racism and sexism in place for many centuries) and repeated practice show that the awkwardness of challenging accepted ways of speaking disappears when a majority of people choose not to use a prevailing habit that misleads so dramatically.

Finding Nonhuman Animals

The beginning of any focus beyond the species line can begin with a most basic question: where do we find them?

One common answer has been alluded to above, namely, the assumption and popular conception that "animals" are represented well by those living beings popularly called "pets" or "companion animals." As already noted, those who travel modern industrialized societies find dogs, cats, and a surprising menagerie of other nonhumans today called "family members." So one answer to this question is, "in our homes."

Another important answer is, of course, that nonhuman animals are food sources, research models, entertainers, property, and much more. These topics are discussed throughout this book, but particularly at the end of this chapter when the companion animal category is examined more closely.

A third popular answer points to those living beings we call by terms like "wild animals," "wildlife" or "free-living animals" who live "out there" beyond our own communities. These living beings are, in fact, ubiquitous, that is, found in so many places that we can fairly say they are everywhere. Organisms of the tiniest kind that we cannot see—and thus cannot even fathom how to treat as individualized living beings—are referred to in this book as microanimals.[16] Microanimals are in us, on us, and around us in our homes, workplaces, and any places nearby. Wherever we are, they are, for as some obscure sciences reveal, there is an unbelievably vast and diverse universe of these microanimals, and they come in almost countless forms. As a famous contemporary scientist suggests, "Five thousand kinds of bacteria might be found in a pinch of soil, and about them we know absolutely nothing."[17] Equally astonishing is the fact that multiple billions of microorganisms exist on the surface of each person's skin and in our guts. This, in turn, suggests that animal communities exist at a level so removed from human abilities that no human individual has ever had day-to-day ethical capacities even remotely capable of dealing with them. In fact, each of the living beings we think of as an individual human, or as an individual dog, cat, wolf, mammal, bird, or reptile of any kind, is a vast, mixed community of micro living beings. Simply said, the world of living beings that we live amid is unfathomably rich and diverse.

Organisms of the larger, more recognizable kind are also, for all practical purposes, everywhere. These living beings can be thought of as "macroanimals" that each of us can notice as individuals and as, possibly, members of their own communities. Evolutionary points of view (whose role in Animal Studies is discussed in chapter 2) suggest that most ancient lineages of macroanimals, which are the kinds of life studied most often in Animal Studies, have been on the earth for perhaps several hundred million years.[18]

Macroanimals, then, are not only companion animals, food animals, research animals, and wildlife (these are the categories of animals mentioned most prominently below). They also include the different kinds of living beings in our backyards, inside the walls of our homes, throughout our cities as domesticated animals gone wild (such as feral cats), and in and passing through or over nearby fields, streams, and indeed our entire local community.

A salient feature of contemporary animal protection efforts around the world is that such efforts are directed at, naturally enough, only those other-than-human animals that we can see easily if they happen to be in our local part of the world. Of these macroanimals that we can actually notice without the help of technology, only some are recognizable as individuals. We are adept at distinguishing one dog or cat from other individuals of the same species. But the vast majority of macroanimals are, even though visible to us as individuals, not easily recognized as specific individuals—in other words, we generally cannot tell one robin from another, one wild mouse or rat from another individual of the same species, and so on. We can, with work, become much better at this exercise, as evidenced by helpful books like Len Howard's *Birds as Individuals* (1953). Yet, even if we hone finely our native abilities to tell one living being from another, there remain many macroanimals that we can see but simply cannot distinguish from one another. Our earth is, then, a shared place of nested communities teeming with countless forms of life found in astonishingly diverse places and in forms that even our fecund human imaginations cannot fathom.

Exploring the Third Issue: Why Is Animal Studies Important?

Beyond the benefits noted above, consider the startling variety of academic fields or disciplines touched upon in one way or another in this introduction to Animal Studies—history, cultural studies, education, natural and social sciences of many kinds, political studies, law, philosophy, critical studies, literature and other arts, comparative religion, ethics, sociology, public policy studies, social psychology, geography, anthropology, archaeology, and criminology, to name only some.

Both a great number and a wide variety of disciplines are needed if Animal Studies is to engage the past, present, and future possibilities of human interactions with living beings outside our own species. There is simply no other way to explore the diversity of other animals, respect the variety in human responses, and describe the peculiar dynamics of human animals.

Thus, for multiple reasons, Animal Studies derives importance from engagement with other disciplines. In addition, by virtue of its range and multidisciplinary scope, Animal Studies confers importance on other disciplines. For example, Animal Studies is important to the integrity of many disciplines whose central work inevitably involves nonhuman animal–related issues, such as veterinary medicine, behavioral studies, ecology, and so much more.

Animal Studies is important as well to the very idea of interdisciplinary work. To think that we could divide our universities up into, on the one hand, "the arts and humanities" and, on the other hand, "the sciences" is to think an impoverished thought. The world is simply more diverse, more populous, and more interwoven than this division suggests. Similarly, Animal Studies gains importance because, as an examination of history will show, our rich abilities with ethics, values, spiritual awareness, and so much else cannot be understood by disciplines that promote a radically human-centered vision of the world.

This introduction can, in fact, be characterized as a sustained argument that Animal Studies needs to be foregrounded whenever some discipline purports to engage the whole world. This is particularly true today because of the number of changes taking place around the world regarding attitudes to other-than-human animals.

Ferment and the Importance of Everyday Life

A key feature of today's Animal Studies is that it is informed by worldwide developments in different societies addressing a surprising range of issues involving other-than-human animals. The introduction suggests that such developments "prompt students of Animal Studies to see the ubiquity and diversity of other animals" as well as to learn about new research findings and our own species' astonishing diversity in cultural attitudes toward other living beings.

The multiple changes one can identify around the world in the last half-century have created a kind of ferment. This ferment, in turn, generates ever more interest, opens up field after field, and catches the attention of diverse secular and religious communities. The fact and pace of change are impacting private decisions and public debates about our capacities to choose a future that we consciously put into place by virtue of the policies, education, and consumer and business practices that individuals and societies support.

The number and variety of fundamental shifts and changes around the world, across cultures and within different domains and academic disciplines, suggest that we now live amid changes in attitude that are still developing. To be sure, this ferment is chaotic, because while discussion of and concern for nonhuman animals appears in many different quarters, it is diverse in both content and intensity such that no single theme or feature dominates. Yet this diversity provides a kind of health, for as information is spread via the twenty-first century's many kinds of media, citizens from around the world have the opportunity to learn about diverse cultural and individual responses to many different forms of nonhuman lives. Implicitly and explicitly, such a range of responses underscores each human's great capacity for caring about other lives. Simply said, our species' overall diversity in finding ways to take other-than-human lives seriously models a wide range of options. Further, because each local world where humans encounter other animals features a unique subset of other-than-human animals, an astonishing variety of perspectives, choices, and ways of talking about many different options are now communicated regularly via the information media.

With other animals noticed more closely, the ubiquitous presence of other-than-human animals prompts adjectives and nouns such as "pervasive," "intertwined," "fecund," and even "pests" or "rivals." These are pertinent to describing well how humans in different situations have in the past acted, now are acting, and can in the future act. It is this combination of history, present practices, and future possibilities of humans in relation to other living beings that is pushing the ferment and, with it, Animal Studies to engage the astonishingly various and deep-rooted features of humans' recognition of other animals. Through such learning, individuals recognize that they and their local communities can, indeed inevitably must, make decisions that impact other-than-human living beings in local communities that humans and nonhumans share with each other.

Animal Studies can affirm, then, the fact that many humans' day-to-day lives bring them into contact with various nonhumans. Noticing this feature of daily life helps individuals be aware of their own local, everyday connections with lives beyond the species line. Such on-the-ground realities have long been an important feature in ordinary people's lives, even in historical eras and places that historians characterize as uncaring about other animals. Addressing a period characterized by some as unresponsive to animal protection concerns, the scholar Brigitte Resl begins her edited volume on medieval views of nonhuman animals with this observation: "The centrality of animals within medieval culture is abundantly reflected in the surviving source material; animal fables and zoological encyclopedias in the broadest sense are among the most widely distributed texts of the period, and hardly any building or illuminated manuscript survives that does not feature animals in its decoration."[19]

Historical work within Animal Studies allows researchers to identify local or familial groups of humans who reflected keen awareness of nonhuman animals in their day-to-day lives even though the larger society's mainline institutions did not recognize, care about, or memorialize such connections (this is pertinent to the idea of "history from below" cited in chapter 2).

The Importance of Animal Studies to Critical Thinking

Because Animal Studies promotes both careful research and a diversity of approaches working together to create an informed view of issues (such as the story of humans' past with other living beings), every subfield of Animal Studies can contribute to the kinds of careful, responsible reflection discussed in chapter 2 as "critical thinking." Such skills reveal not only unreported facts but also new perspectives on problems. They help students and scholars address the multiple vocabularies and jargon that populate any intensely interdisciplinary inquiry, and provide the skills to examine the sources and possible resolutions of existing tensions. Animal Studies, then, can model and even *advance* critical thinking skills because so many areas of inquiry about nonhuman animals are dominated by caricatures anchored in the dismissive attitudes and self-inflicted ignorance that harsh, exclusivist human-centered formulations have sustained and spread.

Exploring the Fourth Issue: The Personal Connection

There is a bottom line or frontier that Animal Studies contemplates as it engages the variety, conceptual complexities, and related intellectual excitement generated by thinking about and studying nonhuman animals—this is another driving element at the heart of humans' interest in other animals, namely, a deeply personal dimension of connection with nonhuman animal individuals themselves. This personal connection is manifested in many ways—throughout this book the one-on-one features of human-to-nonhuman meetings and relationships in modern societies are mentioned, but the connection is apparent in many other ways as well. In the most revered scriptures of many religious traditions, it is easy to find stories of religious figures meeting other animals. Both religious and secular philosophers often wax poetic about humans in wild places. The humanists of Western Europe who in the late thirteenth and early fourteenth centuries became fascinated with Roman and Greek culture and thereby nurtured the historical development commonly called the Renaissance wanted to set humans free by recovering a sense of humans as a part of nature (see chapter 13). Like many ancient religious sources, the humanists recognized within the human spirit a deep fascination with the more-than-human natural world.

It is worth noting that just as the lives of many people prior to the scientific revolution reflected personal fascination with other-than-human animals, so too such a fascination is fully evident after the changes wrought by our species' remarkable scientific developments from the seventeenth century onward. Following the publication in 1859 of Darwin's *Origin of Species by Means of Natural Selection*, other animals continued to rivet people's imagination. Consider just a few of the titles published in the English-speaking world within the six decades that followed, reflecting that authors and readers alike wondered again and again about nonhuman animals:

1871 *The Intelligence of Animals* (Ernest Menault)
1880 *Mind in Animals* (Ludwig Büchner)
1882 *Animal Intelligence* (George Romanes)
1894 *Animals' Rights* (Henry Salt)
1898 *Animal Intelligence* (Wesley Mills)
1901 *Beasts of the Field* (William Long)
1909 *The Place of Animals in Human Thought* (Evelyn Martinengo Cesaresco)
1922 *The Mind and Manners of Wild Animals* (William Hornaday)
1923 *The Animal Mind* (Margaret Washburn)
1927 *The Minds of Animals* (J. Arthur Thompson)
1931 *The Intelligence of Animals* (Frances Pitt)

In scientific circles, too, personal connections and values are powerful factors, even though advocates of science regularly claim that science is value-free. A classic claim of this kind was made by one of the most respected scientist-mathematicians of the twentieth century: "Ethics and science have their own domains, which touch but do not interpenetrate.

The one shows us to what goal we should aspire, the other, given the goal, teaches us how to attain it. So they never conflict since they never meet. There can be no more immoral science than there can be scientific morals."[20]

There are abundant reasons to recognize that while in some contexts the assertion that science and values-based claims, like ethics, never meet, this is only an important partial truth. As a generalization the claim that science is value-free ignores much, including the role of scientists' personal connections and values as factors impacting how science is practiced. These recurring claims are, upon examination, more ideology (in the sense identified above) than a fair description of science as it actually goes forward in one context after another.

Indeed, many who have written about the practice of science have pointed out that scientists can be startlingly unaware of their own values and exclusions as they pursue "science."[21] A noted philosopher of science observed in an award-winning 1986 book that although scientists constantly suggest that science is objective, rational, and value-neutral, this is seriously misleading—scientific work at the theory level and in practice is full of hidden values, unacknowledged interests, and more that control which problems, theories, methods, and interpretations of research prevail.[22]

Among the driving, values-based factors in contemporary science is personal connection with nonhuman animals. This goes well beyond the familiar fact that many individuals choose to pursue a science-based career because they were, as children, fascinated by a particular kind of nonhuman animal. Many scientific textbooks, for example, openly advocate value-driven positions. A leading textbook in the field of conservation biology, which carries a dedication to "those who teach conservation biology, ecology, and environmental sciences, whose efforts will inspire future generations to find the right balance between protecting biological diversity and providing for human needs" includes a section titled "Ethical Arguments Supporting Preservation" and another section describing the inherently ethical issue of restoring damaged ecosystems.[23] In primatology, values have long been part of the discourse. Many readers will, for example, be familiar with Jane Goodall's advocacy for chimpanzees. In marine mammalogy, comparable examples are easy to find. The 2011 volume *The Dolphin in the Mirror* includes a subtitle that makes clear the driving values of scientist-authors: *Exploring Dolphin Minds and Saving Dolphin Lives*.[24] Similar animal-friendly values appear on the opening page of a textbook describing the highly technical scientific work known as passive acoustic monitoring: "For an air-breather, living in the water is a continuous challenge and as such marine mammals deserve our respect and our protection. Consequently, in addition to a pure interest in knowledge, scientific research increasingly studies marine mammals to support their conservation and protection."[25]

From the vantage point of the kinds of critical thinking examined in chapter 2, the author's conclusion about protection does not, as a matter of logical reasoning, follow from the fact that an organism meets a "continuous challenge" (strictly speaking, this conclusion is a non sequitur), but the appearance of the author's personal values in favor of protection surprises very few today because science involving nonhuman animals, like so many human endeavors, is often driven by personal commitments, passions, and connections with the subject matter.

Recall, too, Beston's moving vision of other animals as "other nations, caught with ourselves in the net of life and time, fellow prisoners of the splendour and travail of the earth." Such images of connection can be found in every human era and in many contemporary circles—for example, a prominent late twentieth-century theologian suggests that modern people "consider ourselves blessed, healed in some manner, forgiven and for a moment transported into some other world, when we catch a passing glimpse of an animal in the wild" (see chapter 11).

Such dimensions of personal meaning and connection are often the principal motivation behind student demands for animal law courses (see chapter 4). They are also obvious features of the commitment evident in teachers and scholars who have immersed themselves in one of the many fields exploring issues that are part of the Animal Studies universe. Of particular note for educators is how personal connection issues animate the dynamics of discussions in Animal Studies courses today. These driving forces of personal commitment and vibrant discussion dynamics are important sociological and pedagogical realities in the broad domain of Animal Studies.

Patterns of Presence and Absence

The "everywhereness" of other-than-humans in daily life contrasts greatly with the planned, systematic absence of nonhuman animals in key human endeavors. For practical and health reasons, this absence is to be expected in some business and mainline institutions, but the absence of nonhuman animals in many educational settings is revealing in a number of ways. Certain kinds of nonhuman animals who are present in students' and teachers' lives outside the classroom are typically left behind in educational settings, and this is frequently, though not always, true of present-day Animal Studies courses.

These patterns of presence and absence prompt teaching and learning problems of significant sorts that have their ironies (see chapter 2). For example, when live nonhuman animals are brought into educational or business settings, they are virtually always members of the category we call companion animals. Although it is not uncommon for someone to unreflectively assume that these animal individuals are representative of all nonhuman animals, it is important to ponder in what ways this assumption is true and in what ways it is seriously misleading.

Making Sense of the Category "Companion Animals"

Dogs and cats are the animals usually thought of when the term "companion animal" is used, for these living beings are on many people's minds as modern industrialized societies develop their discussion regarding other-than-human animals. The relationships that humans have with these animals, as well as with other nonhuman animals we place in the companion animal category, are characteristically relationships of dominance, though many animals of these species are "feral" (that is, they are not attached to a particular household and thus roam freely in or near a community). Human dominance over companion animals is often signaled by the term "pet" or "domesticated animal" (the latter term is also used for the fundamentally different kind of domination we have over food animals).

Much attention is given in certain societies today to the particular roles that law and other major public policy mechanisms can play in shaping our choices regarding connections to and relationships with companion animals. In one sense, then, companion animals are, to borrow a famous phrase, "good to think."[26] If one attempts to understand or deconstruct the category "companion animals,"[27] one soon recognizes that exploration of the world makes it apparent that this category has little to do with the animals themselves. The category is an example of what sociologists and philosophers call a constructed category—the living beings in the category are grouped together because of their relationship to humans, not because of their inherent qualities. The category is, then, constructed by humans for human purposes.

Such construction makes the category elastic—whenever a new species of animal becomes widely owned and is called a "pet," then that animal too will be spoken of as a companion animal. Importantly, animals found in this category can easily be found in other categories (such as food or research animals). Dogs and cats, for example, as well as some other animals such as horses and pigs, appear in many other constructed categories commonly found in industrialized societies, such as research animals, food/production animals, and entertainment animals.

Nonetheless, in some circles the "companion animal" construct is so traditional and familiar that some people view it as part of the natural order, that is, a result of biological evolution. Others frame the availability of these animals for humans' personal use as "the way things are supposed to be" because this arrangement was put in place by the command of a creator divinity. Interestingly, it is by no means clear that humans themselves were the agent responsible for the domestication of, say, dogs and horses. Some researchers have suggested, instead, that ancient dogs and humans may have codomesticated each other tens of thousands of years ago, and similar speculation about a codomestication of horses by humans and humans by horses has been published.

If one asks straightforwardly which living beings should be put into the category "companion animals," one could answer, as already noted, that humans can be thought of as fully belonging to this category. This is not, of course, a common way of speaking, for the vast majority of humans have been trained by parents, educators, and other elders in various communities to use "companion animal" only for nonhumans.

In chapter 3, where public policies that mention companion animals are discussed, the observation is made that it is *not* nonhumans that are the principal focus of these public policies—rather, it is the human owners of those dogs and cats deemed "owned animals." This is so because even though some government-promulgated regulations talk about the generic categories of "dogs" or "cats," the real target of such public policies is owned animals, not all members of the nonhuman species involved. The policies have lots of loopholes by which feral dogs and cats, shelter animals, and the animals confined to puppy mills are excluded. So the driving force behind many laws that on their face talk about "dogs" or "cats" is humans' sensibilities about their owned "family members," not the welfare of dogs and cats generally.

The Paradigm Question

Around the world, humans are companioned by between 1 and 2 billion nonhumans, and we spend several hundred billion dollars per year on their food, veterinary care, and more. For this reason, companion animals have become a dominant model or paradigm for talking and thinking about nonhuman animals today. This is evident in many conversations about "animal rights" or "animal protection" where the focus is overwhelmingly owned dogs and cats.

The reason why so many people around the world speak in this way is subject to debate, but a number of features of the human-nonhuman companion animal phenomenon are clear. First, contemporary humans choose to take care of the owned dogs, cats, and other animals we put in the companion category because owners want these animals present in their lives. Second, many (though not all) of us become attached to them when we live with them.[28]

Whatever the source of contemporary humans' fascination with dogs, cats, and the other nonhumans called companion animals, our heavy concentration on dogs and cats has made the category of companion animals a leading edge in contemporary animal law courses, bar organizations, scholarly conferences, publications, and, of course, media. For example, such issues dominate the agenda of the American Bar Association's Animal Law Committee—when this committee meets, the kinds of animals that get attention are, overwhelmingly, companion animals. The participants talk about "animal protection" but what they mean is expanding protections for certain animals, namely, owned dogs and cats and, at times, some of the nonowned (feral) members of these species.

The emergence of companion animals as a dominant model for nonhuman animals is also evident when animal protection is discussed in local communities and in regional government circles, in custody disputes that take place in divorce courts, in the emergence of laws that allow people to take care of "animals" in their wills and trusts, and in proposals to expand the amount of money that a pet owner can recover when her or his companion animal is harmed by someone else.

Problematizing the Paradigm

When the principal basis for understanding nonhuman animals is anchored in a constructed (artificial) category like companion animals, there are significant benefits and risks. Without question, care for these nonhumans opens up the basic issue of connection beyond the species line. Through relationships with companion animals, many humans become personally familiar with the existence of rich and distinctive personalities, the existence of communication, and varied forms of intelligence "out there" in individuals who were born outside our species. These phenomena have informed many animal protection efforts, and thereby "animal" concern has developed as a social issue. Once the possibility of some protections for a group of nonhuman animals becomes familiar, the mentality of animal protection can spread and even foster ever deeper concerns for yet other animals. Thus, because many people easily and naturally think of companion animals as important in and of themselves, not merely because they are owned and belong to a particular human family, discussions of legal protections based on such views naturally lead to serious consideration for such protections for some noncompanion animals as well.

Because many people have insisted on elevated protections for companion animals, then, today it is obvious that, if the political will exists, public policy can establish protections for nonhuman animals. In one sense, then, companion animals can also be seen as ambassadors for the nonhuman beings who do not fit into this constructed, overtly human-centered category. The beings we have slotted into the category "companion animals" can open minds and hearts, as evidenced by the extraordinary number of fact-based accounts and fiction-based works that use companion animal themes to increase awareness of the benefits of living in the presence of these animals. Companion animals, then, create opportunities.

Critical thinking also requires that one notice that these opportunities can involve risks—for example, the risk that the larger category "nonhuman animals" is, at best, partially represented by domesticated animals such as companion animals. Similarly, this category has as a salient feature of humans' domination over these living beings. Companion animals are, by definition, subordinate to humans in humans' households.

A related risk is that *if* the human-animal relationship implicit in the companion animal model becomes the image of what humans' relationship to all animals should be, *then* there is an obvious downside for nonhuman animals who need to be free living (since surprisingly few animals easily fit into human patterns of life). Another risk is that if a key element in our society's notion of all nonhuman animals is their subordination to humans, then even the most compassionate forms of public policy will misfire. Another possible risk that exists when the companion animal paradigm is too dominant is the implicit message that the mere exercise of some care toward one or a few nonhuman animals (for example, one's own dog or cat at home) is sufficient to meet one's moral obligations beyond the species line. Some humans can—and do—rationalize the harms they do to, say, food animals, because they are "animal people" by virtue of the fact that they rescue cats or dogs.

One scholar has observed, "Pet keeping is kindness toward only a few favored animals. The practice of pet keeping operates in a world where other animals are used for work and food."[29] While some would surely reply, "Well, at least some nonhuman animals are getting protected," note what this scholar suggests may be the consequence of the pet-keeping traditions many countries inherited from Victorian England:

> Placed in this different register, pet keeping appears as a phantasmagoria, a fantasy relationship of human and animal most visible in the trope of the animal as child, the pet as a member of the family, which the nineteenth century inaugurates.... The pet who is a child is a de-animalized animal. In this symbolic logic, is not our animal nature, too, denied? That we ourselves might be aggressive and dominating is doubly hidden in the culture of pet keeping.[30]

In a different vein altogether, one of the pioneers of modern animal law who has advocated both legal rights and abolition of many common practices is not at all impressed with today's animal law courses. He criticizes those who seek change through compromises that promote less harsh forms of animal use and domination, suggesting, "Modern animal law, for the most part, promotes traditional welfarist change as a way of modifying [but not abolishing] the property status of animals."[31] Any number of other commentators have also had reservations about what they see as genteel dog and cat protection among the elitist, consumer-oriented segments of society. Such treatment can rightfully be seen as opening a door to a better future for owned animals, but if such treatment makes people complacent or apathetic about all other animals (the free-living ones, and those in humans' research and food production facilities), then the relationship with one's own dogs and cats becomes a substitute for openness to the invitations that many other nonhuman animals extend to humans' abilities to care.

In such a case, the door potentially opened by companion animals leads not to the larger community of other-than-human animals, but to an all-too-human closet in a world completely dominated by humans. Something like this happens in certain veterinary circles where the human-animal bond is touted even as the official institutions of the profession (such as national organizations of veterinarians) promote factory farming and use of animals (including unowned cats and dogs) in research that provides veterinary schools and research laboratories with multimillion-dollar grants. In such cases, the human-animal bond emphasis sits alongside, and is often subordinated to, attitudes that shut down, even betray, a full range of animal protection.[32]

At the very least, deep fascination with only dogs and cats leaves the vast majority of nonhuman animals unexplored. It leaves aside, for example, issues that scientists like Bernd Heinrich raise on the basis of their field research. As part of his account of stories of extraordinary relationships between a bird and a human family who referred to the bird as their "son" and "a true friend" and said they could "not imagine life without him," Heinrich concludes, "There is something unique about ravens that permits or encourages an uncanny closeness to develop with humans."[33] Heinrich suggests that the reason for such attachment is mutual communication: "A raven is expressive, communicates emotions, intentions, and expectations, and acts as though it understands you. This communication is privileged. It occurs when the individual close to the bird is trusted, has earned a trust that is not offered lightly. Given that trust, much is revealed that could otherwise never be seen."[34]

Note how this special relationship is being touted by a scientist who studied these animals in both domesticated and free-living circumstances or, as we often say, "in the wild." The latter kind of work requires great patience and, surely, the willingness to take the free-living birds seriously, much in the spirit of Goodall's decades-long work with chimpanzees. Her work led one major scientist to name her "one of the intellectual heroes of this century."[35]

Work with ravens in domesticated circumstances played a role, too, for it helped Heinrich explore the birds in free-living circumstances, and vice versa. Ravens have long had a special relationship with humans of many cultures and religious traditions. Heinrich's work on human-raven relationships may have opened vistas that our normal companion animal models do not allow us to see well.

Because such observations suggest that there are truly extraordinary relationships "out there" beyond what we consider the standard companion animal relationships, Animal Studies must ask whether making a companion animal model the leading edge of human exploration of other living beings can lead to undesirable consequences. For example, do we, to use Kete's words, risk "de-animalizing" other-than-human animals through our domination of them?

Asking such questions helps us recognize that if Animal Studies goes forward on a companion animal paradigm, then it is at risk of one-dimensionality in thinking about other animals. In particular, the ownership model, which creates allocations of responsibility that are needed if other-than-human animals are to live among humans, fails in many instances. Elephants and chimpanzees, for example, are very dangerous animals when they live among us. They are intelligent and capable of emotions, but they can become psychologically damaged by domination and even unpredictably violent. Animal Studies foregrounds questions about the obvious fact that many nonhuman animals seem to want to choose for themselves something other than human domination.[36]

There are other risks as well when a companion animal paradigm prevails without much reflection. Superficial evaluations of other animals' abilities can occur—companion animals' intelligence and relationship abilities do not exhaust what all other nonhuman animals can be or do. Some animals have entirely different kinds of sensory abilities (bats and dolphins), different kinds of perception and intelligence skills, and, clearly, unfathomably different needs—the problem is encapsulated wonderfully by John Webster in his 1612 play *The White Devil* when Flamineo observes (act 5, scene 4), "We think caged birds sing, when indeed they cry."

Constant repetition of claims of superiority, which echo throughout our educational, religious, legal, and political institutions, lead us to believe that we see and understand other animals' realities well. But most of us know that we often fail to notice other animals' actual realities, and even when we notice them, we fail to take those realities seriously, let alone understand them fully. This is sometimes true of our understanding of companion animals, of course, but it is even truer of less familiar animals like wildlife and the food and research animals hidden away in limited-access situations like factory farms and laboratories.

Our bonds with companion animals reflect, then, but a few of our possibilities with other living beings because the world of other-than-human animals is vastly larger and more complicated than the companion animal world. To be realistic about this, and to prompt open-minded discussion on such matters, Animal Studies must strike a balance—first, it clearly needs to recognize that companion animals open a door, but, second, it must also be realistic that companion animals do not fill all the rooms beyond.

Through the Open Doors

Chapter 2 begins the journey through some of the doors opened by this first chapter. Chapter 3 suggests that the category we call "wildlife" is different from the category of companion animals. Characteristically elusive and mysterious when free of human domination, nonhuman animals are difficult to understand in a variety of ways. Thus even the more familiar free-living animals that exist in urban, suburban, and less human-dominated areas offer a paradigm that is different than that associated with companion animals. The following chapters, then, push the reader to encounter a variety of "traditional," "commonsense," scientific, artistic, or other kinds of claims about our fellow living beings—such claims sometimes illuminate the actual animals themselves, but far more often they obscure, distort, subordinate, imprison, marginalize, and even kill them.

Through Open Doors
The Challenges of History, Culture, and Education

Journeying through doors opened in chapter 1, those who pursue Animal Studies necessarily encounter three fundamental areas of human life—an astonishingly varied past, rich cultural diversity, and multiple educational traditions. Each of these areas offers a range of stories and possibilities that impact the direction of Animal Studies. Some of these stories and possibilities offer breathtaking beauty and eloquent testimony to the breadth and depth of our human spirit. Others, though, suggest that the ugly and baffling can also populate the human-nonhuman intersection.

Consider what humans today encounter behind the door that opens to our species' past. Much of our history is only now being told—scholars in the early twentieth century, for example, inaugurated the "history from below" tradition that expands historical accounts to more than merely a chronicle of rulers, conquerors, and other dominators. This more encompassing, mind-opening view of our past offers, by virtue of the many different stories it uncovers, profoundly important lessons for those who want to comprehend the tasks, shape, and future of Animal Studies.

Through a second door we will find a bewildering variety of human cultures, exploration of which prompts several kinds of awareness that are crucial to Animal Studies. A review of human cultures provides, for example, increased appreciation of humans' extremely diverse sensitivities and claims about the identity of our community and, thus, who we are. Animal Studies offers a surprisingly rich set of insights into what human groups have thought of their place in the more-than-human communities and worlds of which we are an integral, inevitable part.

Passing through a third door, we explore the quintessential human endeavor of education. Sometimes a realm of open-mindedness and awareness about diversity, education has also been a stultifying, even suffocating place for some humans. Shortcomings and even contradictions within formal education have from time to time prompted inquiries about whether humans are "born ignorant, not stupid—they are made stupid only by education."[1] An examination of educational traditions regarding other-than-human animals attempted through the lens of Animal Studies can prompt the noteworthy conclusion that both formal, institutionalized education and informal education have often been far too narrow-minded. In this chapter, variety comes to the fore as both formal and informal means of education regarding other-than-human animals are addressed. By taking into account informal education as well, Animal Studies finds itself examining the importance of family, language, culture, and, especially, the significance of being in the presence of other living beings. Through the education door one is also introduced to profoundly important insights about the seminal contributions of our creative arts to exploration beyond the species line (see chapter 5). The engagement with education started in this chapter continues in many subsequent chapters, for informal education about other living beings cannot be examined well through a concentration on institutionalized education.

Through a First Door: History's Narratives, Ancient and New

Our species' history features what can seem a central paradox—abundant appreciation of other-than-human animals sits alongside wide-ranging denunciations. Our forebears appear to have bequeathed us both long-standing interest in animals of all kinds and aggressive denials that humans should be deeply interested in the world's nonhuman citizens. Sorting through what we can know about, and then telling, the full story of this complex heritage is one of the central tasks of Animal Studies.

As chapter 1 noted, whenever we survey our present physical environment, some other-than-human neighbors are always nearby. Learning this lesson is simply a matter of paying attention to the world into which we have been born—our "neighbors" may be companion animals, or birds traveling the skies above our community, or even small mammals and other life forms in an urban, suburban, or countryside setting. In addition, each of us is richly familiar with myriad images of other animals that populate secular or religious stories. In particular, our creative arts and media provide a steady stream of such images, although some of these are symbols that have little or no connection to real nonhuman animals themselves (see chapter 5).

Each of us is thus regularly surrounded by diverse likenesses, words, and phrases that call to mind other-than-human animals. Importantly, some of the images are fundamentally negative—ordinary language includes animal images that imply debasement, disdain, dismissal, and the like. For example, it is rarely a compliment to be told, "You acted like an animal." In the "carpentered world,"[2] many other-than-human communities and individuals have been viewed as pests or worse, or merely as the equivalent to inanimate resources, and thus controlled or exterminated so as to make this "our world." But in a merely carpentered world, where individuals learn ideas and discuss viewpoints and "thinking" that draw energies only from human interests and viewpoints, learning opportunities are impoverished in both overt and subtle ways.

Chapter 11 considers Richard Louv's observations in his 2005 book *Last Child in the Woods*. The subtitle, the hopeful *Saving Our Children from Nature-Deficit Disorder*, calls out a risk to children who live in impoverished, merely carpentered parts of the larger world.

Children have been removed from the more-than-human world even though the multiple risks of such a strategy, such as impaired development of children's rich cognitive and ethical abilities, have not been studied well. Such impoverishment is only one of the reasons, of course, that activists in social movements as diverse as child protection, environmental protection, and animal protection call upon individual and corporate citizens to contemplate the value of including nonhuman "others" within our moral circle. In effect, such calls plead for each of us to see holistically and thereby anticipate the harms caused by living in a merely carpentered world.

Seeing Our Heritage

Such pleas are made against a background of formidable, well-entrenched dismissals that have intentionally distanced humans from other-than-human animals and, more generally, the natural world. The carpentered world is not merely physical, being composed of right angles and rectangles in buildings. It is also characterized by even more severe angles, as it were, in the exceptionalist tradition that so completely dominates contemporary history books, educational goals, law, and public policy.

One of the most commonly cited developments in the long history of Western European culture as it moved away from viewing humans as an integral part of the natural world is a claim by Immanuel Kant (1724–1804), often described as the most influential thinker in the Western philosophical tradition. Kant argued confidently that only humans could be persons—he made this argument in spite of having had virtually no exposure to complex nonhuman individuals (see chapter 6). Dividing living beings into two categories, Kant's most notorious comment on animals denies that humans as moral beings owe any duties to them: "So far as animals are concerned, we have no direct duties. Animals are not self-conscious and are there merely as a means to an end. That end is man.... Our duties to animals are merely indirect duties to mankind."[3]

When Kant gave the lectures in which he made this claim, the exceptionalist tradition had long been prominent in the official circles of his culture's dominant religious tradition and its mainline institutions. A late nineteenth-century version of this prejudice from a respected theologian is a blunt statement: "Brutes are as *things* in our regard: so far as they are useful for us, they exist for us, not for themselves; and we do right in using them unsparingly for our need and convenience, though not for our wantonness.... We have, then, no duties of charity, nor duties of any kind, to the lower animals, as neither to stocks and stones."[4]

Similar comments today dominate some secular circles, such as modern legal thinking, where the stark dualism "humans and animals" still prevails. The same heartbeat can be detected in the following statement from a deeply respected classical theorist of economics: "Man alone is a person; minerals, plants and animals are things."[5]

The attitude that other-than-human living beings can be treated by humans as mere commodities dominates other circles as well, of course. A number of modern defenders of biomedical experimentation on nonhuman animals use a version of this argument by promoting the exclusivist notion of speciesism (caring for all humans, but only humans): "I am a speciesist. Speciesism is not merely possible, it is essential for right conduct."[6]

These purported justifications of humans' right to dominate any and all nonhuman life forms come from sources that are both mainline and diverse—philosophy, the dominant religious traditions in Western culture, economics, and the science establishment. The prestige and power of these mainline sources help explain why the counterintuitive claim that "other animals are mere things" is well entrenched throughout society—this "business as usual" worldview prevails not only in commerce, but also in many education and the public policy circles described in succeeding chapters.

A Principal Challenge and Its Risks

Animal Studies can, in terms of the realities and ideas it explores, call out the problematic features in our traditions of thinking and speaking about the living beings beyond our species line. Yet note what meeting this challenge requires—negative generalizations that dominate thinking about other-than-human animals must be forthrightly and rigorously described; prevailing human-centerednesses backed by politically powerful elites must be handled in an analytical fashion; prevailing practices must be treated dispassionately and not with a bias in their favor.

It would be disingenuous to ignore the obvious risks taken by those who call out these features of present thinking about other-than-human animals. Proposing alternatives to radical dismissals of nonhuman animals, challenging myopias and self-inflicted ignorance, and forthrightly identifying disingenuous, self-serving denials of other animals' sentience and complexities are risk-prone strategies as politically fraught as they are ethics-laden (see chapter 7).

George Orwell, who cared to make the English language as serviceable as possible, once observed that telling the truth in times of universal deceit can be viewed as a revolutionary act. But telling the truth fully is often impacted by more than conscience—in 1747 the Swedish botanist and zoologist Carl Linnaeus suggested that giving one's full opinion of the truth at times is curtailed because complete frankness has its risks:

> I demand of you, and of the whole world, that you show me a generic character...by which to distinguish between Man and Ape. I myself most assuredly know of none. I wish somebody would indicate one to me. But, if I had called man an ape, or vice versa, I would have fallen under the ban of all ecclesiastics. It may be that as a naturalist I ought to have done so.[7]

The goal of stating the truth plainly will seem to some a commonsense obligation and an obvious goal of history and, of course, education as well—many universities tout truth seeking as their principal purpose, as is implied in Harvard University's one-word motto "Veritas." Yet, as a review of historical accounts attests, frankness about the treatment of nonhuman animals has often been—and still is—wanting.

The development of Animal Studies requires, as argued in chapter 1, that we meet the primary task of telling the entire story about our past with other living beings. Encompassing the choices made in many human cultures as they encountered different kinds of nonhuman animals, the full story reveals again and again how fully human it can be to develop ways of living in which humans and nonhumans coexist in a larger community. When we survey the great variety of approaches across human cultures in different times and places, we readily notice that the versions of history we have inherited are but partial stories of our past. We can also wonder, are such partial accounts serious distortions of the larger story?

In the Beginning: The Deep History of Life

While the vast majority of human history took place prior to what we think of as recorded history, there is an altogether more vast history of life that took place before humans arrived as a species. This deeper history of life itself is a staple of mainline scientific thinking and education, although by no means is it universally acknowledged in all circles today (see chapter 3). But even though not everyone subscribes to the belief that those animals we describe as "the first humans" were direct descendants of other, nonhuman life that had been on the earth for, literally, hundreds of millions of years, there are reasons that Animal Studies is committed to scientific approaches. This commitment follows from commitments to rigor, ethics, and critical thinking because, as chapter 3 suggests, it is impossible to do an informed version of Animal Studies without the help of many sciences that have developed over the last three centuries. The organic, biological connection of humans to other animals is supported by such an overwhelming amount of empirical evidence that Animal Studies must, by virtue of its commitment to rigor and evidence-based approaches, constantly take account of an evolutionary viewpoint.[8]

Because evolutionary points of view are, like all science-based views, not absolute certainties comparable to the certainties found in abstract mathematics or claimed in religious revelation, Animal Studies is properly said to trade in something short of absolute certainties. This is because human explorations of the world, including the elegant work we call science, have long proceeded on the basis of hypotheses, broad generalizations, theories, and, thus, probabilities. These can be—and often have been—revised in major or minor ways. Successful revisions may in fact bring ever greater certitude after further inquiry, but this form of certainty still remains something less than absolute.

Animal Studies, nonetheless, affirms that the evidence of humans' fundamental biological connection to other animals is an essential ingredient of the best science has to offer, and has produced countless further scientific discoveries. This is an important fact of the modern world, for competing, non-evolution-based hypotheses have produced no scientific discoveries. Thus science-based, evolution-informed views of human and nonhuman animals are an essential feature of any robust version of Animal Studies.

A corollary of this kind of realism about scientific findings as essential but not absolute is that there is room for alternative, nonscientific points of view and insights regarding any and all animals. Because any wide-ranging version of Animal Studies soon encounters many explanations of our connection to other animals, keeping an open mind in exploring competing views, as well as our possibilities of understanding some features of some other animals' lives, is both a practical and theoretical requirement. Chapter 1 made a complementary point, namely, that honoring the fundamental need for humility is a sine qua non of mature forms of Animal Studies.

Macroanimals provide the context of our day-to-day awareness because humans simply are often unaware of the bewildering array of unseen microanimals on, in, and around us (see chapter 1). The kind of organisms that humans can recognize in ordinary, daily life first came into existence about 650 million years ago, although it was only much later that the specific kinds of animals recognizable to modern people emerged. For example, high-quality scientific evidence suggests that the first mammals likely appeared somewhere in the range of 200 million years ago, but it was not until perhaps 50–55 million years ago that recognizable herbivores, carnivores (and the first doglike mammals), proto-whales, and bats emerged.

The estimates of our species' age (currently, a few million years) have been repeatedly revised for more than a hundred years. During our species' first several million years, our ancestors learned to survive amid and deal with other social mammals. Archaeological records, such as cave and rock art (see chapter 5), suggest that such macroanimals were frequently the focus of ancient humans.

The fact that humans developed wisdom about survival and community in relation to other social animals accounts for the fact that ancient, prehistoric humans had truly remarkable abilities to be interested in and react to their fellow macroanimals. Another reason is that humans, as but one of several hundred species in the large and eminently social primate group, are biological heirs to primates' special abilities in communication, intelligence, learning by imitation, and general conceptual abilities. This part of our evolutionary heritage is in the range of tens of millions of years old. Other special abilities, such as familial loyalty and caring, predate even our most remote primate ancestors because they are mammalian features developed in the even more remote past. We mobilize these intellectual and emotional abilities inherited from primates and mammals whenever we connect with other mammals such as dogs and horses.

It is also possible that some extraordinarily interesting features in our own lives—such as individualized self-awareness, membership in societies with developed material culture and learning traditions, the rich set of conceptual abilities that are possible with big primate brains—are also older than the human species. Of course, other ancient, perhaps prehuman abilities have made possible the nonmaterial cultural achievements of human societies, as well as the inclination to communicate that prompted the development of human languages and the highly specialized allegiances we know as communal identity.

Once human languages and specific cultural traditions were established, they were no doubt full of references to macroanimals. For this reason, it is likely that virtually every domain of human endeavor has always included recurring references to nonhuman animals—the myriad references one finds today in human arts and sciences are but the latest version of an awareness that stretches back scores of millennia.

Finding the Obvious History: Biological Connections

Given humans' full membership in the broad, diverse animal community, much of what is true of each nonhuman animal on earth is also true of each human animal. For example, each is the latest link in a distinctive chain of life that stretches many hundreds and more millennia, even billions of years, back into the remote past. Given the pace of evolution, only a few hundred generations back in one's own direct family line one can find ancestors that seem altogether different from modern humans. Go further, say, tens of thousands of generations, and the territory is startling. The biologist Richard Dawkins has suggested that all of the great apes (this category includes humans, chimpanzees, bonobos, gorillas, and orangutans) who have ever lived "are linked to one another by an unbroken chain of parent-child bonds."[9] Dawkins explains how immediate and close this connection is with a revealing image.

Imagine a contemporary human on a beach at the edge of a continent holding her mother's hand. Then picture the mother in turn holding the hand of her own mother. Then, with your imagination, extend this hand-holding chain additional generations into the past. When this imagined chain of ancestors reaches 300 miles in length (this takes 500,000 generations), it will have reached the point where the mother at the head of the line resembles the beings we call chimpanzees. Keep in mind that we started with humans, and that, of course, all the mothers and daughters are directly related to one another.

One can discover something fundamental about our remote ancestors by imagining a second chain: assume that the earliest female ancestor in this unbroken chain, whom we would recognize as chimpanzee-like, holds not only her first daughter's hand leading back to the human on the beach but also the hand of a second daughter. Then imagine this second daughter holding her own daughter's hand. Picture this new, second chain standing face-to-face with the individuals in the first chain. Extend the second chain in this fashion so that it parallels the first chain all the way back to the edge of the continent 300 miles away.

Assume, too, that all of the mothers and daughters in the second chain remain chimpanzee-like in appearance, such that the last daughter in this line, who stands opposite the human at the beach's edge, is a cousin who is a chimpanzee alive today. These two lines are composed of individuals who are related biologically—except for the ancient mother and her two daughters at the very beginning of the line, each individual in the two lines is in every instance a true cousin of the individual she faces in the other line.

Keep in mind, too, that the most ancient mother also has her own ancestors who eventually, more and more remotely in time, become altogether unlike chimpanzees. Nonetheless, these, too, are the ancestors of the individuals in both chains.

These chains carry a simple, profoundly important message about both our kin and our history—humans are, relatively speaking, only the rather recent precincts of our ancestors. Each of us is a member of an unbroken chain, as it were, that extends (biologically and historically) back through time and species to, literally, millions of other animals of many different kinds.

The nonhuman animals in these chains are, literally, the extended family of the human standing on the beach. Importantly, this is true even though it is a much broader sense of "family" than we use today to signal but a few generations (perhaps three or four or five) of our ancestors. But we all know that "family" as a biological concept goes much further back into the mists of time.

The individuals in these lines have significance for another reason as well. As this chapter and later ones attest, they are macroanimals who are possible members of our shared community. Our biological connections and our communal possibilities help explain why the primatologist Roger Fouts titled his 1997 book *Next of Kin*. "Kin" here can mean "biologically related," but it can and does also mean "member of my day-to-day family." This is why so many people in industrialized countries refer to their dogs and cats as family members.[10] This now common use of the word "family" reaches far beyond biological parents, brothers, sisters, and immediate cousins. It means more than "lives in the same familial household" because it has an emotional dimension anchored in a shared history of affiliation, loyalty, and, many humans would contend, even love. This is why Fouts observes at the beginning of his book, "This is Washoe's story. I tell it to repay a lifelong debt to her and all the other chimpanzees who have touched my heart and opened my mind."[11]

Ignoring Relatives, Narrowing the Truth

These reflections should make it clear that all Animal Studies must contend with a particularly narrow sense of "history"—the truncated version of the whole story expressed in exclusively human-centered terms. The operative assumption in such versions is that only humans have a history. This narrow story has often been given a positive spin, as the Roman philosopher Marcus Tullius Cicero did: "Who does not know history's first law to be that an author must not dare to tell anything but the truth? And its second that he must make bold to tell the whole truth? That there must be no suggestion of partiality anywhere in his writings?"[12]

The nineteenth-century historian Johan Gustav Droysen made an even more optimistic claim about history, relying on the influential ancient Greek philosophical exhortation "know thyself" chiseled into the wall of the front porch of Apollo's temple at Delphi: "History is the 'know thyself' of humanity—the self-consciousness of mankind."[13] Notice here the exclusively human overtones.

There are, to be sure, less positive views of history. Many people, including historians themselves, have recognized that accounts of the past have often been manufactured in a variety of ways—an example is the common suggestion that the victors write history. In spite of the partial nature of any history written by victors alone, in educational settings the traditional habit has long been, and often still is, to teach history by focusing primarily on "great men and great deeds."

Nonetheless, many humans easily recognize on their own that what passed for history in their education was, in both substance and effect, far narrower than a full, accurate account of what is actually known and knowable of our own species' history. Some historians today make an effort to tell "history from below," that is, from some vantage point other than that of victors, rulers, and other elite segments of human societies. Such inclusive accounts of human history reveal that humans have long led lives intertwined with local nonhuman animals—we examine this larger human story below, but chapter 10 works with history from below to engage the natural question of whether any nonhuman individuals have a history in any meaningful sense.

Earliest Records

Contemporary sources speak of cave paintings, which are thought to be the earliest records of humans, as being 30,000–40,000 years old. Cave paintings in Chauvet, France, which may be the oldest reliably dated ones, feature as their principal theme hundreds of exceptional quality images of nonhuman animals.

The same could be said of other cave paintings from Africa, Australia, the Americas, and other places around the world. Because some of these images were executed with materials that cannot be dated, claims about age are always subject to dispute. The simple facts remain, though—records of human art are available from tens of thousands of years ago, and they very often include nonhuman animal images. Such facts invite anyone interested in humans' relationship to other-than-human animals to reflect on a question that is peculiarly within the province of Animal Studies—why did ancient humans so often picture other-than-human animals?

Another ancient category of human art is petroglyphs (from the Greek *petros*, stone, and *glyphein*, to carve). Found around the world, the oldest petroglyphs are often said to be 10,000–12,000-plus years old. As early as 7,000–9,000 years ago, pictographs and ideograms (precursors of writing) began to appear, but the earliest written human records come from much later—perhaps somewhere around the year 3200 BCE. The emergence of written records is commonly used to mark the end of humans' prehistory, although it is important to keep in mind that this epoch-marking change took place at many different times around the world. Ancient records from both Sumer and Egypt generally date around 3000 BCE, whereas in other places the dates of the first known writing are far later (China, India, Greece, and Israel, for example, date between 1000 and 2000 BCE). Human records of local history obviously began far later for many other societies and cultures.[14]

The Need for Many Disciplines

The scholarship needed to piece together the puzzle of ancient humans' connections to other-than-human animals is drawn from astonishingly diverse fields. Not only were these connections universal and diverse, but they were also invariably imbued with a religious dimension and thoroughly saturated with ethical concern even when the humans were hunting. Evidence of the recurring connections to and concerns with nonhuman animals appear in far more than rock art and written records—they are revealed in burial records and various arts such as music and dance (see chapter 5).

Because exploration of this complex intersection involves studying peoples all over the world as they interact with local nonhuman communities, it takes the work and perspectives of many researchers and scholars to piece the larger story together. One can draw from comparative religion scholars, such as the modern pioneer Mircea Eliade or the respected contemporary scholar Wendy Doniger, art historians such as the eminent H. W. Janson and Kenneth Clark, and veterinarians interested in interdisciplinary work such as Elizabeth Atwood Lawrence, who also had a doctorate in anthropology.[15] In addition, there are today many comparative literature scholars, specialists in single cultural traditions, or comparativists who combine fields such as ethnology, psychology, and archaeology to study the most ancient of human traditions.

The need for such an intensely interdisciplinary approach is common in Animal Studies. Further, because our ability to identify "the earliest recorded history" continues to change from decade to decade (see chapter 9), in a very important sense our take on human history is constantly changing. The task of Animal Studies here is not, of course, to hone specific techniques by which to answer with certainty these issues—rather, that is the task of other fields that can deploy highly specialized expertise and methods. Animal Studies, however, clearly relies on, and thus needs to stay cognizant of, the great wealth of information that has been responsibly developed by a wide variety of sciences and humanities regarding our most ancient ancestors and their diverse interactions with their local worlds.

The First Stories and Myths

Although "myth" is a controversial word in some modern circles, "often tossed around as a casual (but intentional) dismissal of the 'emotional' or 'irrational' views that other people hold,"[16] it is a foundational notion in many crucial disciplines that have made seminal contributions to Animal Studies. Paul Shepard, one of the great minds of the modern reengagement with nonhuman animals, described myth as part of everyone's life:

> Great naturalists and primal peoples were motivated not by the ideal of untouchability but by a cautious willingness to consume and be consumed, both literally and in a mythic sense. *Everyone lives in a mythic world, however ignorant of it they may be.* The most revealing source of information about how people conceive of themselves in relation to the nonhuman world is myth....All myths operate on three levels: one deeply personal, concerning an inner, unconscious life; another the social and ecological milieu; and third, the society of spiritual and eternal things in tales of creation.[17]

"Myth" is, then, a valuable tool when used as a synonym for "a powerful story" of the kind valued by a community whose members take the myth as a way of speaking about the group's identity or explaining some feature of how the world works or came into being. As the Canadian scholar Robertson Davies suggested, "History and myth are two aspects of a kind of grand pattern in human destiny: history is the mass of observable or recorded fact, but myth is the abstract or essence of it."[18] In this sense, everyone has myths, for such stories orient and integrate people into the community—in effect, telling these stories creates and maintains community.

If one analyzes some of the "humans-only" versions of history now prevailing in so many contemporary circles, it is possible to see these accounts as having some features of myth. Yet in fact these versions of history fail to explain any feature of how the real world works. Further, while "humans-only" myths may confer identity on some groups, they are versions of history that risk the dysfunctions promoted by the exceptionalist tradition (see chapter 1).

We recognize easily, of course, that while some myths are positive, others can have negative effects—the modern world is only too familiar with examples of myths of racial, national, or gender superiority. Knowing how a community's stories—whether in the form of myth or a conventional historical account—impact people is important, especially because a community's stories give its humans some sense of possibilities. Of great importance in Animal Studies is that a community's collection of myths and other stories always includes fundamental notions about other local animals, whether they are humans or not.

Human groups have often couched their claims about the world's living beings in story form—for example, in religious texts we can find views of other living beings (see chapter 7).

Some of these views hold, upon close examination, important fact-based information. But such examination also reveals that texts held to be "revelation" can include fanciful elaborations. Some also include dismissals and exclusions of certain other living beings, human and nonhuman alike.

Myths can work as psychologically significant exclusions, educating those invested in the myths to ignore and even actively exclude certain living beings. Similarly, narrow versions of history that talk only about humans can, like some myths, cause people to miss altogether noteworthy complexities of the world we inhabit. Just as clearly, however, myths that affirm can educate in ways that open minds and hearts. They can supply profound insights on matters such as the limits on human analytical abilities or common flaws in our character. In important ways, the very universality of myths, as well as the prevalence of stories about other animals, suggests how our human minds and hearts contain much more than conceptual and calculating abilities. To be sure, our analytical abilities on their own can be powerful for some purposes, but the persistence of myth testifies in countless ways to how multivalent humans' meaning-making minds and hearts are, as when myths deal with origins, kindredness, and other meaningful connections. In this sense, myths testify to how we make meaning—each human group clearly lives amid an immediate randomness that engulfs us as a form of animal life, and yet we clearly want to share with each other something satisfying about our origins, larger community, and penchant for meaning and significance. Myths offer a kind of order, even though they do not exhaustively describe all of the realities that we experience. They give us a story about why this world is as it is, even as it continues to challenge and baffle.

From another angle, of course, the prevalence and universality of these powerful stories testify to some of the limits on our thinking. The extraordinary roles that myths have played in our ancient past, our more recent history, and our present-day framing of the world suggest indirectly that our important and powerful analytical abilities are bounded, limited, and even fragile. As individuals we often experience the obvious limits of individuality, but our social dimensions regularly cause whole groups to struggle with the humility of the limits of our intelligence, wisdom, and place in the world. In our social dimensions—in groups, organizations, nations, cultures, and religious traditions—the limitations that individuals know so well are often ignored in tragic ways. So while we know that, at times, our individual limitations are surmounted by our social ability to work together, nonetheless social processes create their own new, debilitating uncertainties that impact our "knowledge" claims dramatically (see chapter 8).

Animal Studies and Myths

Because myths, for all these reasons, have great psychological power as they provide a group and its individuals with a commanding sense of the surrounding world and our possibilities in it, one of the key challenges of Animal Studies is to point out how some myths work well while others simply fail us when it comes to illuminating the inevitable human-nonhuman intersection. For example, even as specific myths function to help groups make sense of certain features of their local world, such accounts can be manipulated. As noted above, dysfunctional myths—such as those suggesting that women were made for men, that people of one race are naturally the slaves of another race, that humans alone are intelligent, or that all nonhumans were designed for the benefit of humans—may create cohesion for a group or justify an existing practice that harms the subordinated group. But even as they function in this way, they fall short if measured against other myths that enable more realistic engagement with the empirically verifiable facts of the surrounding more-than-human world.

Much is at stake, to be sure, in the evaluation of a myth that is, on the one hand, psychologically significant for a group's identify even as it is, on the other hand, dysfunctional on some other key feature of life. Some myths distort the realities of some other living beings, whether human or nonhuman, and some play into certain cultures' insistence on human domination and a related refusal to consider the whole story of the breadth and depth of human-nonhuman connections.

It will surprise no one that there are many disputes over how such myths function and mislead. Animal Studies is peculiarly well suited to noticing how often myth-framed accounts of the world employ animal images and how such accounts fairly reflect or unfairly distort the biological beings who, apart from the myth, inhabit the same ecosystems as do the mythmakers.

Finding the functioning, healthy side of certain myths even as we call out the dysfunctional features of other myths is, then, no simple challenge, especially with regard to those myths that each of us hold and live by. But such analysis and description of the roles myths play in developing or retarding basic ideas about other animals are an important collective function that Animal Studies is capable of carrying out through its myriad connections with many other disciplines.

Myths and Real Animals

Many myths show that certain biological animals themselves (that is, not just their images) have had a major presence in human cultures. A revealing set of comments by the scholar Karen Armstrong reflects the ancient nature of humans' recognition that real animals have a major role in our choices, actions, and cherished notions of identity and ultimate reality. Armstrong underscores that pivotal developments in our religious past and its concern for true, ultimate reality serve to focus each believer on the local world. The "Axial Age" referred to by Armstrong has been given various dates by different scholars, but generally is said to have occurred during the six or seven centuries in the middle of the first millennium BCE.

> The [Axial Age] sages certainly did not seek to impose their own view of this ultimate reality on other people.... What mattered most was not what you believed but how you behaved. Religion was about doing things that changed you at a profound level.... *First you must commit yourself to the ethical life*; then disciplined and habitual benevolence, not metaphysical conviction, would give you intimations of the transcendence you sought.... Your concern must somehow extend to the entire world.... Each tradition developed its own formulation of the Golden Rule: do not do to others what you would not have done to you. As far as the Axial sages were concerned, *respect for the sacred rights of all beings*—not orthodox belief—was religion.[19]

The religious traditions to which Armstrong refers—those of ancient India, the Hebrew Bible prophets, ancient China, and the earliest Greek religious thinkers—are the foundation of religious traditions that today are followed by well over half the human race. The fact that these traditions share a breakthrough insight about the importance of all living beings—namely, true spirituality is constituted not by orthodox belief but instead by respect for the sacred rights of all beings—is important to Animal Studies for any number of reasons. Religious traditions that honor this insight attain an ethics-intensive form that focuses on empathy and compassion, neither of which can be, according to Axial Age religious sages, confined to one's own people or species.

Taking Responsibility for How We Have Recently Told This Story

More recent evaluations regarding the moral dimensions of acting responsibly toward other living beings are, ironically, narrower. This provides yet another reason why an essential task for Animal Studies is development of a realistic account of humans' long and diverse interactions with other animals.

Many modern histories of animal protection tend to focus on efforts in the nineteenth and twentieth centuries in Europe and North America, thereby treating concern for other-than-human animals as a recent, often merely secular phenomenon. In this, they reflect only the recent ferment, not the fuller, more deeply culturally rooted, millennia-long human story from which the ferment draws so much sustenance. The associations called to mind by the titles of three respected historical accounts of the animal protection movement in one country (the United States) are freighted with overtones of morality and spirituality:

> *The Animal Rights Crusade: The Growth of a Moral Protest*
> *The Animal Rights Movement in America: From Compassion to Respect*
> *For the Prevention of Cruelty: The History and Legacy of Animal Rights*
> *Activism in the United States*[20]

The invocation of "crusade," originally a Latin-based word tied to the religious notion of marking with a cross (*crux*), "moral protest" and "compassion" (both with long religious histories), and "respect" and "prevention of cruelty" (the latter an unalloyed evil condemned in no uncertain terms by every human cultural tradition) reveals how much the authors think is at stake and how concerns for other living beings parallel concerns that have long been deemed spiritual and ethical.

Because most past accounts regarding our dealings with other living beings are truncated, even biased versions told from clearly human-centered vantage points (the animals protected are often only those treasured by an upper class, or the stories involve one segment of society trying to control another), providing the entire story continues to be a profound challenge. Today, that task falls not only to the specific disciplines involved in the study of nonhuman animal issues but to Animal Studies generally.

History from Below

In the last century, the move to tell history from the vantage point of ordinary people opened many people's eyes to the tremendous bias in our received historical accounts. The early twentieth-century historian Marc Bloch is sometimes credited with pushing the field of history to tell a fuller, history-from-below story. Such an approach was taken by Howard Zinn when he published the widely read *A People's History of the United States: 1492–Present* in 1980. In this book, Zinn frames the story in terms of people other than the victors, rulers, and social elite who dominate popular but superficial textbooks. Zinn's work was made possible by pioneer researchers who sought out information from those who gave attention to marginalized and forgotten human groups—a good example is a 1925 article in the then-popular magazine *Survey Graphic*, "The Negro Digs Up His Past." In this article, Arturo Alfonso Schomburg gave details refuting a claim made by one of his elementary school teachers that "black people have no history."

Zinn's widely read and admired book, which expressly reiterated Albert Camus's observation that the job of thinking people is not to be on the side of executioners (invoked by Zinn in his chapter "Columbus, the Indians and Human Progress"), has prompted many other historians to provide astonishingly rich accounts of ordinary people's struggles of all kinds. One of the many implications of history from below is that prior histories are severely deficient. Importantly, the problems caused by such narrowness are significant—telling one-dimensional accounts of human history does more than cause groups to disappear from history. It also contributes to the marginalization of people in contemporary society, which is why Schomburg worked to refute the ignorance-driven claim that "black people have no history." Women, disenfranchised males, the politically oppressed, the poor, nonconformists, and "the subaltern" (the Marxist Gramsci's term for those humans who are socially, politically, and otherwise excluded by an oppressive power structure) are no longer remembered.

Education premised on narrow-minded, exclusivist versions of history has great impacts, too, for such education exacerbates political disadvantages—those learning narrow versions of history are not alerted to the full range of problems and suffering confronting a society. Even when marginalized groups are discussed, "those people" are often so marginalized as to be effectively erased. Research about them is disfavored or worse; caricatures of the marginalized prevail because they are not mentioned in research topic lists, indexes, style manuals, and the like. They are not well described and thus rarely appear in book titles, dissertations, or articles in scholarly journals. New academic positions and tenure possibilities favor those who study canonical problems and peoples. Programs end up revolving around only mainline issues and recognized peoples.

The upshot is that lack of awareness of those who have been marginalized is a manufactured problem, yet another form of self-inflicted ignorance. The limited information that is available often is mere caricature, misleading in other ways, or outright false. Even the descendants of marginalized people often do not know them.

The result is obvious—human history about some humans is incomplete—and hence the need for history from below so that the whole human story can be engaged. Just as the tradition of narrow, human-centered history reveals that important chapters of human history remain to be written, it also reveals that other parts of history, such as our relationship with beings deemed inferior to even the most marginalized of humans, have been radically ignored as well.

Today as History

Beyond the themes of history from below, there remains the obviously important theme of what humans are doing today to nonhuman animals. A wide range of actions—from consumer choices to business practices to religious rituals to educational emphases—are impacting an astonishing number of other-than-human animals today. Even as the ferment in contemporary human-nonhuman connections reflects the growth of protection-oriented sentiments in many human cultures and subcultures, today's politics, public policy, and law also shape the world in which our children will live. Thus, just as our historical past awaits sensitive, respectful examination on the issue of human societies' engagement with the more-than-human world, so too is such an examination needed for today's contemporary practices and cultural diversity.

Through a Second Door: Depth and Breadth in Cultural Diversity

Once one learns that the discipline of history has often supplied a decidedly narrow version of humans' own story regarding both human and nonhuman animals and that such limitations continue to prevail in many important circles, it is worth assessing whether other narrownesses also hold sway. Consider one such narrowness that long prevailed in spite of the fact that other cultures produced great variety in the ways humans thought about and related to other living beings. Anthropology, the discipline classically associated with the study of cultures which today offers astonishingly rich contributions to Animal Studies, in its early phases was extremely dismissive of non-European attitudes toward lives outside the human species (see chapter 9). This was the case in part because many other cultures exhibited different attitudes toward nonhuman animals and thus did not promote the forms of human-centeredness that early European anthropologists felt essential to "civilization."

Because dismissive attitudes regarding many non-European cultures and religious traditions have long been influential, some people around the world remain unaware of the daunting diversities that quickly become apparent when one looks at "the animal issue" across many different cultures. Yet given the combination of ubiquity and diversity of other life forms, it is not surprising that human curiosity has made a marked fascination with some other-than-human living beings a cultural universal. This fact is often obscured altogether or distorted in circles where thinking is not only constricted by traditional human-centerednesses but actually impoverished because such thinking misses the diversity of views toward nonhuman animals in a wide range of human cultures. This diversity suggests something extraordinarily positive—we are, as a species, culturally rich beyond any one individual's imagination when it comes to stories about and perspectives on other living beings.

This diversity has multiple implications that Animal Studies necessarily explores. At the very least, the great diversity suggests that humans' abilities to care are broad, our imaginations are fecund, and both of these abilities are central to human life. Further, the diversity itself begs questions about the nature and limits of any one culture's claims to have definitive, total "knowledge" about certain animals or the entire collection of life beyond our species line, because different societies around the globe experience different segments of the other-than-human citizens of this far-more-than-human world.

This fact, in turn, suggests that a certain humility is needed, for the great differences in views about other living beings hint at the nature and limits of any single culture's claims about any and all nonhuman animals. Our species' cultural diversity, then, provides extraordinary opportunities to study the phenomena of how humans learn and treat other living beings.

As Animal Studies explores diverse views found in different human societies, it must contend with some baffling features or tensions in group approaches to the more-than-human world. Even though many humans have a deep fascination with other living beings, they nonetheless also display a recurring willingness to adhere to majority views even when such views are demonstrably inaccurate. For this reason, the important problems of group-level knowledge being distorted by social factors are addressed in much greater detail in chapter 8.

The Risks of Generalizations

Diversity lures description, which in turn employs generalization to frame the underlying subject matter's complexities. Cultural views of other animals reflect two different but intersecting diversities. First, human cultures are, as already noted, often quite distinct from one another. Second, the diversity of life itself has the greatest importance in animal studies. The variety among the individuals and communities that make up the millions of different species on earth is so unfathomably vast, regularly featuring extraordinary complexities and mysteries, that without question this second diversity is orders of magnitude more complex than human cultural diversity.

Animal Studies must assess the strengths and shortcomings of the generalizations used to make these diversities comprehensible to human minds. For example, there are obvious risks when one generalizes about a subject with broad statements that are underdetermined by actual facts—generalizations that purport to describe our many human cultures, then, can sometimes fail to represent well special or unique features of views and values that play crucial roles in a specific culture, subculture, or community.

A parallel problem exists for those who want to talk about other-than-human animals. Here the risks are great because, as noted throughout this book, generalizations underdetermined by discernible facts have often prevailed such that incredibly diverse nonhuman animals are often lumped into an amorphous category that ignores fine-grained distinctions based on empirical evidence. This is the problem that bedevils the phrase "humans and animals." This dualism and the many flawed habits of speech, thinking, and generalization that it supports are the principal form of generalizing about other living beings in many circles. Animal Studies must work hard to catalog the problems and misunderstanding that this habit has created. For example, in succeeding chapters some of the complications created by this generalization in central human endeavors—politics, science, ethics, religion, language, and so much more—are unpacked. Similarly, as noted below, they have profoundly negative impacts on educational content and techniques in modern communities.

Animal Studies itself involves broad generalizations, of course—such as the notion of "animals." Staying aware of the implications of employing this and any other generalization is important—for example, it helps to recognize that the border between the living beings we call "plants" and those we call "animals" is not nearly as precise as the phrase "plants and animals" might imply. There are, in fact, living beings that straddle this border in a number of ways. Staying cognizant of the risks of using generalizations is an important part of the task of critical thinking described below.

Of particular relevance to industrialized societies, however, is the set of problems created by the dominant generalizations that create barriers of many kinds—political, legal, ethical, theological, scientific—between humans and all other animals. Thus, even as students may at one turn be exhorted to "seek the truth," they are trained to use dualisms such as "humans and animals" that are based not on careful, open-minded assessment of facts but instead on preexisting ideology and bias that preclude any possibility of more fact-sensitive accounts.

Frankness about Dismissals of Some Cultures

Among the most unconscionably inaccurate and unfair generalizations are those regarding small-scale cultures generally. The following passage reveals how arrogance was once part of many European and American scholars' interpretation of the views of nonhuman animals found in small-scale societies around the world:

> Civilization, or perhaps education, has brought with it a sense of the great gulf that exists between man and the lower animals. . . . In the lower stages of culture, whether they be found in races which are, as a whole, below the European level, or in the uncultured portion of civilized communities, the distinction between men and animals is not adequately, if at all, recognized. . . . The savage . . . attributes to the animal a vastly more complex set of thoughts and feelings, and a much greater range of knowledge and power, than it actually possesses. . . . It is therefore small wonder that his attitude towards the animal creation is one of reverence rather than superiority.[21]

In this single passage, both nonhumans and humans ("the savage," that is, those "below the European level") are deprecated. Animal Studies can reveal the arrogance of this passage because of its basic commitments to seeing the realities of both nonhuman and human animals.

The passage also reveals other risks—for example, those who assume without careful argument that their own "common sense" should control the thinking of all humans (as the author of this passage does regarding "the distinction between men and animals") fail to appreciate the complexity, variability, and frailty of human thinking. The author's views also lack the humility needed to describe nonhuman others. In effect, because this author mixes two broad and enormously complex generalizations about, first, cultural views and, second, nonhuman animals, the passage offers a paradigmatic example of the risks of superficial thinking—it misses both the sheer number of different views found in cultures regarding who or what "an animal" is and the complexity and diversity of many creatures found in the more-than-human world we inhabit.[22] If, however, one takes into account both the complexity and diversity of the nonhuman world and the remarkably fecund abilities of human groups to create unique cultural views, then it is hard to subscribe to such broadly dismissive generalizations unless one wishes to promote extreme bias and ignorance.

A second example, involving opinions about wolves, reveals that it is often the allegedly careful or clear thinkers who pass along views that are radically underdetermined by other animals' (human or nonhuman) actual realities. There is a remarkable shortfall between research on the daily lives of wolves and the image of wolves that dominates certain circles. The English philosopher Mary Midgley, after reviewing the current state of knowledge about the lives of real wolves, suggested how remarkably different the actual biological beings are from the image of a wolf "as he appears to the shepherd at the moment of seizing a lamb from the fold."[23] Midgley noted that those who have taken the time to watch wolves "have found them to be, by human standard, paragons of steadiness and good conduct." Summarizing, she takes philosophers in particular to task for their misleading use of inherited caricatures: "Actual wolves, then, are not much like the folk-figure of the wolf, and the same is true for apes and other creatures. But it is the folk-figure that has been popular with philosophers. They have usually taken over the popular notion of lawless cruelty which underlies such terms as 'brutal,' 'bestial,' 'beastly,' 'animal desires,' and so on, and have used it uncriticized, as a contrast to illuminate the nature of man."

What makes use of inherited caricatures ironic is that philosophers are usually held up as seeing "the truth," or at least carefully analyzing our claims to have "knowledge." But in this matter, Midgley suggests the real result has been a perpetuation of what is no less than self-inflicted ignorance.

As suggested by the passages above, the tendency to caricature nonhuman animals has become a prominent feature of some of the most influential human cultural traditions in today's world. This tendency is particularly evident in certain humans' steadfast refusal to employ constructively those very abilities that distinguish humans as excellent discoverers of the realities surrounding us.

Risk upon Risk: Key Opportunities in Realism about Generalizations

Animal Studies goes forward, then, in a context in which discussing the views of one dis-favored group (other animals) as they have been developed by another disfavored group (marginalized cultures) can involve piling one set of risks upon another. But there are a number of great advantages to studying the diversity of views of other animals found in small-scale cultures and religious subtraditions around the world. These worldviews reveal, for example, astonishingly diverse thinking about other-than-human animals. Through studying a range of different views, then, it is possible to recognize that while any single group might have been fascinated with only a few dozen different kinds of animals, the human species as a whole has found literally thousands of nonhuman animal species to be of great interest. These are but a few of the many reasons it is crucial that Animal Studies engage such diversity (indeed, diversity upon diversity) with as much responsibility as possible.

A number of tools can facilitate attempts to learn views of other animals developed in a wide range of cultures. Humility about one's own preconceptions is indispensable, as is realism about one's experiences in day-to-day life. A willingness to explore artistic sensibilities in other cultures helps, too, for literature, dance, and other arts can provide important insights when one's goal is to use limited human abilities to identify the actual realities of nonhuman animals (see chapter 5).

Animal Studies must engage the fact that long-standing biases and practices, official policy, and even educational institutions' infrastructure can become impediments to open inquiry because they are so often invested with the force of tradition and authority. This explains continuing resistance to and skepticism about inquiries at the heart of Animal Studies, even as it also suggests the possibility that we may need to unlearn many traditional claims that anchor wide-ranging dismissals of nonhuman animals.

Good and Bad Generalizations

Those who pursue Animal Studies, then, must strike a balance, underscoring that some generalizations can be immensely helpful even as others are misleading and harmful. Some generalizations about other animals, such as those of the scientific classification schemes known generally as taxonomy or more technically as "systematics," supply extremely valuable insights about many forms of life. Helpful insights can arise from generalizations across cultures about what appear to be common phenomena, even cultural universals, like fascination with at least some forms of local life. Many cultures have developed what can only be called respect for certain other-than-human forms of life.

Similarly, it is valuable to notice—and even generalize about—the fact that humans have often engaged intelligence and other complexities evident in the lives of certain other-than-human animals. Many peoples around the world have used common sense again and again to notice certain behavioral or physical traits of animal lives that fit into our modern category "natural history."

But realism cannot stop at the positive value of some generalizations. Many of the most common generalizations about other-than-human animals are clearly misleading or clumsy. In the spirit of an influential twentieth-century definition of philosophy as "a battle against the bewitchment of our intelligence by means of language," Animal Studies needs to un-bewitch language about nonhuman animals as much as possible and, instead, apply a variety of critical thinking skills such as those discussed at the end of this chapter.[24]

The Recurring Heartbeat: Noticing, Taking Seriously, Caring

Because humans are, as chapter 8 reveals, prone to rationalize their failures to engage the realities of other animals and, *at the same time*, fully capable of caring beyond the species line, Animal Studies deals with some tensions in humans' approach to the human-nonhuman intersection. Even as some people and our sciences insist that we seek out and explore other living beings' realities, others attempt to control, even shut off, inquiries that open up what we can know and do in relationship to other living beings. Sometimes simple things, such as economic considerations rule, for, as Upton Sinclair observed, "It is difficult to get a man to understand something when his salary depends upon his not understanding it."[25] There are, to be sure, many other reasons that individuals in industrialized societies today fail to notice or take seriously nonhuman animals' abilities or moral significance—some of these are anchored in tradition, while others may be the function of human limits or simple selfishness.

One can, however, find in *any* culture a variety of indications that *some* humans in their day-to-day lives do in fact recognize the frailty of inherited views and the importance of observation, inquiry, and engagement in one's local world. One can also recognize that individuals become aware of the suffering caused by uncritical uses of inherited biases, short-sightedness, selfishness, and various myopias that impact human claims about other living beings. Thus, even though public discourse in modern societies as a whole may often promote views that are demonstrably wrong or otherwise radically dismissive of all nonhuman animals, many individuals nonetheless take responsibility for their own views and thereby choose to foreground compassion, humility, and curiosity (chapter 3). That some individuals are led in a variety of ways to encounter, then engage and care about, other macroanimals is a simple, recurring phenomenon and one of the key drivers or heartbeats of contemporary Animal Studies.

Through a Third Door: Education and How We Learn about Other Animals

At the education door, one encounters astounding variety in approaches to both teaching and learning about the more-than-human world. What is most relevant to animal studies as it explores our species' past, present, and future intersection with other-than-human animals is that some kinds of education on these issues operate in tension with one another. Some forms of education beckon us to learn openly about other animals, while others close the process down because they are not invested in avoiding self-inflicted ignorance regarding other beings. Thus, if the suggestion that "humans are born ignorant, but made stupid by education" is true in any domain, it is true about the formal education that many modern humans receive regarding other-than-human animals. As Andrew Knoll, a professor of natural history at Harvard's Earth and Planetary Sciences Department, has suggested, "The average adult American today knows less about biology than the average ten-year-old living in the Amazon, or than the average American of two hundred years ago."[26]

The issue can be framed, then, as good and bad education—some formal education is both fact intensive and imaginative, and thereby enabling. But of course one may be tempted to frame the issue as "formal education versus other, informal, more effective ways of learning about other animals." "Versus" here is meant to be provocative, that is, to call forth a fundamental tension—some views seem superficially reasonable because they are widely held, but when subjected to critical thinking they fail because they are as overdetermined by group dynamics as they are underdetermined by facts.

Such is often the case with views of the natural world passed along in the education systems of industrialized societies. Andrew Knoll also observed, "Through the fruits of science, ironically enough, we've managed to insulate people from the need to know about science and nature."[27] Decades ago the English commentator C. P. Snow said, "Technology...is a queer thing. It brings you great gifts with one hand, and it stabs you in the back with the other."[28]

Obviously, formal education is important, and thus it needs to function well if industrialized societies are to responsibly identify and address the problems they cause and otherwise face. As the US Supreme Court observed in its famous 1954 decision *Brown v Board of Education*, "Today, education is perhaps the most important function of state and local governments."[29] Government involvement has a variety of implications for solving human problems, of course, but when it comes to the impacts of human actions beyond the species line, government-based education is often dominated by agendas that make such "education" dramatically inadequate.

This is not surprising from certain vantage points—education promoted by societies that have pulled away from nature can be extraordinarily harsh on many different living beings. Historically, for example, government-run education has often implicated political elites in demeaning people or subcultures whose views are different from those of the elite. Two scholars in a 1965 paper titled "American Indian Education for What?" observed that "many western reformers have viewed formal education as a benevolent instrument of social change and social uplift—the principal and ideal technique for developing the underdeveloped." But social scientists, these scholars suggest, "when functioning as scientists trying to discover why education fails to move some [indigenous] peoples...have stumbled upon the fact that education does not look the same from the bottom as from the top."[30]

Education fails when it is seen primarily as an abstract, culture-inculcating activity that prompts students and even members of another culture to conform to an ideology—this is, in fact, what passes as "education" about the proprieties and dismissals that undergird the "humans and animals" approach to living beings.

Considering Informal Education

Contemporary English employs two important but distinguishable senses of the word "educate." In ordinary conversation, we say that someone becomes "educated" about other-than-human animals through day-to-day encounters. We also speak of people as "educated" because they have spent time in, perhaps graduated from, formal educational institutions. Distinguishing these two types of education on animal issues helps one see that there are countless ways in which each of us learns about animals (this is equally true of our education about human animals).

From birth onward, a variety of informal educational processes, such as learning how our family and community talk about living beings, are of paramount importance for any child. As very young children, we are often told stories about and shown images of certain animals, or hear adults or older children talking about other-than-human beings. We also experience other living beings, of course, and in this sense perhaps "educate ourselves," but more commonly our views of other living beings are deeply impacted by what we are led by other humans to expect. Through such processes and eventually our own exploration of the different ways in which ideas and images are conveyed to us, we become aware of constructed images of a number of the living beings outside our own species. We learn language-based traditions eventually, becoming users of phrases such as "humans and animals" that teach us to divide the world up into categories. In the end, we appropriate a series of fundamental notions regarding living beings.

Informal learning expands into awareness of the local world's shape as we eat, play, travel, and grow into our family and respond to the psychological dimensions of group life. Pioneering researchers in cognitive psychology have discovered that children develop early in their lives a "naive biology," a core domain of knowledge about living things. Its first glimmers are discernible in infancy, and by the preschool years, far earlier than Piaget thought, this knowledge base, particularly about animals, already is well established.[31]

So one answer to questions such as "How do we really learn about other animals?" is, "in many different ways." In fact, what industrialized societies think of as formal education relies in countless ways on many of the processes that comprise what here is called informal education. An implication of this fact is that, even when formal education takes human-centered forms that create imbalances and lack of realism about other animals, many humans nonetheless learn in other ways and contexts about the more-than-human world. Even in the face of the exceptionalist tradition advanced by institutions, then, individuals' awareness, curiosity, compassion, and connection to other-than-human animals may continue to develop.

Whatever conclusion one reaches, then, regarding the proper focus of institutionalized, formal education, there are other, informal means of education with important consequences for nonhumans, ranging across a broad continuum from highly beneficial to extremely harmful. This book argues that today's education about nonhuman animals is beset by a combination of, on the one hand, institutional failures in formal education and, on the other hand, failures of the human spirit reflected in informal education mechanisms dominated by dysfunctional human-centerednesses. This combination produces, first, devastating impacts on other-than-human animals and, second, an uninformed public. As the educator David Orr has suggested, "The truth is that without significant precautions, education can equip people merely to be more effective vandals of the earth."[32]

These failures, in turn, have had profound implications for all forms of life, whether human or otherwise. Suffice it to say, then, that education is critical because when this "most important function of state and local governments" goes wrong, it causes very severe problems. These problems, in turn, explain why a noticeable heartbeat of Animal Studies is a very specific question—Which forms of education help us notice and take seriously actual biological creatures and their realities? This question has power in part because other animals' realities provide crucial truths about the nonhumans who populate our shared world.

Yet the question also has power that goes beyond good answers of this kind, for it has the special power possessed by any good question which opens minds to complex realities. This is the power to open up hearers to unimagined problems and possibilities. The more specific question "Which forms of formal education help us notice and take seriously the actual biological creatures?" opens the listener to the idea that the values driving formal education may be other than a search for truth. Having heard the question, the listener can decide whether the views of nonhuman animals at issue are driven by the animals' actual realities or some other values, such as a tradition-anchored ignorance or apathy, a self-interested motive like making money, or an ideological bias such as those that characterize the exceptionalist tradition.

For some people, of course, education, like history, is only rightly focused when its dominant theme is some form of human-centeredness. When such people dominate education, questions about which forms of education help us find the truth or think better can create political risks for those who explicitly or even implicitly question refusals to notice other animals' discernible realities.

Rarely a Place of Daring? Animal Studies as Response

Narrowness in formal education is nothing new, to be sure. Thus, while virtually everyone is enamored of the possibilities of education, many prominent voices have, like Helvetius, been critical of what has actually passed as education. Further, even though many people think of formal education as a domain of liberal thinking and values, some have suggested other problems. The social critic Theodore Roszak, for example, once observed, "Let us admit that the academy has very rarely been a place of daring."[33] Such concerns abound in the present era, when many books published every year lament the state of education in contemporary society.[34]

Animal Studies has developed as a response to the inadequacies of both formal and informal education. It can be seen as a true example of daring in an educational system, for it is attempting to go forward in an environment that is overwhelmingly focused on human issues. It can hardly be denied that much of the academic world has long been dismissive of the idea that protecting nonhuman animals is a social issue that deserves to be mentioned in the same breath (or classroom) with protecting humans. Thus Animal Studies must regularly confront the limits and foibles of modern education precisely because it is dominated by traditions and people so used to human-centered ideals that they assume an exclusive focus on humans to be natural, just, and without any moral dimensions.

If, however, the prevailing forms of human-centeredness are seriously dysfunctional, then Animal Studies can offer balance. Further, even educators who subscribe consciously or unconsciously to the exceptionalist tradition might consider the value of Animal Studies if corresponding benefits for humans can be identified. Above all, if Animal Studies can foster key skills, like critical thinking and a maturation of humans' moral possibilities, then remedying a wide range of present dysfunctions may well be possible and thereby benefit the human community.

Parsing All of Education

The well-known division in "higher education" between "the humanities" (sometimes "humanities and the arts"), on the one hand, and "the sciences," on the other, suggests how diverse human inquiries about the world can be. In the humanities, students explore in ways that are overtly qualitative, conceptual, ethics-laden, or "creative" (here meaning "imaginative" in a very rich sense). Sometimes, of course, the approach is merely observation based.

The sciences also use multiple approaches—some sciences utilize patient, observation-based approaches even as others are highly theoretical and quantification based. Yet other sciences are aggressively experimental, manipulative, and invasive as they probe and isolate various phenomena. The many different scientific methods have, in combination and in isolation, produced impressive and extremely powerful analytical tools.

While in both humanities and sciences many techniques, such as observation or trial and error, are akin to skills we use in ordinary, day-to-day life, nonetheless both of these megafields use many additional methods. Thus these megafields have often achieved intellectual feats of understanding that greatly expand human consciousness. Both megafields may utilize methods and vocabularies, perhaps even overly intellectualized pyrotechnics, that can seem irrelevant to daily life—consequently, some fields may seem to lack substance because discussions baffle all but insiders deeply committed to a highly specialized vocabulary that seems like mere jargon or trendy fashion to outsiders (see chapter 7).

Getting to the Basics Informally

Whether one finds wonderful or wanting the different kinds of learning that happen in the humanities and sciences of formal education, it is important to recognize just how true it is, especially in Animal Studies, that formal education in big institutions is by no means all of education. Thus even if it is true that "education is perhaps the most important function of state and local governments" (and, one should add, private educational institutions), myriad forms of informal education are as or even more important when it comes to realistic, effective education about nonhuman animals. Indeed, Animal Studies would be deeply impoverished if it had to go forward primarily on the radically inadequate formal "education" about other-than-human animals that state and local governments or private institutions offer.

Such informal education can be identified by looking at what happens before, outside, and after one's formal education. There is stunning variety at each of these stages since human learning skills generally are very diverse. Further, individuals' peculiar talents for learning are often idiosyncratic, that is, they can be unique to each individual and highly dependent upon the specific context (such as family, religious community, and local ecological niche) in which each individual lives. Further, since some individuals are far less impacted than others by social psychologies and pathologies (chapter 8), informal education for humans is characteristically highly individualized.

The net result is that individuals and communities create awareness of other living beings of a kind and to an extent that cannot be matched by institutional means. Given that formal education so often involves sitting in classrooms isolated from the more-than-human world (even when computers are available), formal education is relatively impoverished in terms of learning opportunities that involve real-life contexts where nonhuman animals exist.

The upshot is that formal education is truncated relative to day-to-day realities, and comes in forms—such as the narrow versions of history discussed above—that feature one-dimensional learning of the human-centered ilk. Thus, even though much educational rhetoric claims that education is about both the truth and the future, formal education is highly invested in a "truth" and "future" that are centered so exclusively on humans (or some powerful and privileged fraction of a human society) that Animal Studies is a mere sideshow or afterthought.

But change is afoot, as the rapid expansion of "animal law" courses in American law schools confirms (see chapter 4). This example strongly suggests that the future of Animal

Studies is likely to be robust as multitudes of primary, secondary, university, graduate, and professional students petition for Animal Studies courses. Already, students in hundreds of courses each year are being taught Animal Studies themes and thereby learning a powerful means of opening up our engagement with the world in which we live. Further, since many students assess the present harms to nonhuman animals as ethically questionable, such courses implicitly or explicitly raise questions about whether our society's social policies governing the protection of nonhuman animals need fundamental changes.

Good, Better, Best Education

Engaging other-than-human animals well, that is, in terms of their realities and with an open mind about humans' rich ethical abilities requires education to address the full range of human intelligence and heart, as it were. How this might be done at the early, elementary levels of education is different from how it might be done in higher education. Chapter 11 explores the possibility that children offer all of us surprising opportunities to see clearly some basic connections that humans can have to other animals.

Here, the focus remains on higher education because in so many contemporary societies it is universities and colleges, as well as professional schools, that continue to be dominated by the deeply entrenched exceptionalist tradition that anchors formal education's radical dismissal of other-than-human animals.

Perhaps Thinking Begins Here

Chapter 1's list of various benefits from recognizing other animals includes developed compassion, strengthened character, enrichment of mind and imagination, and enhanced critical thinking skills. It also includes opportunities for self-actualization through self-transcendence, and connection to our larger community. Both of these also have the potential for the developed ethical and even spiritual sense sought by the Axial Age sages when they advocated, in Armstrong's words quoted above, respect for the sacred rights of all beings.

These benefits on the human side of the ledger are integrally tied to a commitment to seek out the truth about other living beings' actual realities. Such a commitment is often associated with scientific pursuits but is just as crucial to ethical inquiries (see chapters 3 and 6). One additional benefit of attention to the realities of other animals is that, as Lévi-Strauss said, other animals are "good to think" (chapter 1). The philosopher Jacques Derrida said, "The animal looks at us, and we are naked before it. Thinking perhaps begins there."[35] Decades earlier (1978), the ecologist Paul Shepard observed, "Animals are among the first inhabitants of the mind's eye. They are basic to the development of speech and thought. Because of their part in the growth of consciousness, they are inseparable from a series of events in each human life, indispensable to our becoming human in the fullest sense."[36]

Although the quotes of Lévi-Strauss and Derrida above use the word "think," what is at issue (and what these seminal thinkers are intimating) is more fully conveyed by words such as "become aware" and "open up." Other living beings invite us in special ways to be more aware than we otherwise would be. This is why they have long been associated with signs, symbols, omens, auguries, and much more. Many peoples have understood other animals to teach humans in ways that are, to use Shepard's language, "indispensable to our becoming fully human" for which there is "no substitute" because "there is a profound, inescapable need for animals that is in all people everywhere."[37]

These themes constantly resurface, as in the following book titles used by contemporaries from widely separated cultures: Debra Rose Bird's *Dingo Makes Us Human* (1992) about Australian aboriginal culture, and Temple Grandin's *Animals Make Us Human: Creating the Best Life for Animals* (2009). These books are but two among many available today that help flesh out what it means to note that other animals are "good to think" such that "thinking perhaps begins there." Chapter 6 explores various implications of claims that multifaceted human abilities like thinking, mind, speech, and consciousness have roots in humans' connections to other living beings—these observations invite exploration of the ways in which Animal Studies foregrounds the central role of critical thinking.

Critical Thinking

"Critical thinking" is a series of processes and tasks that many educators and others have advocated as a way of investing our mental processes with responsibility and humility— thereby increasing the breadth and depth of human reflection. Critical thinking prompts abundant questions in order to increase the chance that our encounters with and reasoning about the realities surrounding us will reflect both features of the real world and our awareness that human thinking has limits even as it is wonderfully powerful at times.

Critical thinking employs multiple techniques ranging from simple, commonsense approaches to more abstract notions. The former include honoring the intuition that each and every human needs to consult the world in which he or she lives as carefully as possible when purporting to describe that world. One method of doing this is patiently observing, when possible, the actual realities about which one desires to make claims. An example of a useful abstract or theoretical notion that critical thinking can employ is Bayes's theorem, which was formulated to deal with ways of wondering about uncertainty and which requires multiple steps and a mathematical formula deriving from a kind of common sense about the relationship between evidence and certainty.[38]

Because critical thinking employs a wide variety of techniques and is open to new insights that help make human contemplation of the world as responsible and fair as possible, it leaves room for flexibility. It is capable of recognizing multiple sources of knowledge and it provides space for multiple forms of reasoning that take a thinker from evidence to a conclusion. Critical thinking also helps immensely when a claim may have compelling psychological value even though its relationship to the truth is uncertain or, worse, nonexistent. Recognizing such problems is important because self-deception can plague both individuals and groups (see chapter 8).

Critical thinking is, in summary, multifaceted, careful reflection about human thinking, with all the humility that that implies. Such attempts at responsible reflection are important for both obvious and subtle reasons. For example, each human's thinking has obvious limits. Similarly, our communal efforts at thinking, while surely intriguing and powerful for us as individual humans, have their definite limits as well, although sometimes the social dimensions of our knowledge claims can be even harder to see than the limits on our individual thinking.

Further, because humans think and are intelligent in many different ways (see chapter 5) and thus use a great variety of processes and patterns in thinking, the simplest definition of critical thinking is, in a way, regularly thinking about our thinking processes. Such an approach requires constant acknowledgment that the manner in which any human might apply multiple abilities and patterns of thought is not mechanical, but rather something of an art.

Consider, for example, what has been suggested so far about the range of tasks to which critical thinking can contribute:

- Recognizing what we really know as opposed to what operates as wishful thinking, bias, or self-inflicted ignorance
- Seeing the range of human abilities to notice, take seriously, and even care about others
- Staying aware of the risks of generalizations and language habits
- Noticing how some views are overdetermined by group dynamics even as they are underdetermined by actual realities of the world

In order to elaborate more fully how critical thinking is useful and must play a role in animal studies, some other meanings of "critical" are examined here, for these, too, help one understand and distinguish the special role that critical thinking plays in Animal Studies.

Acknowledging "Critical" as a Contested Term

The term "critical" has many meanings in ordinary language—including "crucial" or "offered many criticisms." But "critical" has a special subset of meanings when it comes to discussions about human thinking and knowledge that are particularly pertinent to education in general and, most specifically, the quality of human thinking about ourselves and other living beings. For this reason, chapters 4 and 7 address the set of highly intellectualized, mostly academic inquiries known as "Critical Studies,"[39] because the word "critical" has in such instances a special range of meanings that are illuminating in contemporary Animal Studies.

In many cases, "critical" is a synonym for terms like "reflective" and "reflexive" in the sense of "thinking back upon thinking" or "being aware of one's own assumptions." As any human who has tried this important task knows, "thinking about thinking" is a humbling exercise. It requires that the thinker recognize (and this feature is truly crucial in all critical thinking) that this exercise is self-referential. In other words, to assess our human thinking, we have to use our human thinking abilities. As we use the very processes that we are evaluating in order to decide whether the processes are valid, there is no avoiding the sense that circular or self-affirming thought is taking place. Such a problem is, in a very basic sense, unavoidable. Arguments are circular when their conclusions appear among their premises. Another kind of circularity occurs, however, when principles of reasoning are supported by arguments that employ them.[40]

Here is an example that makes this conceptually complex issue easier to follow—if someone uses logic to argue that logical arguments are valid, while illogical arguments are not, the very principles of reasoning one is hoping to justify are being employed to justify their own validity. Similarly, at the level of our attempts to think and inquire carefully or critically, we have to use careful thinking to assess what careful thinking is—and here our human minds are at the end of the line, as it were. There simply are no other resources by which this endeavor is possible.

In formal philosophical circles, the most famous use of the term "critical" is that of Kant, whose fame is related to the fact that he focused a great deal on what it means to think as carefully as possible about thinking. Kant's preeminence has prompted widespread use of the term "critical" within and beyond philosophy. Earlier philosophers had been interested in the process of thinking about thinking, often explaining this with terms like "critical," which comes from the Greek word *kritikos*, meaning "relating to judging, fit for judging, skilled in judging." For Kant, however, the most basic task for philosophy was to judge whether and how knowledge is possible. He thus understood the philosopher's primary role not as proposing theories, but rather as subjecting all theories and knowledge claims to a rigorous examination of the very possibility of knowledge.

In Kant's work, then, "critical" is a reference to a kind of judging or reflective review of the deepest foundations and limits of our claims to know. Because of Kant's immense influence, the term "critical philosophy" is most commonly a name for Kant's philosophical approach in which one first judges how human thinking ("reason" for Kant) works and, second, discovers thinking's limits so that one can, finally, apply the findings to one's own life.

Terms like "critical study" and "critical studies" have long been widely used for many different things. For example, *Contemporary Buddhist Ethics* (2000), edited by the Buddhist scholar Damien Keown, is part of an entire series that carries the title Curzon Critical Studies in Buddhism. A quarter of century earlier (1975), the philosopher Norman Daniels published an edited collection titled *Reading Rawls: Critical Studies on John Rawls' "A Theory of Justice."* In such volumes, what passes as "critical studies" is diverse, but the approaches share much with the general spirit of what Kant was attempting, namely, to think as responsibly and carefully as possible about how we make claims about a particular subject.

Another, even more commonsense use of "critical studies" is something like "careful, rigorous, evidence-based analysis of the subject." For example, the respected American commentator on modern law and society, Judge Richard Posner, addressed the subject of popular biographies about certain judges: "These traditional biographies are pointless if you're interested in understanding the significance of the judge as a judge.... What are needed are critical studies, as opposed to biographies."[41]

It is important, then, to acknowledge two aspects of the word "critical." It is an adjective that many people like to use, and it is called upon to do many different kinds of work in diverse contexts. Further, since in the history of human thinking there have been many attempts to invest our reflections with features of responsibility, discipline, rigor, humility, and the like, any number of terms like "critical thinking" might also do the same work.[42] In other words, the term "critical thinking" is by no means the only one that could be used to describe the care with which Animal Studies must be pursued.

In summary, whether one prefers "critical thinking" or some other way of describing the important task of aspiring to the best in human thought, the issue is to find ways to do careful thinking that connects us with realities (our own, those of other beings, and those of the universe of which we are an integral part) rather than caricature and distort these realities. The two words "critical thinking," then, by no means have an automatic, definitive range of meaning that everyone will agree on. This means that what is claimed below in the name of "critical thinking" must be taken as involving an inherent flexibility and open-mindedness— said another way, what is aimed at here is a willingness to reflect on our thinking processes, to learn about learning, and at times to acknowledge the need for rethinking and even unlearning. Through all this, it is important to avoid the impression that critical thinking is a simple process—in fact, it is a rich, versatile series of processes that help us see our own thinking better. Without such flexibility, even the term "critical thinking" can be used uncritically, that is, in undisciplined and sloppy ways.

Whose Critical Thinking?

"Many teachers who don't have a deep appreciation of science present it as a set of facts," said David Stevenson, a planetary scientist at Caltech, quoted in a 2007 book dedicated to increasing scientific literacy among the general populace. "What's often missing is the idea of critical thinking, how you assess which ideas are reasonable and which are not. Even more than the testimonials to the fun of science, I heard the earnest affidavit that science is not a body of facts, it is a way of thinking. I heard these lines so often they began to take on a bodily existence of their own."[43]

The confidence with which this scientist employs the notion of critical thinking is noteworthy—the plain implication is that "the idea of critical thinking" is an essential tool for assessing "which ideas are reasonable" in the domain of science. Critical thinking applies, of course, to more than a single domain, although there is no consensus as to which domain offers the paradigmatic version of critical thinking by which all others might be measured.

Some researchers, educators, scholars, and professionals, however, talk as if the form of critical thinking employed in their own field is paradigmatic. One can find narrow accounts of critical thinking among very specialized thinkers of diverse kinds—among natural scientists, as above, but also among philosophers, theologians, and ethicists. Economists and other social scientists also may tout, in a fundamentalist-like fashion, their own way of thinking as the essential measure for assessing "which ideas are reasonable."

Such thinking about thinking is obviously full of challenges. By one practical measure, what it means to be "reasonable" is to give reasons that others can scrutinize and assess as fair-minded in the situation and for the subject one is trying to illuminate. In such cases, "reasonable" ends up meaning well balanced, empirically based, and nondogmatic. Stevenson also suggested that critical thinking in science leads to humility: "Part of critical thinking includes the understanding that science doesn't deal with absolutes."[44] The humility of this understanding in no way handicaps science; instead, it is this very feature that gives researchers confidence in the breadth and depth of science enriched by critical thinking—Stevenson immediately added, "Nonetheless, we can make statements that are quite powerful and that have a high probability of being correct."

In such contexts where humility is allowed to play out alongside careful thinking, human reflection has deep prospects. Our thinking is, as we all know from experience, an enterprise as complicated as it is rich—humans' multifaceted abilities and multiple kinds of intelligence can contend with one another. When humility is foregrounded as we work with such various abilities, it facilitates communal sharing, which in turn enhances the potential range and power of thinking.

Science, at its best, does this well. But at its worst (and, as chapter 3 suggests, this happens regularly), science can be shaped by researchers and administrators of institutions who violate even the simplest canons of critical thinking when it is to their advantage. At its best, science promotes open-minded asking of questions by anyone who would like to inquire. This commitment to public verification or testing of claims has prospects of increasing everyone's understanding of the universe we inhabit.

In this regard, sciences can model for all disciplines the value of foregrounding openminded, good-faith questioning of any claim. It is especially claims about the significance and meanings associated with nonhuman animals that need such a model. For many humans, answers to fundamental issues involving nonhuman animals were given long ago, fixed by human authorities, and therefore resolved. Today Animal Studies employs a wide range of critical thinking skills to assess both past authorities and present claims about animals.

Some Relevant History

The history of critical thinking as both a term and concept is summarized in the second edition of Matthew Lipman's *Thinking in Education*. It is noteworthy—and eye-opening about the continuing evolution of the contemporary meaning of "critical thinking"—that the newer edition includes a number of additional penetrating questions that reflect how broadly the notion of critical thinking can be used in education. Lipman observes at the beginning of this new edition, "Parts Three and Four are almost completely new. What these new parts offer is a view of education at a more comprehensive level of effectiveness than critical thinking by itself could ever hope to achieve. Some components new to the elementary school level of education have been introduced: emotions, caring thinking, mental acts, and informal fallacies."[45]

All four of these elements are pertinent to careful thinking by humans about any subject, but especially so in nonhuman animal matters. The issue of emotions is, for example, pertinent to the polarized environment in which much discussion of animal protection proceeds, as well as the question of emotions as integral parts of the life of many mammals (including humans, of course). Caring thinking and mental acts are, as noted throughout this book, an essential part of humans' ethical capabilities. Informal fallacies are also crucial to the convoluted reasoning by which humans' obvious power over other animals is converted into a justification for human traditions and privileges. Animal Studies can contribute to the continuing development of critical thinking based on these and other important features of human thinking and valuing because, first, human thinking about other-than-human animals is no easy matter for human minds in general, and, second, the intersection of humans and nonhumans presents both formidable challenges and wonderful opportunities.

The Importance of Critical Thinking to Animal Studies

An even more obvious point, of course, is that Animal Studies as a new discipline in the academic world needs to foreground critical thinking at every turn. Both the field and education more generally benefit when everyone is invited to address how we think as we assess the past, present, and future of the human-nonhuman intersection. Indeed, a willingness to examine, question, redo, and even unlearn is essential for a variety of reasons. Traditional dismissals of nonhuman animals are institutionally entrenched, and of course they are psychologically and religiously significant for many people. Further, the frailty of human thinking in general has led to a very specific, ethics-fraught problem—learning about other animals is complicated because they (other animals) are often impacted by the very attempt to know them in any detail. Everyone knows, for example, that one cannot study certain wild animals simply by planting oneself squarely in their midst—some animals are shy or fearful in ways that are exacerbated in the presence of outsiders. Beyond the harms that human presence may cause, there is the additional problem of humans being misled by distorted behavior.

Critical thinking offers key tools for those who want to recognize such problems as well as features of their own thinking about animals. It is also crucial to each person remaining as aware as possible of the values and hidden assumptions driving research, education, and advocacy. It is precisely by resorting to practiced awareness about the complexity of our thinking and valuing processes—which is the central task of critical thinking—that we can see and work at minimizing avoidable distortions. If and when these tasks are accomplished, then our thinking about any subject, but in particular difficult topics like trying to know other animals' realities, can be invested with the best possible qualities.

Summarizing Central Tasks

Given that critical thinking is constantly developing additional methods by which to assess human thinking, no single list of its central tasks is likely to be exhaustive. But at the very least, the following tasks are important as critical thinking is applied to issues arising at the human-nonhuman intersection.

1. Give questioning a central role. It is suggested at various points in this book that questions can have more power than their answers. In Elie Wiesel's autobiographical novel *Night*, Moche the Beadle observes that every question possesses a power that does not lie in the answer. This is particularly true when questions prompt us to think about other beings that have previously been unimportant to us, maybe even unknown because our culture or our own personal actions have marginalized these beings in one way or another.

Such an attitude opens up much—an ethic of inquiry prompts one to do science-like work about one's surrounding world, just as it prompts one to ask ethics-intensive questions about the consequences of one's own choices. It also prompts issue and information literacy, the latter enabling a person "to recognize when information is needed and [to have] the ability to locate, evaluate, and use effectively the needed information."[46]

2. Reflect regularly on one's own thinking and claims. Readers will notice that reflexive thinking, an ancient hallmark of critical thinking, is invoked in any number of chapters in this book. Such a task requires a great more than the skill of remaining open to a variety of questions and answers—it requires, for example, a comparably deep and honest commitment to ascertain as best we can how we are going about efforts to collect and think about what we colloquially call "the facts." This remains an especially central challenge when what might

be learned disfavors one's own preconceptions, privileges, and heritage. It has already been suggested that humans are in peril when they are either mindlessly subservient to the cultural and religious heritage or to the familial, social, and political traditions into which they are born. Honesty about one's heritage—its strengths and weaknesses—is mandated by the fact that one's actions and ethics are always about choices in the present. Those who purport to follow the past meticulously fail to acknowledge that such an approach to decision making and value choices requires constant choices each new day and thus inevitable interpretation. Cultural and religious heritages are not simple but cumulative—in a word, complex. They often change dramatically over the centuries, such that what is now taking place differs dramatically from the choices and values made in the past. So following them blindly puts one at risk of being completely dysfunctional in today's world.

Critical thinking inquires about such changes and factors them into the simple fact that each human makes key choices in daily life. Further, just as critical thinking prompts each individual to be honest about the complexities of his or her heritage, it also keeps in the foreground the possibility of challenging even its own traditions and methods, for careful thinking implies that we must constantly think and talk about the very functions of critical thinking.

3. Set an open table. Critical thinking requires many different kinds of expertise to be consulted, worked through, and used when fitting. This sort of approach has been, like critical thinking itself, named variously—in this book, the term "interdisciplinary" is meant to do this work. In addition, the corollary task of being comparative, that is, of consciously comparing different areas of human endeavor that can be seen as alike in some ways even as they are distinctive in other ways, is called out often in this book. Yet another corollary of such work is the possibility of weaving together, when possible, multiple ways of talking about issues (sometimes called traditions of discourse, specialized vocabularies, or even jargon). These exist in abundance and can be, if worked with respectfully, helpful in identifying various subject areas' complexities.

4. Foreground a developed sense of humility when in pursuit of "the facts." Critical thinking pursued regularly by means of (1) through (3) above balances the search for reality with humble acknowledgments that what any one of us, or all of us as a group, can know and therefore call "the facts" is no simple matter. This is so for a great variety of reasons, not least of which are the diverse psychological and social processes that help to build and shape what we claim to know.

In one sense, it is both common sense and the spirit of critical thinking that push Animal Studies to acknowledge at every turn the importance of other animals' realities. When seeking such realities, critical thinking also leads one to keep in mind the advice "seek simplicity, and mistrust it."[47] Simple explanations have an allure because they give us comfort that we fathom other animals, but such explanations can be in tension with more thoughtful analysis—they can, for example, slide almost imperceptibly into oversimplification. When this happens, the power of questions can introduce a useful mistrust that prompts one to ask if an oversimplifying generalization falls short of rigor and analysis.

Oversimplification, nonetheless, often prevails in our notion of the facts about other animals—recall the caricature of wolves that Midgley challenged. Almost everyone entertains ideas about other animals that go well beyond what is described in our sciences—at issue here is, overall, a certain fairness to the realities around us, which this book suggests humility mandates. But even if we must again and again acknowledge limits in what our fascinating but finite human abilities can perceive, we confidently surmise that we are capable of knowing some aspects of the world "out there." As we humbly and fairly try to work out what it is that we in fact know, we can celebrate "what a thing is the interested mind with the disinterested motive." There may be disputes about whether such a claim applies to any other-than-human minds, but the remark is surely true of human minds.

5. Stay aware of social psychology and pathologies. What often makes critical thinking necessary and valuable are some peculiar features that regularly show up in the history of human thought. These powerful factors in our identity and awareness need first to be seen, then considered, if we are to think carefully (chapter 8). Our minds are not mirrors of the world, but active producers of meaning. Since a major goal of critical thinking is to eliminate as many forms of mere wishful thinking, unfair bias, prejudice, and self-inflicted ignorance as we can, the tools of critical thinking necessarily prompt us to engage how we make meaning. Thereby we can ferret out the phenomena called, among other things, "social construction" (described in chapter 8) that are inevitable in our thinking and which, therefore, play particularly important roles in the matters Animal Studies engages.

6. Give a place to nonanalytical thinking and ethics. As the foregoing suggests, it is not merely analytical but also both nonanalytical thinking and what might even be called meta-analytical thinking and caring about the world that contribute to our understanding of animals' lives. Humans feature multiple kinds of intelligence (chapter 5)—this is almost surely true of some other animals as well, but the specific relevance of this observation to human thinking is that we think in a rich variety of ways that form the tapestry of our understanding and interaction with the world. One of the more complex tasks of critical thinking is to provide existential space, as it were, for the many rich forms of human thinking and awareness of the world and its communities of living beings.

7. Recognize multiple approaches as part of human understanding of the world. As noted in the introduction, students of Animal Studies face multilayered complexities as they address the human-nonhuman intersection amid the world's "buzzing, blooming confusion." While some difficulties are connected to the overwhelmingly diverse and often mysterious lives beyond the human species line, other difficulties exist because of inherent limitations in human knowledge and the complexities of our communication and social realities. Humans need great flexibility as they address these kaleidoscopic complexities. We as a community need not only to work through multiple disciplines but also to recognize that knowledge comes in many different ways. For example, critical thinking prompts one to recognize that what it means to "know" goes far beyond mere working out of ideas—"knowledge" includes existential, psychological, and even bodily features. In his 2012 book *The Great Animal Orchestra*, Bernard Krause suggests that awareness of sound can be an indispensable tool of knowledge because humans evolved amid a raucous "biophony," which he defines as "the sounds of living organisms." Humans have, by virtue of their evolutionary heritage, deep capacities to become attuned to "the many subtleties of untamed natural environments," although the world's biophony has been diminished by human-caused changes in the natural

world.[48] Human artistic abilities can produce valuable insights, just as the realities of caring can open up modes of knowing that differ from quantification-based modes (chapter 5). Critical thinking prompts caution about one-dimensional approaches that may be insufficiently critical even as it encourages exploration of the possibilities of the many different disciplines that can assist one in coming to "know" the "truth" about the world's diverse features.

"Education Perhaps Begins There"

Beyond the obvious problems of access and fairness to humans that educational systems generally face, many find formal education to have questionable features of other kinds. This book suggests that the matter of nonhuman animals is a particularly serious challenge for human-centered versions of science and the humanities. Chapter 3 addresses ways in which the science establishment closes off questions, and additional chapters address how some precincts of the humanities choke off questions about nonhuman animals that touch on essential features of education, history, ethics, policy, and more.

The value of Animal Studies for education is hinted at in Armstrong's observations about Axial Age sages, Lévi-Strauss's observation that other animals are "good to think," and Derrida's more recent comment, "The animal looks at us. . . . Thinking perhaps begins there." Given that Animal Studies provides a steep learning curve, urges the uncovering of marginalized truths, and vibrantly questions complacency and wishful thinking, perhaps not only thinking, but education, too, "begins there." At the very least, Animal Studies has the potential to help rework and improve fragile human understanding.

Animal Studies also has prospects of helping education more generally. For example, gains in critical thinking can be achieved through Animal Studies in each of the six areas listed above, which is important in education generally given that "an astounding proportion of students are progressing through [American] higher education without . . . improving their skills in critical thinking, complex reasoning, and writing."[49]

In the next chapter, we move from these preliminary encounters with history, cultural diversity, education, and critical thinking to the profoundly influential endeavors of science and politics as they impact other animals and Animal Studies.

Science, Politics, and Other Animals

The perspectives on nonhuman animals in, first, our sciences and, second, our political systems and policy discussion circles can be profitably compared. Many individual sciences are crucial to the development of a robust form of Animal Studies because of their commitment to exploring other animals' realities as fully as possible, that is, on their own terms rather than on terms dictated solely by human interests and biases. Detailed information about many nonhuman animal individuals and communities has been developed over many decades now through various sciences, and media of different kinds have made much of this information widely available. Such developments clearly have raised awareness of basic aspects of certain nonhuman animals' lives—for example, elephants' matriarchal social organization, dolphins' intelligence and playfulness, chimpanzees' friendships and political intrigues, and on and on. Increased awareness of such realities is today a recognizable force both in the worldwide animal protection movement and in the crystallization of demand for better education-based offerings on nonhuman animal issues.[1]

From the science vantage point, commitments to discover other animals' actual realities are fundamental to the very enterprise of science. But as discussed below, this ideal and the commitments it generates can be overridden in a variety of ways in modern scientific circles, thereby offering the chance to compare views of nonhuman animals found in the practice of science with views that prevail in politics. In contrast to the ideal of science, politics and policy provide the paradigmatic example of an arena of human life in which our human realities, including our power relations with each other and the more-than-human world, are worked out in very species-centered ways.

In this chapter, then, both science and politics are examined. Each of these major human endeavors is a central concern in Animal Studies. At times in completely different ways, at times in surprisingly similar ways, each impacts other-than-human animals greatly through holding harms or fundamental protections in place.

The Question of Their Realities

Inquiry about the actual realities of other animals as individual members of their societies is a driving issue in Animal Studies. This commitment is easily recognized in various sciences, but also exists widely outside the realm of science—in many individuals, small-scale societies, and even major religious traditions (see chapter 7). Importantly, though, human learning about other animals' real lives and more complicated dimensions (such as personality, social interactions, emotions, communications, and intelligence) involves, as experience readily confirms, great challenges. This book suggests in several ways that it takes all of humans' abilities worked out in healthy communities using interdisciplinary forms of communication to get even a partial picture of other animals' realities.

Speaking of one of the great modern figures who illuminated many basic issues for Animal Studies, a scholar observed, "The Paul Shepard I knew...knew firsthand (as an academic himself) that intellectual culture is insecure, isolated from the biophysical context of life."[2] Some academic discussion of other living beings is surprisingly removed from the beings themselves and any context in which they might be fully and fairly understood.

This book identifies important limits in any individual's ability, not merely those of academic scholars, to learn and speak of the actual realities of other animals. These limits may be practical, scientific, philosophical, ethical, or ecological. Still other limits may take personal forms, or be created by political, religious, and cultural factors. Often, attempting to understand other animals is so challenging that the very attempt launches us on a journey of self-exploration about our own way of understanding our local world.[3]

Importantly, a contrast between sciences, on the one hand, and politics and policy discussion, on the other, reveals how various limits play out in our species' interactions with other animals. These different spheres of human life overlap, to be sure, since in both scientific and political circles there are inquiries that are thoroughly values driven in ways that raise ethically charged issues. This may seem controversial given the recurring claim that science is either value free or value neutral. The contrasts and comparisons in this chapter about the handling of nonhuman animal issues reveal, however, that much science falls far short of this ideal. In summary, in both science and politics, human creativity has prompted both beautiful ideals and human-centered hubris of debilitating sorts.

Science and Other Animals

A principal reason for the importance of scientific perspectives is this simple, straightforward fact—without highly specific details about actual animals, Animal Studies risks being irrelevant or empty. Science has within its very heart an important set of commitments to seeking animals' realities, which have prompted development of a variety of methods because, despite the common phrase "the scientific method," there is by no means a single scientific method. Rather, different sciences use a great variety of methods to ensure a fundamental openness among those who seek to learn what they can of the actual realities of living beings.

There is a balance to strike, however. Beyond observation and data collection, one needs the creativity of broad ideas or theories about other animals that open up inquiry. Human attempts to get details about specific animals can miss the mark, if researchers remain unaware of broader issues of the kind that Animal Studies raises (such as the distortions of traditional caricatures, the dominance of the exceptionalist tradition, the inevitability of ethical issues in invasive research, etc.). To be sure, one needs to question closely forms of thinking that are merely theoretical, that is, thinking that pays scant attention to actual realities and thereby runs the risk of being completely unrelated to real-life issues. In summary, forms of Animal Studies done without reference to biological animals risk being both irrelevant and empty.

There are other risks, of course—empirical inquiries about nonhumans without reference to ethical and personal dimensions risk creating significant problems, including blindness to nonhuman animals' realities and needs, which has in many instances led to notorious failures to recognize, let alone honor, other animals' realities and suffering, or even humans' most basic ethical abilities.

Animal Studies is, by virtue of its commitment to the central place of other animals' realities, an extension of the spirit of the scientific revolution. It is another step in the journey away from the fantasy that humans are the center of the universe. More particularly, it is a refusal to believe that humans are the raison d'être of the world, or that humans are so qualitatively superior to any and all other animals that the harms humans do to other living beings have no moral implications. Such wishful thinking and prejudice have no evidentiary basis— further, they narrow humans' ethical abilities and thereby drive the exceptionalist tradition's justifications of human domination of the more-than-human world.

Frankness about the Practice and Power of Science

As discussed in the second half of this chapter and then throughout this book, the exceptionalist tradition's justifications hold sway in political, educational, legal, and institutional circles of many kinds. The practice of science, too, can feature human-centered biases that rival those of politics. Thus, the achievements of science notwithstanding, no discussion of the importance of science is complete without a frank appraisal of the difference between the ideals and the actual practices of science in the real world. These problems are one reason Animal Studies has a special role to play in using other-than-human animals' actual realities as a lens through which to view humans' treatment of their fellow animals.

Humans, of course, value science for reasons other than its power to elucidate realities, for science has given humans extraordinary power. That power includes the astonishing and varied ways humans have to dominate, harm, and subordinate other animals, only some of which are discussed in this book. Large food production industries (also known as agribusiness) have promoted the academic field of animal science, which now has hundreds of departments granting undergraduate and graduate degrees. This field, which is a major part of veterinary schools and university science departments, is a narrow, production-oriented enterprise that uses much technical science. In practice, however, animal science circles have features that suggest that many of its practitioners are not at all interested in broader issues of science, but instead play down many actual realities such as the suffering of those nonhuman animals treated as mere resources by for-profit industries and many research laboratories.

When one reflects on which "discoveries" and "breakthroughs" interest the large of majority of scientists in animal science, one notices that these scientists focus their attention, by and large, on research opportunities that either create grant opportunities or promote greater profits for industries by creating additional "efficiencies" in the use of other-than-human animals. Science-based findings that illuminate these nonhuman animals' sentience, intelligence, and cognitive abilities, all of which are pertinent to the possibility of animals' suffering in modern production facilities, are at best of secondary interest and, at worst, of no interest at all to the scientists and educators employed in animal science. Because those who promote animal science are predisposed to challenging any such findings as not scientifically certain (which means, of course, that they need not take such a finding seriously), proponents of animal science exemplify Upton Sinclair's quip about the difficulty of getting "a man to understand something when his salary depends upon his not understanding it." Academic departments that go by the name animal science would, in fact, be less misleadingly named if they advertised themselves as "food animal engineering."

This phenomenon is not confined to science regarding nonhumans—what has been called "corrupted science" takes place in many contexts that involve humans, as evidenced by the astounding harms done to humans by the marketing of drugs, tobacco, and asbestos.[4] But the extent to which profit-oriented manipulations of science harm humans is minor compared to the harms done to nonhuman animals. It surely cannot be denied that animal science endeavors are immersed in some science, although, as noted above, animal science programs ignore other science relevant to the medical and psychological problems that industrial efficiencies impose on the nonhuman animals used as mere resources. Animal science as a field, then, is a blinkered approach to science pursued to justify policies whose purpose is advancement of the exceptionalist tradition.

What reveals how unscientific animal science has become is the treatment of new students and dissident faculty members. If a student taking an animal science course insists on using strictly scientific terminology (by speaking of "nonhuman animals" and "human animals"), the student risks ridicule and even poor grades. Those students or faculty members who persist in asking ethics-focused questions about modern practices risk marginalization as well. Such questions have power beyond any answer that might be offered, for the question itself implicitly reminds every hearer of the antiscientific denial that is the heartbeat of common phrases such as "humans and animals" (chapter 2).

Two examples help reveal the stark contradictions that animal science entails because its narrow focus is profit-oriented practices squarely in the exceptionalist tradition. The following comes from a speech delivered by a veterinary ethicist and later published in the *Journal of Animal Science*:

> One of my animal scientist colleagues related to me that his son-in-law was an employee in a large, total-confinement swine operation. As a young man he had raised and shown pigs. . . . One day, he detected a disease among the feeder pigs in the confinement facility where he worked, which necessitated killing them with a blow to the head, since this operation did not treat individual animals, their profit margin being allegedly too low. Out of his long-established husbandry ethic, he came in on his own time with his own medicine to treat the animals. He cured them, but management's response was to fire him on the spot for violating company policy. He kept his job and escaped with a reprimand only when he was able to prove that he had expended his own—not the company's—resources.[5]

A passage from a best-selling book published in 2006 further describes exceptionalist efficiencies actively promoted within animal science circles:

> Piglets in these CAFOs [a US government term that means "confined animal feed-ing operations"] are weaned from their mothers ten days after birth (compared with thirteen weeks in nature) because they gain weight faster on their drug-fortified feed than on sow's milk. But this premature weaning leaves the pigs with a lifelong craving to suck and chew, a need they gratify in confinement by biting the tail of the animal in front of them. A normal pig would fight off his molester, but a demoralized pig has stopped caring. "Learned helplessness" is the psychological term, and it's not uncom-mon in CAFOs, where tens of thousands of hogs spend their entire lives ignorant of earth or straw or sunshine, crowded together beneath a metal roof standing on metal slats suspended over a septic tank. It's not surprising that an animal as intelligent as a pig would get depressed under these circumstances, and a depressed pig will allow his tail to be chewed on to the point of infection. Since treating sick pigs is not economi-cally efficient, these underperforming production units are typically clubbed to death on the spot.[6]

It is not uncommon for students to ask why "an animal as intelligent as a pig" would be treated in this manner. The prominent American political commentator Matthew Scully, who was the senior speechwriter of President George W. Bush, observed at the beginning of a bestselling 2002 book, "no age has ever inflicted upon animals such massive punishments with such complete disregard, as witness scenes to be found on any given day at any modern industrial farm."[7]

Later in the same book, Scully describes his tour of a modern slaughter facility with the president of the largest pork producer in the world. He reveals why a veterinary student whose goal is to help heal animals might question the *actual* role played by industry-paid vet-erinarians who oversee prevailing food animal practices: "Some [industry-hired] shill of a vet comes by every few days to check on the stock. But for the vets, too, they are not even animals any more. They're piglet machines. And tumors, fractured bones, festering sores, whatever, none of these receive serious medical attention anymore."[8]

Because this high-profile conservative political commentator sees intensive food produc-tion systems as promoting suffering, and because he understands veterinarians as obliged by their professional oath to minimize suffering, he challenges any veterinarian who enables harm, worrying openly about the "profound betrayal of veterinary ethics everywhere around us—the sworn obligation of every veterinarian 'to protect animal health [and] relieve animal suffering.'"

The vision driving Scully's critique is simple—veterinarians are supposed to be leaders in animal protection. Their familiarity with science is important, but when this familiarity is used only to increase production without regard to increased suffering, rather than to heal as a veterinarian's oath requires, questions need to be asked. Many individual veterinarians share this vision of the primary and fundamental purpose of the veterinary profession, but the official positions of some national veterinary associations unequivocally support present agribusiness practices.

In many countries, then, the veterinary medicine establishment and animal science often follow rather than lead industry. Both yield to political realities, regularly opposing animal protection efforts by ordinary citizens, animal law developments, and certain protections often called "animal rights." In some countries, though not in the United States, Canada, or Australia, the veterinary profession takes a leadership role in discussions about alternatives to the harsh "efficiencies" now promoted by animal science.

Similar observations could be made about the way researchers in laboratories follow rather than lead. Because so much scientific research is government funded, the exceptionalist values of public policy and law completely dominate scientists' research choices. The fact that government-funded research is often extremely harsh on nonhuman animals has drawn many challenges, but those challenges only rarely come from within the laboratories. There is in laboratories, as in animal science courses, a pronounced effort to discourage any criticism of existing practices. When the sociologist Arnold Arluke assessed the ethical socialization of workers in laboratories that conducted research on nonhuman animals, he found that laboratory managers attempted to control how employees spoke about the experimental subjects. Because he found that words like "sacrifice" were mandated replacements for more literally correct words such as "kill," Arluke observed, "Unstated rules dictated how people interacted with laboratory animals. Social norms stipulated that they were objects and not pets, and sanctions supported this definition."[9]

While industry representatives rarely describe their intentional killing of other living beings as ethically charged, this reality is more readily called out in veterinary circles. As Adrian Morrison, a veterinarian who for decades has been among the most ardent advocates of using nonhuman animals in research, suggests in his 2009 defense of animal-based research, such work is difficult on the researchers and clearly comes out of a past that was "without doubt terribly cruel" and "brutal."[10]

While candor about such harms is often curtailed by those who benefit from the profits generated by this form of scientific practice, science as a larger human enterprise has its own power and logic that go well beyond profit motivations, funding priorities of governments, or desire of individuals for financial reward or political power.

Humility and Inquiry as the Heart of Science

Whenever work within individual sciences slips into exceptionalist manipulations because researchers fail to elucidate all relevant realities, Animal Studies can underscore that the overall enterprise is nonetheless an astonishingly broad and powerful tool for discovering a wide range of realities. Science can, if scientists choose to do so, explore even those realities being ignored in industry or animal science, such as the actual realities and suffering of the food animals treated as food production machines.

Animal Studies, in fact, has abundant resources to contribute to the debate about how our collected sciences must, if they are to realize their full genius, be integrally tied to self-imposed humilities of several kinds. These humilities apply at multiple levels—they include the importance of individual humility, but, more importantly, they also require species-level humility. Each of these humilities is a form of the self-transcendence described in chapter 1 that prompts forms of community—recall Frankl's observation that "self-actualization is possible only as a side-effect of self-transcendence."

The respected historian of science Steven Shapin puts the issue of personal or individual humility in this way—"the irrelevance of the personal in scientific knowledge-making has been vigorously asserted at least since the seventeenth century."[11] Personal dimensions, as Shapin points out in great detail, nonetheless play extremely important roles in science. But in order to meet the goals of doing good science and educating a future generation in the mentality needed to do good science, personal biases and preferences cannot be allowed to control. The same must be said for species-level biases. In a theoretical sense, then, there is no place in science whatsoever for bias in favor of one person or one species—the canons of science are decidedly neutral at both levels.

A corollary of these first two humilities is the hallmark of science—the commitment to seek the whole truth even when it is not self-serving. This is a hallmark because a self-contradiction prevails when bias for a particular finding, rather than an open-minded, question-welcoming search for the actual truth, prevails in any field of science. It is, in fact, this inclusive openness that makes so plain the antiscientific features of the agenda-driven phrase "humans and animals."

Similarly, this inclusive openness reveals how bankrupt it is to discourage student questions in animal science courses or to control vocabulary in laboratory settings. The Western scientific tradition emerged, and then came into its own, through a series of revolutions against a restrictive worldview from the sixteenth to the twentieth century. That restrictive worldview insisted upon extraordinary illusions about humans' central place in the entire universe. Arguably, science loses its genius when it loses a willingness to tolerate questions or frank ways of speaking about prevailing illusions or other problems that block the vigorous, untrammeled pursuit of truth. The exceptionalist agenda of agribusiness, establishment veterinary medicine as it caters to agribusiness and government funders, laboratory directors who mandate how their workers speak, or educators who insist their students share the educator's bias for human advantage make one thing clear—the practice of science can violate the open-minded spirit of modern science's revolutionary turn of mind.

Science's Integrities

Along with the modern scientific tradition's humilities, a constellation of commitments or integrities functions as the heartbeat and engine of the scientific worldview. The most obvious integrity of science is the search for the truth, which Animal Studies appropriates as a fundamental common sense. Humans in a group can readily be dominated by bias, love of fantasies about our self-importance, and recurring inclinations to be political rather than moral or principled in pursuit of the truth (chapter 8). For these reasons, a successful exploration of the realities amid which we find ourselves requires a powerful commitment to the truth—this is why development of critical thinking skills is vital to the emergence of healthy thinking in any field but especially in Animal Studies in a human-centered era.

Related commitments or integrities of science are its structured commitment to open-mindedness and a disposition to explore. These are corollaries of any foundational, organizing commitment to the search for truth, and together they amount to an ethic of inquiry of the kind described in chapter 2. These related commitments pay homage to reality by prompting simple, humble questions such as, What is our world like? and How can we describe it best given our limited abilities? A succinct summary of how these integrities work in combination is the motto of the Science Channel, "Question everything," which also displays prominently another key idea: "When you ask questions about the world around us, that's when it happens. That's how the revolution starts."[12] A fine example of how science is not only honored but marketed as a respected, even privileged way of speaking about the world around us, the Science Channel's motto embodies how a structured commitment to open-mindedness and a disposition to explore bring out the importance of a frank, basic commitment to explore wherever that effort takes us and our thinking.

A further integrity of science is a sustained dedication to transparency. This is, in effect, the public or communal version of a commitment to truth. This integrity, which helps ensure that others are able to assess how any one person or group arrived at conclusions regarding their claims about reality, is anchored in a form of skepticism about individual experiences even as it reveals confidence in communal work. As every reader will have experienced, an individual human may be confident, even certain, of what he or she perceives, but end up being wrong. Transparency is one way of pushing science to less guessing and more knowing, as it were—there are, to be sure, forms of guessing at the very heart of science (prediction and statistics and approximations are forms of guessing). These are not, however, mere guessing, but rather a kind of principled guessing. Transparency in scientific work is one way of getting all of us to assess science at work.

An open mind regarding the role that unlearning plays in the pursuit of truth is yet another integrity that follows from the individual- and species-level humilities needed to do science. This particular integrity of science can seem irreverent to those who believe they already possess absolute truth. But, as noted below, sciences have often been subversive in the sense of challenging widely accepted views that many hold to be a matter of traditional authority or divine revelation. This open-mindedness arises out of an important feature of the history of science—the scientific revolution was driven by more than efforts to discover what we might know about our surrounding worlds. It was also driven by reactions to traditional explanations and claims that exponents of tradition insisted were the absolute truth about the world in which all live. The purported "knowledge" of traditionalists, which so often in reality amounted to nothing more than requiring conformity to some authority's approved viewpoint, was often merely inherited custom. Since many traditional claims were obviously unverifiable in any meaningful sense, adhering to the mainline form of traditional beliefs often came to be what twentieth-century citizens termed "politically correct." Disbelievers were often disadvantaged, persecuted, ostracized, even executed simply because they openly refused to subscribe to one feature or another of a dominant view. In response, science over time created its structured commitment to open-mindedness and the role that unlearning sometimes must play if we are to have less guessing and more principled knowing.

One additional integrity of modern science stems from our biological origins and ecological realities—this is a willingness to explore the inevitable connections that each human has to "others" in the world. This includes, of course, connections to other humans, but it also pertains to nonhuman animals, ecosystems, and the inorganic world as well. As Carl Sagan famously summarized in *Cosmos*, "The nitrogen in our DNA, the calcium in our teeth, the iron in our blood, the carbon in our apple pies were made in the interiors of collapsing stars. We are made of star stuff."[13]

Animal Studies works constantly with, and is enriched by, each of these integrities—the search for the truth, a structured commitment to open-mindedness, a readiness to explore, a sustained dedication to transparency, an open mind regarding the importance of unlearning, and a willingness to explore connections.

One Implication of Species Neutrality

While science can, as a whole, fairly be said to be neutral on the issue of both humans and nonhumans, it is common to hear the claim that using certain nonhuman animals as subjects in invasive, harmful experiments is "scientifically valuable." There is a commonsense truth in such claims, for given all animals' interrelatedness, experiments on one sort of animal have prospects of delivering information that helps one understand other sorts of animals. Thus, entirely apart from the moral issues raised by such practices, critical thinking suggests that it is useful to wonder if experimenting on certain nonhuman animals can produce some information that informs humans about the illnesses and other biological realities humans share with the experimental nonhumans.

While critical thinking mandates that moral issues be dealt with frankly, it is possible to set them aside temporarily in order to get the clearest possible view of the line of reasoning one pursues about the issue of obtaining "scientifically valuable" information through experimentation on nonhuman animals. Critical thinking also pushes those who advocate using nonhuman animals in this way to ask if there is an even greater truth, so to speak, in the claim that experimenting on humans helps humans with their own biological problems. Using scientific criteria, humans are likely to be the best experimental subjects by far if the goal of the experimenter is to identify and solve human problems. Thus it is impossible to avoid the conclusion that in whatever sense it is true that using nonhuman animals as experimental subjects is supported by science, it is even truer that (1) science supports using humans, and (2) science leads to the conclusion that, for the stated purpose, using humans is superior to using nonhuman subjects. It follows from these observations that those who claim the needs of science favor using any and all nonhumans for humans' benefit will be inconsistent if they fail to acknowledge that their own argument implies an even more powerful argument that the needs of science support the use of humans as superior experimental models.

When the goal is "scientific value," but only nonhumans are to be considered as experimental tools, then some other value is controlling the conclusion that bars use of the superior scientific tool. Critical thinking allows one to identify such incomplete patterns of argument, and also to identify the values that are bootlegged into arguments of the kind set out above. In this regard, critical thinking can illuminate the ethical dimensions of scientific practice—those who advocate the use of only nonhuman animals as tools to develop scientific findings expressly meant to benefit humans are clearly not reasoning in a value-neutral manner.

Critical thinking can, of course, go far beyond approaches that examine only isolated lines of reasoning and their shortcomings. It also permits one to view science as a whole, especially the claim often made by scientists that their work is value free. A truly value-neutral version of science would be silent on the question of which animals to favor through scientific experimentation. This is the point made by a Nobel laureate in physics, Erwin Schrödinger: "The image of the world around us that science provides is highly deficient. It supplies a lot of factual information, and puts all our experience in magnificently coherent order, but keeps terribly silent about everything close to our hearts, everything that really counts."[14]

Science is, then, ideally neutral in an ethically important sense—it works in ways that fail to favor any one position, person, or species. An upshot of this observation is that the prevailing conclusion in today's scientific establishment that researchers can, without moral qualm, use nonhumans in harmful experiments reveals a value judgment that obviously favors humans and ignores the science-relevant point that humans are in fact superior experimental subjects when the goal is truth about humans' biologically based problems.

The practice of using only nonhumans, but not humans, as experimental subjects, then, obviously contradicts any claim that science is completely value free. It is just as obvious that many scientists are not distressed by this, for most express no concern whatsoever that established practices and policy in matters of experimentation are anchored in non-science-based values and thereby, in this case at least, support the exceptionalist tradition. Nonetheless, several things are clear—this human-favoring policy, while consciously chosen, is a decision made in precincts where citizens of any society with this policy can, if they choose, also push for a different policy that refrains from using some or all nonhumans in harmful ways.

Animal Studies, Biology, and Complexity

Critical thinking also raises important queries about whether it makes any sense for Animal Studies to stay focused so heavily on the animal world rather than broader topics that include plants, ecosystems, and the inanimate world. One obvious reason that Animal Studies stays focused primarily on animals is that, as animals, humans are particularly fascinated by living beings who are noticeable to unaided human perception.

But there are reasonable and even powerful counterarguments by which one can push Animal Studies well beyond macroanimals—it can be said, for example, that living beings made visible with aids such as microscopes draw our interest as well. Further, staying solely within animal-defined parameters is not really possible, for those who study animal lives of all kinds must also study the extended environments where animals live in order to grasp any meaningful features in animals' lives. Because there are, not surprisingly, any number of reasons for humans to look with responsibility and imagination beyond animal issues, the reach of Animal Studies naturally includes careful examination of the more-than-animate features of ecological life.

At the same time, there are powerful reasons that Animal Studies keeps animals as its heartbeat issue. Four reasons are listed below, although there are surely more that human imagination can develop. The first argument was already stated by Shepard regarding other animals as "indispensable to our becoming human in the fullest sense" (chapter 2). The second, third, and fourth arguments involve, respectively, a science-based reason, an ethics-based issue, and an education-based practicality.

The science-based reason Animal Studies stays focused on animals is related to a special feature of complexity in the world around us. A commonly stated paradigm for science in general is drawn not from the biological sciences, but instead from nonbiological, more exclusively physical sciences sometimes thought of as purer forms of science, such as physics and astronomy, as well as the fascinating science of mathematics.

But consider the claim that it is not these sciences that are the most complex, but rather biological sciences. Biology, this argument goes, is more interesting to us because it is inherently more complex in an important sense. It includes the complexities of the merely physical world and more. The reasoning behind this claim has been explained by one of the twentieth century's leading evolutionary biologists:

> Insistence that the study of organisms requires principles additional to those of the physical sciences does not imply a dualistic or vitalistic view of nature. Life…is not thereby necessarily considered nonphysical or nonmaterial. It is just that living beings have been affected for…billions of years by historical processes.…The results of those processes are systems different in kind from any nonliving systems and almost incomparably more complicated. They are not for that reason necessarily any less material or less physical in nature. The point is that all known material processes and explanatory principles apply to organisms, while only a limited number of them apply to nonliving systems.…Biology, then, is the science that stands at the center of all science.…And it is here, in the field where all the principles of all the sciences are embodied, that science can truly become unified.[15]

Said more simply, the biological arena is one in which all of the material processes analyzed by physics and chemistry are evident. These material processes, however, have undergone a process of historical development (evolution) that has produced processes that are qualitatively more complex than the physical realities studied by physics, chemistry, astronomy, geology, or mathematics. There are, to be sure, bridge sciences, as it were—a relevant example is biochemistry, where a science dealing with the building blocks of reality (in this case, chemistry's stunning sophistication in molecular and atomic matters) illuminates biology, and does so in ways that teach humans much more about life than they can observe by simply watching the other macroanimals and plants in their environment.

Still, Simpson's point remains an important reason that nonhuman animals incite particular fascination in human animals. Said simply, animals' obvious biological complexities intrigue us, often more so than the processes that we notice in the parts of our world that are inanimate. Biological phenomena are, then, particularly fascinating to us because they alone seem to have a particularly rich and inviting complexity.[16] This deep fascination on our part provides a first reason that explains why Animal Studies keeps a focus on other animals as its heartbeat.

To the extent this first reason is explained solely in terms of science-mediated knowledge about the world around us, it does not fully explain our fascination with other animals. A second, ethics-based reason helps fill out why other animals are so compelling to human animals. Our interest in other living beings is anchored in our own special abilities—our ethical abilities that make us capable of recognizing how to treat another individual as a morally important "other." In summary, this ethics-based reason for Animal Studies concentrating on animals more than on inanimate objects relies on the undeniably rich individual-to-individual skills that each of us has. As noted below, these skills prompt caring within and across the species line.

This undeniable feature of our lives creates our special genius for one-on-one connections. This set of abilities is, however, not exclusively human, for our abilities to care have deep and wide mammalian roots. To make this point about our close mammalian cousins is not to deny that humans' version of ethical abilities is unique and seemingly the most capacious. Yet critical thinking requires that while this conclusion does seem reasonable, clearly we do not know yet that comparably rich and interesting abilities are absent from any or all nonhuman animal communities. This is so because we have only begun to explore which features are found in those societies composed of nonhuman individuals who have large brains, social skills of great complexity, and distinctive personalities (all of which are true of cetacean, nonhuman great ape, and elephant societies). This humility-focused observation suggests we do well to heed advice given by one of the twentieth century's most famous philosophers in 1922: "whereof one cannot speak[,] thereof one must be silent."[17]

What we can say with great confidence is that it is within encounters with other individual living beings that our own ethical abilities begin and then occur on a daily basis. These "others" are sometimes human, sometimes nonhuman, for while it is true that our earliest encounters are usually with parents and siblings, very young children also engage companion animals, backyard wildlife, and sometimes other domesticated animals. The "others" we encounter in childhood do not exhaust our abilities to care any more than our membership in a particular group of humans exhausts our abilities to care. Beyond immediate family members (which, again, these days are commonly thought of expansively as including nonhuman members as well) and certainly beyond our local human community, we eventually meet many other, non–family members of both the human and nonhuman ilk.

It is at this basic level of individuality where we inevitably learn and play out the full range of our ethical abilities. Just as early encounters give birth to ethical possibilities and abilities, encounters later in life actualize our abilities to care, helping them mature to their fullest forms. One-on-one caring opportunities, first in the family and home, then beyond, are our earliest moral opportunities, and they invite us constantly throughout life to go further. As we mature and notice this, we recognize that using these abilities beyond the species line is well within our ethical capabilities.

Such encounters, especially with non–family members who do not protect our interests, are also where our ethical abilities can atrophy. We can be trained to ignore encounters with nonhumans, just as we can be trained to ignore encounters with other humans. But it remains true that in any one-on-one encounter, the connection possibilities are one of the keys to our ethical character. For this reason, caring one-on-one, that is, caring in relationship with an "other," is the door that opens onto caring about multiple others.

In one sense, then, caring for others is a chrysalis of ethical abilities, as ethicists have known for a long time. Insights about the importance of instilling care-based virtues early in life appear, for example, in Aristotle's ethics, modern virtue ethics, the tradition of feminist care ethics, and ecofeminism. The upshot is this—if one fosters caring abilities in a child early and often, then the child stands a much better chance of actualizing these abilities throughout life. The converse applies, too—retard caring about others early, and the child is at risk of losing such abilities for life. Retard these abilities in adolescence, and the potential for becoming a responsible adult can also be lost. Nurture these abilities during adolescence and young adulthood, and the prospects of responsible adulthood are more likely to be realized.

Animal Studies takes this key insight across the species line. It provides reasons to believe that training humans to care only for family or local community or one's own culture or nation or species misses a key opportunity to develop our human capacities for one-on-one caring about others who come from outside family, local community, society, or species. Given that the ferment in our society on animal issues is driven in part by a renewed sense of the importance of a truly broad notion of communal caring, Animal Studies in a number of ways features an unmistakable ethical cast as its draws energy from the fact that the complexities of animal life intrigue us in very special ways.

There is, relatedly, one final, education-based reason that Animal Studies keeps its primary focus on animal issues—this reason is a practical version of our abilities in one-on-one caring of the kind discussed above. Practically speaking, each of us as an individual animal can manage thinking about other animal individuals. This insight plays a role in an interesting observation by an environmental studies professor from Massachusetts Institute of Technology named Steve Meyer. As an adjunct lecturer at Tufts' Center for Animals and Public Policy, Meyer for several years taught graduate students pursuing a master of science degree.[18] In 2004, he commented that while he knew MIT students pursuing environmental studies were excellent students, he consistently detected greater focus, direction, and pursuit in the Tufts graduates studying animal-related issues.

Professor Meyer's observation prompts one to speculate how focusing on nonhuman animals might produce educational benefits. One possibility is that such a focus offers an immediacy and involvement with identifiable nonhuman individuals that appeals to, even tugs at, each individual human's existing ethical abilities in ways that the elegant but abstract ideas of environmental protection cannot. Students at Tufts clearly arrived at their graduate program with interests that led them often to heavy involvement in specific animal protection causes such as improving adoptions from shelters, stopping cruelty, or raising awareness of some specific category of food animals, research animals, or wildlife. Said another way, the Tufts students arrived with developed abilities in individual-to-individual concerns typical of the animal protection movement. They were already convinced that specific acts of compassion can easily and regularly make a difference in real lives. This realization helped foster keen inquiry about the potential reach of ethical consciousness and action, as well as pursuit of specific ways to alleviate problems. In this respect, the Tufts students seemed to Dr. Meyer to have more specific forms of inquiry and pursuit than did the MIT students studying higher-level generalities like the extinction of species, habitat destruction, and the environment. To be sure, these more general issues are crucial, but their very breadth and generality are harder to grasp and less immediate and personal than is stopping harms to familiar and treasured individuals like dogs, cats, horses, and similarly familiar "neighbors."

These observations are by no means the final word, for they are tentative generalities based on limited exposure to only certain kinds of students in highly specialized circumstances. But Meyer's observations suggest this important possibility—focusing on actual individual nonhumans—calls upon basic skills that humans learn early and then use often.

As importantly, it is through focusing on individuals that one grasps why it is important to avoid harms—not only does each of us as an individual know the risk of harm, but additionally each of us readily can see and feel the harm to another individual in one's presence. A related insight soon follows—what matters is individual acts impacting real living beings. This realization is easily understood and remembered, which means it can anchor learning in ways that generalities like "the welfare of a species" and "the good of the environment" often do not. One-on-one relationships, because they are something each of us intuitively grasps, offer the collateral benefit of opening each individual up to the importance of ethics, which in turn allows commitment and "pursuit" to become part of one's life. Further, such experiences may even open us to early consideration of more general issues such as how to refrain from harming abstractions like "a species" or even more generally "the environment."

History of Science as a Key Field in Animal Studies

The dual features of science addressed above—the key integrities or commitments of science as a whole versus the politicized practices of many contemporary research institutions, organizations, and individual scientists—are largely based on the Western scientific tradition that is commonly understood to be rooted in an important period in Western cultural history usually called the scientific revolution. Interestingly, the idea of a single scientific revolution is misleading for two reasons. First, there are other science traditions that humans have developed which, though less well known, have importance in their own right.[19] Second, historians of science who focus on the powerful Western scientific tradition for decades have regularly pointed out that while science has caused revolutionary changes in Western culture, those changes took place in fits and starts and in so many areas for so many reasons that one cannot talk of a scientific revolution. This is one of the reasons that Shapin opens his 1996 book *The Scientific Revolution* by stating, "There was no such thing as the Scientific Revolution, and this is a book about it."

Whatever position one holds on scientific revolutions, or on the achievements of one scientific tradition versus others, Animal Studies is an extension of the spirit of this kind of human work. This field embodies and extends the mentality that led to the changes we think of as "the scientific revolution," as well as the general spirit of inquiry and concern for the truth that drive the scientific spirit. This general spirit is the source of the commitment within Animal Studies that requires careful study of other-than-human animals' actual realities. Careful, empirically based work about the realities that surround each of us is surely not the sole province of science, but such work is nonetheless done well in scientific traditions when they hold true to their basic humilities, commitments, and integrities.

Balancing Discoveries and Intrigue

Science-based discoveries about other animals' lives give those who pursue Animal Studies strong reasons to infuse even nonscientific work—such as doing history, honoring cultural diversity, or assessing education—with science-informed views of other animals. Further, the basic humilities and integrities of science, particularly as they coincide with and thereby reinforce commitments to see other animals as they actually are, help anyone explore the many diverse forms of life and awareness outside our own species. In effect, the findings of science and the larger commitments of the entire scientific enterprise help individuals and entire academic fields get beyond the exceptionalist tradition.

The history of science, though, has many episodes in which this major human enterprise was not dominated by humilities and its commendable integrities, but instead fraught with intrigue and constant battles. In tragic ways, such problems share some features of the better-known battles within religious traditions or political systems. One source of such intrigue has been human limitations—scientists are, after all, humans prone to biases, limited vision, jealousies, conspiratorial motives, and assorted other human frailties. These all-too-human traits subvert the very integrities that give science its power, thereby undoing the humilities that science at its best prompts so well.

Another source of intrigue is the fact that some who wrap themselves in science conspicuously fail to honor its basic ideals. Sometimes this is due to dishonesty and fraud, when research is falsified for financial advantage. But this is a relatively minor problem because of the sustained commitment within scientific circles to transparency. More commonly, problems arise because science is pursued in ways that serve human biases of one kind or another—such as reluctance to discuss other animals' cognitive abilities, described as "mentophobia" driven by an altogether unscientific "paralytic perfectionism" (chapter 6).

A third source of various intrigues that have marred the history of science is the fact that scientific method, even when pursued as fully and ideally as possible, has inherent limits. Some try to build from the fact that science done through precise measurements and quantification can be remarkably powerful to an altogether more far-reaching suggestion, namely, that measurement-based science is the only valid form of knowledge and can tell us all we want to know. Without question, science-based quantification is a powerful tool for those who seek to peer into nonhuman animals' lives. It can overcome what many have long taken to be insurmountable obstacles—for example, the techniques of passive acoustic monitoring can be used to learn where cetaceans swim and even with whom.[20] This fascinating science on its own, of course, hardly exhausts questions about cetaceans' daily lives, let alone their inner lives. What is learned from the quantification-based techniques of passive acoustic monitoring can be connected with what is learned in other science-based work, giving us a much better picture. Such combinations can even help us peer into other animals themselves and features of their ancestors' remote past. Below the mid-twentieth-century development called "neo-Darwinism" or "the evolutionary synthesis" is mentioned because this powerful set of scientific tools using genetics-based information can help us understand a great deal about any living being's basic abilities as inherited from ancestors.[21]

Quantification-based science is, thus, obviously powerful and important—sometimes it may even supply an essential piece for solving certain puzzles. But it remains partial in a most fundamental sense—for example, in the case of cetaceans, even multiple sciences in combination do not exhaust questions about cetaceans' lives. In effect, quantification-based science falls short of the Schrödinger test, for it "keeps terribly silent about" so much that is "close to our hearts." Using individual sciences' evident power to illuminate some aspects of the world to justify the far-reaching claim that scientific methods or even all sciences together are the only basis of knowledge is to leap far further than logic can carry us.

Challenging one-dimensional claims focused on but one method, or one science, or one suite of natural sciences is crucial. There are not only many productive scientific methods, but also valid sources of knowledge outside of science altogether, as implied by Schrödinger's lament. As this book points out, even when science is done as well as possible, it turns out to be just one of the major ways of exploring other animals. In fact, before modern sciences were developed, there were abundant observation-based explorations that remain just as relevant today as they were in the past.

Intrigue also is evident when one group of scientists privilege their own field by deriding others. A now-classic example of this problem is the dismissive view of less theoretical and less mathematics-reliant sciences embodied in the quip of a Nobel Prize–winning physicist: "all science is either physics or stamp collecting."[22] Animal Studies is particularly adept at pointing out the advantages of "stamp collecting" (observational work that helps humans recognize so much about other-than-human animals)—such efforts have made vital contributions to learning and thus clearly qualify as an essential form of inquiry about the truths and realities foregrounded by both science generally and Animal Studies more particularly.

Perhaps the greatest intrigue in the history of science, however, has been resistance to new ideas, for this human tendency has led to much strife, polarization, and many other problems. Many examples from geology, anthropology, physiology, astronomy, and countless other sciences illustrate how ideas now widely considered common sense were long repudiated in scientific circles. One of the most dramatic examples, to be sure, is the debate over Darwin's work about the relationship of humans to other animals. Before discussing how Darwin's achievements were received, it should be acknowledged that even when scheming, bias, and resistance to new ideas have produced tension in the history of science, these eminently human failures in no way eclipsed the value and power of science. Animal Studies has been, and will continue to be, a beneficiary of an astonishing range of science-based work done in terms of the most basic commitments and humility that drive the scientific tradition generally. So although this introduction to Animal Studies at times points out that some work within science—such as veterinary medicine, zoos, and animal science—continues to go forward with forms of bias that foster the exceptionalist tradition and serious harms to many nonhuman animals, contrasting such problems with the astonishing successes evident throughout science is important to providing a full and fair account of this powerful human enterprise. Similarly, Darwin's work readily reveals how science in both theory and practice is an astonishing cumulative achievement of humans across many centuries and cultures.

Darwin's Evolution and the Larger Community

The work of Charles Darwin is pivotal not only in the history of science and the development of Western culture, but also for humanity and the world generally. Darwin's great contributions are well known—his insights about evolution, natural selection, and common descent, including his detailed documentation of forces affecting the survival and proliferation of species, produced deep changes of many kinds in and beyond Western culture—and worldviews have shifted to include much more awareness of the fundamental continuity between humans and other species.

All of these developments are of the greatest relevance to Animal Studies, but, in particular, Darwin advanced the processes by which humans and nonhumans alike have been demythologized. His powerful synthesis of perspectives on the biological bases of intelligence, emotions, cooperation, communication, and competition over time opened up many people to connections which, though obvious, had been played down greatly. Historically, Darwin's argument was only the latest in a long series—many different individuals had suggested that people already knew a basic reality, namely, that one had, as suggested by the Roman writer Lucretius in the century before Jesus was born, "only to look attentively at the world around us to grasp that many of the most intense and poignant experiences of our lives are not exclusive to our species."[23] Scores of science-based researchers had for centuries before Darwin's *Origin of Species by Means of Natural Selection* (1859) provided countless observations about the obvious similarities between humans and some nonhumans. The famous French philosopher Jean-Jacques Rousseau had even asserted in 1753 that humans and chimpanzees were members of the same species, and this view was again asserted in 1770 by the almost equally famous Lord Monboddo. In 1781, Charles Bonnet, comparing apes and humans in *Contemplation de la Nature*, stated plainly, "We are astonished to see how slight and how few are the differences, and how manifold and how marked are the resemblances."[24]

The obviousness of the overlaps between humans and some other animals notwithstanding, similarities had been played down in many influential circles. This is in part because the powerful European tradition had been exposed to very few nonhuman great apes before the nineteenth century, and very little to elephants, cetaceans, or other nonhumans that feature large brains, relatively complex communications, and even cultures. Recall the 1747 comment by Linnaeus, who created biology's basic classification scheme still in use today— he cited nonscientific forces as playing a crucial role when he made his decision to "distinguish between Man and Ape": "if I had called man an ape, or vice versa, I would have fallen under the ban of all ecclesiastics. It may be that as a naturalist I ought to have done so."

Darwin's work famously challenged very powerful religious and secular dismissals of humans' organic relationship with other-than-human animals. It is telling that although his explanation of the connection between human animals and the rest of life has the kind of elegance that appeals to many people today, in the second half of the nineteenth century many people thought it ugly. Many versions of one story in particular reveal that Darwin's ideas were thought by some to be unseemly—the wife of the bishop of Worcester is reported to have said in 1860, "Let us hope that what Mr. Darwin says is not true, but if it is true, let us hope that it will not become generally known!"[25]

In a way, the bishop's wife got her wish—at least for a while. Darwin's ideas did not carry the day in scientific circles until the 1930s (half a century after his death) when a new era in Darwin-influenced thinking produced a wide-ranging synthesis of his insights and important discoveries, like population genetics, known generally as "neo-Darwinism" or "the evolutionary synthesis."[26] Through a combination of mathematical techniques and genetics-based discoveries, highly technical and powerful predictions became possible regarding gene frequencies in populations over generations.

Even if, scientifically speaking, this important set of ideas about animals prevails today, Animal Studies must still contend with two important points of resistance. One is the culturally and politically powerful opposition to evolution from many religiously inspired sources. In 2004, the widely respected publication *National Geographic* ran a story under the headline "Was Darwin Wrong?" which included this passage:

> The most startling thing about these poll numbers is not that so many Americans reject evolution, but that the statistical breakdown hasn't changed much in two decades. Gallup interviewers posed exactly the same choices in 1982, 1993, 1997, and 1999. The creationist conviction—that God alone, and not evolution, produced humans—has never drawn less than 44 percent. In other words, nearly half the American populace prefers to believe that Charles Darwin was wrong where it mattered most.[27]

Thus, although Darwin's ideas have shifted many humans' understanding of our own animality and rich connections with other animals, the debate over Darwin's legacy remains fierce. This can be seen not only in the continuing resistance to Darwin's ideas in many spheres but also in the resilience of language-based habits like "humans and animals" that predate Darwin by centuries.

Yet while there will almost certainly continue to be challenges and adjustments to Darwin's evolutionary insights, today his views are a widely respected component of modern science and a key to understanding, then telling, the story of humans' past and present thinking about animals. Thereby, Animal Studies today is one of Darwin's greatest beneficiaries because his work fully exemplifies the ways in which mature, scientifically focused thinking can prompt recognition of connections and overlaps between human and other-than-human animals.

Darwin-Inspired Maps of Life

The continuing resistance to Darwin is telling in modern society and the basis of some claims that the public is not well versed in science. More troubling, however, is the prevalence of dualistic language like "humans and animals" in scientific circles. That antiscientific habits of mind prevail in, sometimes are even promoted by, the scientific establishment is more than a challenge to the scientific spirit—it is also a revelation regarding how much is still thought to be at stake. In a very real way, when scientists perpetuate the dualism "humans and animals" (as when, for example, they teach their students to talk in this way), they exhibit either a lack of conviction about the truth of scientific categories or a form of cowardice in relation to reality.

Despite this lingering resistance, science circles have influenced Animal Studies greatly in one very structured and specific manner—they have created exhaustive maps or taxonomies of life that reflect Darwin's central insights. These are the most complete taxonomies available today, providing a basic inventory of life by which Animal Studies can measure its own work (in the sense of noting which nonhuman animals are spoken about compared to which ones are known to exist).

The development of science-based taxonomies has by no means eclipsed the prevalence of folk taxonomies—modern categories such as "companion animal," "food animal," and so on are a kind of folk taxonomy, and such nonscientific categories remain extremely important. Yet, revealingly, it is scientific taxonomy that is invariably called upon whenever a reliable taxonomy is desired. Using the great breadth of scientific taxonomies, scholars and students can shed light on other cultures' construction of maps or taxonomies of life. Most cultures offer a map of life or folk taxonomy that seems, relative to today's scientific schemes, simple and easy to follow. To any person with a late twentieth-century education, it is nonetheless obvious that most folk taxonomies leave out many animals. This is not surprising in one sense—any small-scale society or local community occupies territory that has but a small percentage of the earth's different forms of life, and thus its citizens have little or no exposure to the vast diversity of the larger world's animal life. Citizens in modern industrialized societies are the beneficiaries of globally shared knowledge about other living creatures, and on that basis alone, but especially relative to science-developed taxonomies, it is easy to recognize that nonscientific classification schemes list only some of the earth's animals.

Caution about the Best Maps of Life

Even if one is familiar with the formidable Darwinian synthesis and the most modern taxonomies that offer surprisingly specific information about the interconnections of different kinds of life, one must not confuse that feature of scientific taxonomies with truly knowing other living beings in context. With that qualification in mind, one can reflect on what one of the most widely used, though by no means universally accepted, maps of life suggests.[28] Known as the five kingdoms classification (the image draws on the old patriarchal notion of kingship), this map of life divides living beings into fundamentally different kinds of living organisms. It is based on the astonishingly detailed information about the diversity of life beyond the human species that our life sciences have provided through much cumulative work in the twentieth century. The simplest forms of life are microscopic single-celled living creatures that scientists call eubacteria (eu is a Greek word that here means "true") and cyanobacteria or blue-green algae. This invisible class of beings contains millions of different kinds of unicellular life that scientists call the Monera kingdom.

Modern taxonomy sets these simple, microscopic forms of life apart from four other, altogether distinct kingdoms of living beings. Each of the other kingdoms (some schemes list four additional kingdoms, while others list five) contains almost countless additional forms of life. As importantly, each is a kingdom because its life forms comprise a dramatically distinct form of life. The first of these additional kingdoms, named Protista by scientists, includes organisms described by technical, unfamiliar phrases like "unicellular protozoans" or "multicellular algae."

Beyond these first two kingdoms are three more whose names will sound more familiar, namely, the kingdoms Fungi, Plantae, and the familiar-sounding Animalia. If this fivefold division is not daunting enough, keep in mind that from a scientific standpoint, these five kingdoms do not comprise the whole of life, for outside of them is yet another important form of beings that have a few features that seem to most of us to deserve the name "living"— these are viruses, but they are outside the normal scheme of life because they are even simpler than the single-celled bacteria.

One implication of this science-based inventory of life is that the vast majority of living beings are not in the scientific category Animalia, but instead in the other kingdoms. Even within Animalia itself one finds overwhelmingly complex diversity because this kingdom features an astonishing twenty-one subdivisions or phyla. Most of these twenty-one phyla include living beings that we would colloquially call "worms," but which are, scientifically, of many different phyla. The very last of the twenty-one phyla is Chordata, which will seem familiar-sounding territory because many of us recall that humans are, scientifically, chordates, that is, possessing a backbone.

The Chordata are, nonetheless, further subdivided into three major subphyla which contain beings that are still not "animals" by common definition. This includes living beings in the Subphylum Urochordata (sometimes called "tunicates") and Subphylum Cephalochordata (or "lancelets"). Finally we get to the Subphylum Vertebrata, which is, one might assume, at last a familiar neighborhood. These subdivisions of life include many living beings that we might call "animals," but which even animal advocates characteristically ignore when they talk about animal protection issues. This fact helps one see an important implication of the scientific classification of life—the vast majority of living beings are dramatically different from the few tens of thousands of species of living beings that humans are used to calling "animals."

Thus even the familiar-sounding category "vertebrates" is subdivided into classes that only rarely, if at all, receive any mention in animal protection discussions. Here are the subdivisions (or classes) of Subphylum Vertebrata:

Class Agnatha (jawless fishes)
Class Placoderms (armored fishes)
Class Chondrichthyes (cartilaginous fishes, such as sharks and rays)
Class Osteichthyes (bony fishes)
Class Amphibia (amphibians)
Class Reptilia (reptiles)
Class Aves (birds)
Class Mammalia (mammals)

In this list one finally encounters the mammals, birds, reptiles, amphibians, and fish that populate folk taxonomies. Among these, too, are the macroanimals that different peoples have historically recognized as potential others. These living beings, but particularly the community of mammals, birds, amphibians, and reptiles, are the focus of most Animal Studies discussions, courses, and, especially, programs and scholarship.

Keep in mind that scientific taxonomies, so important because they reveal that nonhuman living beings are more diverse than any culture ever imagined (in part because they list so many micro forms of life beyond the range of humans' eyes, ears, and touch), by no stretch of the imagination confirm that humans today know a great deal about other animals. In fact, the diversity that scientific taxonomies memorialize only testifies to how impoverished are many contemporary concepts of "animals."

Deeper into Science: The Problem of "Animals in and of Themselves"

Identifying and putting into full context the actual realities of other animals is a goal that makes sense from many vantage points—the search for truth as it is found in the sciences and in ethics, the interests of the curious, and even the simplest forms of common sense. This goal of first attempting to discover, then contextualizing what one thinks one has learned about other animals' realities will, no doubt, seem to many an intuitively obvious objective of Animal Studies. It is, for example, the kind of intuitively appealing approach to other animals that is the backbone of the insights by the zoo expert quoted in chapter 1 about life in the wild being the measure of zoos.

Realizing this aim, however, poses very difficult challenges for humans. This is one reason Animal Studies must work with and through many sciences and other fields. In one sense, efforts to accomplish the task of noticing other animals' realities can start in a familiar place, for each of us, as an embodied animal, knows a great deal about certain dimensions of animal life. Talking of ancient human hunters' awareness of macroanimals and then contemporary humans' attitude about "the obvious similarities of sexual acts and bodily functions," one scholar suggests,

> We are so very squeamish and silly about these today because our culture operates to distance ourselves from, and deny our biological kinship with, other animals. Our early ancestors, however, were intimate with the living world, not alienated from and hostile to it. When they followed a bison herd and watched a bull, penis red and dripping with semen, mount and move his loins against a cow, they would surely vividly recall their own sexual experiences. On their daily foraging rounds, they were likely to see animals eating, drinking, defecating, and urinating—acts that are daily human experiences as well.[29]

Each of us also knows, from our own experience, that at least one macroanimal (our own self) has a point of view, personal interests, and awareness of other macroanimals. Further, we often have a profound certainty that the familiar macroanimals in our lives have their own experiences of the world and it matters little whether these individuals are human, nonhuman mammals, birds in the household, or local wildlife.

Some philosophers, of course, aggressively deny that we can know much about any other being's actual realities—for millennia, various philosophers have also extended this skepticism to even our certainty about what any other human being is actually experiencing. But most of us are willing to admit that while such philosophical challenges are humbling (because they speak to our obvious limits,), they are not complete roadblocks. Our imaginations are able to handle notions such as, "I cannot know precisely what this other macroanimal's life is like, but I can guess realistically and successfully at some features of that life." We know intuitively about such things as pain, dislike, fear, and anticipation. We also guess confidently that various experiences and loyalties in other animal individuals are important to these others in some way. We can even sense that other lives are radically different from our own, an idea that may perhaps silence us about other lives except for very general kinds of statements like those made in this paragraph.

At the very least, critical thinking suggests that we be cautious when attempting to describe the point of view of these others. In such descriptions we may in fact admirably elucidate externals (as do scientific taxonomies), but still fail to go deeply into the actual realities of other animals' individual or communal intelligence, their emotions, and on and on. Very often the obvious truth is that we do not really "know," that is, have any reasonable certainty, about the details of the experiences of familiar others. In other words, we are guessing, even if in an informed way. So, plainly stated, the elusive facts of other animals' individual and social lives are hard to ascertain.

One possibility is spending time in the presence of other living beings—let us call this "the Goodall principle" to honor the great scientific contributions of Jane Goodall, whose legendary observational persistence led to discoveries that made her "one of the intellectual heroes of this century." But even if one is persistent and in the right place to observe some nonhuman animals living undisturbed lives, one needs a talent for observation. One also must engage in dialogue with others who have tried to address the range of issues in trying to know these alien beings' realities.

Since, as already noted in a variety of contexts, humans share many features with some other animals, the shared features might provide some basis for understanding other animals in their communities. But humans share only some features with other primates and fewer features with other mammals. We are extremely dissimilar from the vast majority of non-mammalian life, which comprises easily 99-plus percent of all living beings.

These facts suggest a very clear bottom line—identifying and confirming actual realities in context (where those realities play out in natural ways that are different from the distorted kinds of behavior that human presence or domination produces) is exceedingly difficult. There is an important insight to be learned, then, from the philosophers and other reflective thinkers who are skeptical of facile claims about truly knowing in any detail what life is like for another living being. That insight is that we must be cautious in our claims.

Yet if humans work humbly and together to achieve the best possible description given our limitations, we can make meaningful claims about other animals' realities. We know, for example, that what takes place for them is not controlled by our preconceptions of their world. In other words, we know their actual, day-to-day realities are independent of us. About those realities we can describe some basics (see next section). Such a viewpoint, which is sometimes referred to as an "objectivist claim," can be found in both naive and more realistic forms of human thought—the key is to be careful, heeding advice of the kind offered by the early scientist Francis Bacon: "God forbid that we should give out a dream of our own imagination for a pattern of the world." Such problems also prompted Bacon, in the opening lines of his influential *Novum Organum*, to warn, "They who have presumed to dogmatize on nature ... have inflicted the greatest injury on philosophy and learning."[30]

Sciences' Obvious Relevance

Science has multiple ways to confirm what we detect in ordinary life—for example, that some other mammals react to pain in ways identical to our human mammal reactions to pain. Such basic overlaps between science and common sense undergird Voltaire's famous reply to Descartes's mechanistic view of animals: "Barbarians seize this dog.... They nail it on a table, and they dissect it alive in order to show the mesenteric veins. You discover in it all the same organs of feeling that are in yourself. Answer me, machinist, has nature arranged all the means of feeling in this animal, so that it may not feel?"[31]

We do not need science to discover that some other animals experience physical pain as we do, for of course people from time immemorial have been confident that an individual human can observe when another animal (human and nonhuman alike) is in pain. It is our absolute confidence in this intuition that is the basis of every culture's development of provisions for protecting other animals in a variety of ways—all of this is informed by confidence in our perception of who and what they are.

Beyond providing means of confirming some basics we know from daily life, science by virtue of a series of different methods has additional prospects for identifying some features of other animals' realities—as Voltaire's quote above reveals so plainly, one method is to confirm through dissection that the mechanisms and organs of other animals are like those we possess. Dissection of live animals, of course, contends with humans' ethical dimensions. Less contentious methods include dissection of animals that died a natural death or patient observation of living beings by which we accumulate information.

But even in the face of a great deal of detail (the size of a nonhuman community, which mothers rear which offspring, what is eaten and when, and which tools, if any, are used in what circumstances), we often can say relatively little about many aspects of other animals' lives. Almost everyone still recognizes about even the most familiar nonhumans, such as dogs, that many features of their lives are fundamentally elusive or even unknowable. Science itself makes clear that we have reasons for such cautious doubt, for one science after another reveals that there can be no understanding of individual, real nonhumans without seeing their larger world, their place and role in a community that is a part of an ecosystem, or what play out as life-and-death issues, loyalties and fears, and more.

Take the example of the largest dolphins in the world, which are commonly known as orcas. Because these animals are exceptionally social animals, humans who seek out their realities cannot understand either an individual orca or an orca community without learning much about their water world, their social lives in their specific groups in their specific water-based territories, or their intelligences and communications. Orcas reveal that, to even make the claim that one knows something about another being's realities, one must engage that living being's fuller world to begin the process of coming to know that individual's life in any detail.

Claims made about individuals as each functions in their society, within their econiche, on the basis of their abilities are, then, multilevel claims that any human is ill equipped to judge exhaustively unless much patient, sensitive, humility-driven observation has taken place. Barry Lopez has suggested, on the basis of his study of wolves, "The animal's environment, the background against which we see it, can be rendered as something like the animal itself—partly unchartable. And to try to understand the animal apart from its background, except as an imaginative exercise, is to risk the collapse of both. To be what they are they require each other."[32]

Science produces many well-articulated statements of this kind about the intertwining of each living being with its local world on the basis of that individual's abilities. Trying to assess the complexities of this individual-amid-its-world has obvious complications and risks, all of which are seen better if one inquires as carefully and persistently as possible and works diligently to identify one's own assumptions. In such a way, one takes seriously the attempt to see nonhumans' world and realities from their viewpoint.

Sciences without Humility or Qualm

Sorting out the realities of other animals, then, is difficult for many reasons. Not only are the realities often inaccessible, but human abilities to understand what other living beings might experience are subject to obvious limitations given that humans are sight-dominant primates, and in many other animals, other sensory abilities are dominant (such as smell). Scientific capabilities can also be quite limited because there may be no way to gain access to the realities one is trying to learn—sperm whales have the largest brains on earth, but how individuals live and play in sperm whale social groupings is not at all easy for humans to explore because sperm whales live in the deep ocean and dive to great depths.

Apart from these limits, of course, there is the problem of preconceptions anchored in cultural distortions that may curtail questions, answers, and opportunities for research. The example that follows is taken from early modern medicine; it suggests how both preconceptions and a strong resistance to change impact scientific views. Alan Cutler tells the story of Nicolas Steno, the "humble genius" who after extraordinary success in medicine later went on to found modern geology:

> Steno was unusual among his colleagues not only for his skill at the dissecting table, but for the fact that he dissected at all. Most anatomists were unwilling to bloody their own hands and left the work to an assistant. In fact, at most medical schools dissection was more like an academic ritual than a method of scientific research. The ancient texts of the Greek physician Galen had been the primary source of anatomical knowledge for nearly fifteen hundred years. Galen's authority often trumped that of actual cadavers. Dissection was the art of opening up the flesh to reveal what Galen said was supposed to be there. If what was found did not match the text, it was an embarrassment to the dissector, not to Galen.[33]

Cutler points out that Steno found existing scientists' "stubborn adherence to tradition" a serious obstacle to progress in anatomy. Incisions were made according to prescribed rules, and the organs were to be examined in a prescribed order. Such rigidity was completely contrary to genuine scientific research, said Steno, which "does not admit of any set method, but must be attempted in every way possible."

Part of the problem stemmed from the delicate tissue of the brain itself. "Every anatomist who has been concerned with dissecting the brain can demonstrate everything he says about it," he said. "Because its substance is soft and so compliant that his hands, without his thinking about it, shape the parts as he envisaged them beforehand."

Steno challenged Descartes, whose *On Man* had just been published posthumously— in this book, Descartes declared, in Cutler's words, "the pineal gland, a small nut-shaped gland in the center of the brain, to be the crucial link [of soul to body]. Twisting and turning in response to the soul's demands, it literally pulled the strings that controlled the body." Descartes had used deduction alone, and not observation, to come to his conclusion about the central role of the pineal gland. Steno as a deft anatomist was able to show that the pineal gland was held fast and could not move as Descartes described. Still, some of Descartes's followers refused to accept the visual evidence provided by Steno's dissection of the pineal gland.

This refusal is telling. It is precisely this kind of psychologically grounded certainty that one already knows "the truth" that permits many to dismiss contrary evidence plainly before their eyes. If the situation being addressed is complicated, it is all that much easier to deny evidence that one is wrong, for the latter requires one to rethink the prevailing view.

Yet Another Revolution

Although science itself is the result of a series of strong reactions to overbearing authoritarianism and human-centeredness, the scientific establishment has nonetheless tended to become rigid and reactionary on any number of issues. Some of these have great bearing on Animal Studies, for the scientific establishment developed a double standard by which the lives of animals were judged (chapter 6). Human abilities were broadly studied while the evaluation of the abilities of any other-than-human animals was ignored due to an "insidious barrier to scientific investigation" anchored in a "paralytic perfectionism" regarding nonhuman animals' experiences and "minds."

Through the efforts of many inside and outside science, a major change in attitude—sometimes referred to as the "cognitive revolution"—developed from the 1960s onward. This development is both cause and effect of the ferment discussed at the beginning of this book. Scientists have now provided wide-ranging research on other animals' mental, communicative, and emotional or affective abilities.[34] Further, researchers have more recently focused their efforts on understanding stress and trauma in other animals that are often captivity induced.[35]

The upshot of such developments is that complexities in a wide range of nonhuman animals are now being explored in detail. The results often astonish those whose learning was nurtured solely with the exceptionalist tradition of modern education. For example, for decades primatologists have talked about cultures based on a close examination of our cousin great apes' actual realities in their natural contexts (chapter 7).

Science and the Search for Realities

There are several ways in which different sciences provide pieces of the picture by which other animals' realities might be identified and put into meaningful context. Some individual sciences take forms that are surprisingly limited even as they provide important puzzle pieces needed to see the larger issues. Modern introductory courses in biology at the university level are often focused not on whole organisms but instead on molecular-level realities. Such courses can deaden some students' sensibilities to the different kinds of issues that arise in studying nonhumans as whole individuals or, on another front altogether, the complex scientific and ethical issues that arise in the area of animals and public policy. Such courses also do not teach nonreductive methods well—skills crucial to seeing nonhumans well, such as patient observation and a consideration of holistic concerns, are not covered.

Similarly, courses in genetics and scientific taxonomy provide little information about the animals as individual living beings. They produce information that is very relevant to understanding individuals, of course, but any picture drawn solely on the basis of molecular-level realities is likely to be inadequate for understanding nonhuman animals in context. Such an approach would fail the Schrödinger test in the same way that a stick figure conveying a rough idea of human body structure utterly fails to provide insights needed for understanding the whole human.

Questions about what real animals are like are in part addressed by the field called ethology, often defined as studying other animals in their environment. The Greek root of the word, *ethos*, is translated with different English words such as "character," "morals," "principles," and even the cognate "ethos." Sometimes described as a part of zoology, ethology is also viewed as a separate scientific field with its own subfields. Many ethologists commonly divide the field between "behavioral ecology," which explores the association between behavior and ecological conditions, and "cognitive ethology," which examines the subjective side of animal beings studied through ideas like information processing, consciousness, intentionality, intelligence, and subjective experiences.

For those interested in the actual realities of other animals, both subfields have obvious promise. Yet many find that publications in behavioral ecology can be surprisingly lifeless—many reports are stiflingly antiseptic, focused heavily on use of specific, often highly technical words and information presented as "data sets." The latter term is used in both nonbiological and biological situations—it usually refers to a record of actual observations obtained through sampling what is technically known as "a statistical population." In studies of other living beings, the observations are characteristically of very minor behaviors isolated for the purpose of the study. These observations are reduced to a number or, sometimes, a short label drawn from an agreed-upon list, and together these numbers or labels comprise the data set.

While in an important sense meeting the specific goal of attaining rigorous data sets is crucial to identifying many features of a living being's life—how and in what way an animal moves, where this animal goes and how long it remains in specific places, with whom this animal interacts, and on and on—there is a limit to what can be pieced together of a whole animal's world from atomized information recorded in this way. Thus, although such information is invaluable, it provides at best a partial view of an animal as it lives its life amid the complexities of communal and ecological contexts. Ironically, such reports can seem lifeless to outsiders who have not been initiated into the highly technical vocabulary and conceptual subtleties of behavioral ecology.

It is thus common even for those who have great interest in the nonhuman animals that are the subject of such data to conclude that ethologists use words and measurements that empty out, not fill up, the search for other animals' realities. It is as if technicality is foregrounded in order to mimic other sciences dominated by mere quantification, like physics, chemistry, and astronomy. These fields have, in fact, gained great respect because their discoveries have been made through quantification-dominated methods. Newton, for example, had a powerful inclination to measurement-based precision and uniformity in his thinking about space and time (also in his less well-known work on alchemy and Christian theology).

Quantification works in ethology as a means by which the field and its individual scientists can gain credibility in those scientific circles that insist on numbers, quantification, reduction, and precise prediction primarily in terms of numbers. But from the standpoint of a robust, interdisciplinary form of Animal Studies that engages the whole, embodied animal in its full context, such a ploy creates problems. Arguing that it is "quite certain that neither Tinbergen nor Lorenz" (the founders of ethology) "wanted to 'desubjectify' animals," one scholar suggests that the field of ethology nonetheless is full of deadening language:

> Despite their intellectual continuity, there is a great disparity between ethologists and naturalists with respect to their uses of language. In contrast to the naturalists' language of the lifeworld, ethologists use a technical vocabulary, in part constructed by themselves and in part appropriated from behaviorist psychology. The linguistic and argumentative edifice created by the pioneer ethologists led to the representation of animals as natural objects.... The inexorable if unwitting consequence of applying a technical language was the epistemological objectification of animals and ultimately the mechanomorphic portrayal. Mechanomorphism was the price of the idiom that the ethologists opted for; it did not involve the ethologists' deliberate endorsement of a mechanistic view of animals, but was an effect of the representational medium that they elaborated.[36]

Another commentator observes that the result is not a rich understanding but, instead, a debilitating distance: "The objectifying language they employed distances them from their animal 'subjects' and seems to deny those subjects volition or intention. For instance, ethologists speak of 'innate releasing mechanisms' within the animals that 'release' behavioral responses, rather than of animals doing something for their own purposes."[37]

Beyond behavioral ecology is an even newer, still controversial field known as cognitive ethology that covers some of the most widespread questions that emerge from humans' inevitable intersection with other-than-human animals. For example, from time immemorial humans have wondered, which of our fellow animals are intelligent or self-aware? Which communicate within their own communities or across species lines, as we do? Which have emotions? Which have abilities that we do not have?

Cognitive ethology typifies the interdisciplinary features of many fields that are potential contributors to Animal Studies. The authors of a leading textbook in the field suggest in their preface that "cognitive ethology refers to the comparative, evolutionary, and ecological study of animal thought processes, beliefs, rationality, information processing, and consciousness."[38]

The interdisciplinary features of such work include a heavy emphasis on the relevance of the realities of the actual animals themselves to theorizing: "The importance of interdisciplinary discussion means that philosophers who would like their theorizing to appeal and be relevant to scientific colleagues must spend an increasing amount of time keeping up with the empirical literature, perhaps even going out to gain firsthand experience of the ordeals of fieldwork. And scientists who have not read technically difficult philosophical papers and books must do so if they are to stay abreast of developments."[39]

Cognitive ethology's questions reflect not only that humans broadly treasure cognitive complexities but also that we pay attention to sentience, as the universality of humans' compassion impulse shows. This universality extends not only across cultures, but also within and throughout individual cultures in the special sense that inclinations to and even traditions of kindness and compassion survive in social circles and subcultures even when the mainline institutions of a culture promote radical dismissals of all members of certain groups. This phenomenon is a by-product of our one-on-one ethical abilities as discussed above, and is what grounds the sense of personal connection with other animals that one often finds at the level of individual humans' day-to-day interactions. Because Animal Studies constantly refers to this level (and, as chapter 1 suggests, is in many ways driven by this personal level), the concern of Animal Studies is far more than the claims made in official histories, established political and policy positions, or prevailing theoretical analyses.

Science as Subversive: Ecology and Other Challenges

In 1964 Paul Sears defined ecology as "the subversive science" because it can "endanger the assumptions and practices accepted by modern societies, whatever their doctrinal commitments."[40] In fact, many scientific fields other than ecology have subverted humans' penchant for interpreting the world in overtly human-centered ways. Astronomy from the sixteenth through the eighteenth centuries succeeded in making it clear that the solar system was not centered on the earth, and biology in the nineteenth and twentieth century returned humans once again to full membership in earth's animal community. Sciences have come away from the exceptionalist tradition and other human-centerednesses in different ways and at different times. Some fields have continued to be persistently human-centered—after describing psychology's "myopically anthropocentric nature" and "inability to free itself from a fixation on human pathologies and abilities, to the detriment of general scientific issues," one critic suggested the relevance of a non-human-centered version of psychology: "Psychology should be the study of intelligence, of adaptive and complex behavior, wherever it is to be found, in animals, people, or even machines."[41] The emergence of cognitive studies in, to name just a few fields, primatology, bird studies, elephant studies, marine mammalogy, and companion animal studies suggests that even psychology has the capacity to move well beyond the exceptionalist tradition.

The fact that science after science has subverted the exceptionalist tradition reveals that scientific inquiry in general is more than a natural ally of Animal Studies—it is, in fact, an essential component because of the integral role sciences play in the development of basic facts about the similar and different abilities that human and other-than-human animals possess. Sciences provide key perspectives on the profound truths of our membership in the interconnected, inviting universe all animals occupy—what Berry called "the larger community."

Note, then, yet another implication of the science enterprise—the dominance of perspectives of the exceptionalist tradition that promote the reign of a sparse and inadequate dualism such as "humans and animals" is easily understood to be not only antiscientific, but also antiecological. Science's dispassionate commitments to the truth about humans and other animals as linked through common histories and shared ecosystems also prompt one to see the radical inadequacy of other terms and concepts as well. Discussions of broad-sounding topics like "sustainability" often remain so insistently human-centered that they amount to what can be thought of as "environmental speciesism," that is, a framing of environmental and conservation programs solely in favor of the human species.[42]

The implications of many sciences' subversion of the exceptionalist tradition's vision of reality pertain also to the humanities and, of course, to Animal Studies. For example, the subversion causes one to look carefully at the way exceptionalist versions of ethics have dominated Western culture's philosophical, theological, and ecological thinking (this approach is sometimes referred to as "ethical anthropocentrism"). The subversive implications of science help one see why thinking that takes its cue from the exceptionalist tradition is inadequate and parochial, falling far short of much broader human visions that clearly recognize humans' membership in the larger community.

The holistic features of ecology and environmental protection also cause them to be integral parts of many visions and organizations that focus heavily on nonhuman animal issues. The example of orcas above was mentioned because no one could possibly grasp the actual realities of orcas outside their existence in the ocean niches where they live, socialize, and have thrived for millions of years. Similar claims are often made regarding indigenous tribes of humans who recognize themselves as part of the land on which they have long lived. Knowing specific beings in any robust sense means seeing them and their community in all the fullness of their actual environment and "home."

Animal Protection in Several Keys

For reasons of this kind (and others as well), any number of environmental groups and animal protection organizations seamlessly present their causes in ways that recognize both habitat protection and the importance of protecting individual members of a community. Similarly, in highly specialized sciences one finds both of these elements. Marine mammalogy, primatology, and elephant studies, for example, look at their subject species as both individual animals and members of social groups featuring complex behaviors that need to be understood in context. It remains true, to be sure, that our understanding of both individuality in these animals and the social dimensions of their lives is rudimentary, all of which mandates that we remain cautious when claiming to know details of these animals' realities. In any number of ways, our sciences are far better at high-level generalities than fine-grained evaluations of other animals' individual realities. The generalities can, however, be fascinating, as evidenced by the following science-based description of the lives of spinner dolphins studied for decades by a pioneer researcher: "The dolphin's echolocation shield is its own special defense. With it these mammals buy an advantage in the costs of predation over their silent antagonists.... Given that seemingly insignificant advantage, they can then afford to express all the complexity and individuality of their mammalian heritage.... They can let down the school's shield long enough to afford nurture, instruction, tradition, and even culture."[43]

Animal Studies looks at how such a passage suggests several different features. There are generalities like ecological dimensions impacting survival skills and the existence of group advantages over predators because of complex communications that are still mysterious to human observers. This description also suggests a rich set of individual-based complexities with the words "nurture, instruction, tradition, and even culture" even though these elusive, important realities are not described in any detail.

Marine mammalogy clearly provides abundant confirmation of the fact that some cetaceans' social realities are very unlike the general realities evident in primates' social lives. But similarities, too, can be postulated, for much knowledge has been hard won by many different researchers over decades. Norris comments, for example, "It took my entire two decades with spinners to formulate a theory for how they worked." Another researcher who has also conducted a long-term study in the wild of a community of spotted dolphins (*Stenella frontalis*) commented in 1995 that "our knowledge is broad enough to know that dolphins are long-lived social mammals, that they form long-term bonds, and that they learn and grow in their multi-generational societies and use many senses to communicate, especially sound."[44]

The accumulation of science-based research on cetaceans has prompted discussion about not only these living beings' abilities, but also humans' obligation to them. A good example is Thomas White's *In Defense of Dolphins: The New Moral Frontier* (2007).[45] Animal Studies is well situated to facilitate an interdisciplinary engagement with White's use of the terms "defense" and "moral frontier," especially as they are driven by humans' ethical dimensions responding to dolphins' special abilities.

Elephant studies is another specialized science-based inquiry that has produced information about a startlingly complex and fascinating nonhuman community. It further confirms that in no meaningful sense are humans "alone" on the earth. Elephants have had such a long and extraordinarily complex interaction with humans that this particular area of human-nonhuman interaction is surely among the richest and most complicated, as suggested by a leading expert on Indian elephants:

> An object of worship, a target of hunters, a beast of burden, a burden to the people, gentle in captivity, dangerous in the wild, the pride of kings, the companion of mahouts, a machine of war, an envoy of peace, loved, feared, hated, the elephant has had a glorious and an infamous association with man in Asia. For its sheer contrast and splendour, this association is unequalled by any other interaction between animal and man in the world.[46]

This description of the diversity of human interaction and fascination with elephants explains much that is of great importance in Animal Studies, but it opens up many questions about contemporary issues. The positive aspects of this appraisal push Animal Studies students to ask, are elephants' actual lives in fact such that we should be paying attention to them? Elephants are major factors in much of the ferment over humans' relationships beyond the species line—they are cultural icons in places, major draws in zoos. Yet we continue to impact them in ways that prompt science-based research into stress and trauma induced by even well-intentioned zoo-based captivity and far more impactful interactions in the wild. For example, a surge of unusually aggressive and violent behavior, including attacks on humans and even rape of rhinoceroses by young elephants in both Africa and India is described in the October 8, 2006, *New York Times Magazine* cover story, "An Elephant Crackup?" The violence has changed elephant-human relationships in many communities, according to one researcher: "Everybody pretty much agrees that the relationship between elephants and people has dramatically changed.... What we are seeing today is extraordinary. Where for centuries humans and elephants lived in relatively peaceful coexistence, there is now hostility and violence. Now, I use the term 'violence' because of the intentionality associated with it, both in the aggression of humans and, at times, the recently observed behavior of elephants."[47]

The cause is not known, but the article focuses heavily on a respected researcher who suggests that the problem is chronic stress suffered by the elephants. These animals are extraordinary individuals by any measure, as was clearly suggested by the distinguished Dame Daphne Sheldrick, whose cutting-edge work on rescuing orphan elephants has been widely lauded:

> Elephants are emotionally very "human" animals, sharing with us the same emotions that govern our own lives, plus an identical age progression, the same sense of family, sense of death, loves and loyalties that span a lifetime, and many other very "human" traits, including compassion. They have also been endowed with other attributes we humans do not possess, such as innate knowledge in a genetic memory.... In such a long-lived species, there is also a lifetime of learning through experience, just as there is for humans.[48]

Animal Studies in particular has the capacity not only to note the science about elephants but also to ask questions about how elephants' remarkable abilities relate to queries about humans' role in the decline of elephant populations and psychological trauma created by elephants' living conditions, including conflict with humans.

Animal Studies beyond Science

This example of different kinds of inquiries about elephants (and other animals as well, of course) reveals how important it is for students of Animal Studies to engage careful, rigorous science as often as possible. Perspectives grounded in science provide, in turn, information needed for another important task, namely, identifying both possibilities and problems in human actions. When taking this additional step, Animal Studies faces the task of analyzing various possibilities and problems, such as the dilemmas inherent in political and moral choices.

Consider, for example, the possibilities and problems related to our human penchant for political intrigue. We are the successors to Machiavellian ancestors, as both our own human history and primatology make only too clear. Books with titles such as *Chimpanzee Politics* and *Machiavellian Intelligence* inform us that scientists have identified how richly political many of our closest primate cousins are.[49] It is great apes in particular (chimpanzees, bonobos, gorillas, orangutans, and humans, of course) who are subtle in their use of social manipulation through deception or cooperation. Such actions create advantages both for social groups as a whole and for those individuals sufficiently intelligent to employ "whatever mechanisms enable an individual to take into account the complexities of social or other life and devise appropriate responses."[50] The net benefit creates selective pressures for the development of more individual intelligence and more group intelligence, all of which causes intelligence to spiral upward over time. While there are many different explanations of what is involved in such interactions, "all these hypotheses share one thing in common: the implication that possession of the cognitive ability we call 'intelligence' is linked with social living and the problems of complexity it can pose."[51]

Importantly, primate researchers also confirm that our close cousins also have morality-like rules on their minds and care in their hearts. De Waal summarizes his views of primates, and for good measure adds observations about dogs as well: "I've argued that many of what philosophers call moral sentiments can be seen in other species. In chimpanzees and other animals, you see examples of sympathy, empathy, reciprocity, a willingness to follow social rules. Dogs are a good example of a species that have and obey social rules; that's why we like them so much, even though they're large carnivores."[52]

Historically, many studies on primates were pursued because researchers wanted to illuminate human origins. But such research has revealed in almost countless ways that our close nonhuman cousins are truly complex individuals in their own right and form connections that astonish those who look. These connections include not only morality-like dimensions but also richly interpersonal, social, political, and emotional lives. Such discoveries give Animal Studies reason to ask about how human political systems impact the world's nonhuman others and their communities.

Politics and Other Animals

"Politics" and "political" are part of the impressive array of words, including "police," "policy," and "polite," rooted in the Greek word *polis* (city). Most relevant to Animal Studies is that while the word "politics" can operate as a synonym for the well-developed academic enterprise known as "political science," far more often this word is a reference to the complex bargaining and power struggles we also call "political life." One of the most famous definitions, encapsulated as a book title, reveals that power is indeed part of the very soul of this realm—*Politics: Who Gets What, When, How*.[53] The power that defines the essence of political realities and characterizes humans as Machiavellian has profound impacts on many human domains, including the content of education and the practice of science. Further, given this central preoccupation with humans' power relations with and over each other, it will seem natural to many that politics is overwhelmingly focused on the exclusively human. That the roots of politics are in city life and power relations means that the exceptionalist tradition is a pervasive, controlling presence in politics.

Politics and the Exceptionalist Tradition's Ideology

Compared to the species-neutral ideals at the heart of the idealized image of science, the ideals and aspirations that dominate politics are paradigmatic examples of the exceptionalist tradition as it has promoted all aspects of human life at the expense of nonhuman lives. Since political ideals track the special importance of human life called out by religious figures, poets, philosophers, and many others, it seems normal that in politics, to use Rachels's language, "the central concern" of humans' multifaceted moral abilities "must be the protection and care of human beings."

On-the-ground realities, of course, have fallen tragically short of such ideals. Thus, just as the practice of science does not measure up perfectly to the ideals of science as an elegant enterprise that foregrounds a commitment to truth, actual political realities have rarely, if ever, honored the importance of each and every human as envisioned in the core ideals of politics, religion, ethics, and education. For millennia, one group of humans after another has used political means to dominate other groups of humans. The vehicles of harm have been diverse, ranging from direct killing of individuals to tragically virulent wars, class exclusions, enslavement, ethnic mass murders, religious strife, racial and sexual discrimination, and countless other forms of ostracism and denial.

Animal Studies examines humans' complex political realities from several different angles. A principal preoccupation of Animal Studies is frank exploration of the fact that in most societies today, the domination of other-than-human animals through politics and public policy is merely business as usual rationalized in terms of the exceptionalist tradition. Animal Studies also explores how different nonhuman animal groups are impacted when one group of humans dominates another group of humans. Yet other issues arise as some nonhuman animals are treated well by human societies, others are marked as vermin, some have their "welfare" (however one defines this key term) protected, and others are ignored altogether.

Since Animal Studies seeks to understand how and why the exceptionalist tradition is a decisive factor in politics, there are several important anomalies to address. No educated person today is likely to be unaware that political decisions have harmed billions of innocent humans by intentionally denying them lives of dignity and opportunity, self-determination, freedom of movement, religious toleration, free expression, and access to mechanisms that can address obvious imbalances and injustice. Given this important reality, how might anyone suggest the exceptionalist tradition is a force in politics? Since this tradition holds that members of the human species alone are sacred beings or, in the secular version, so important that they are (in Rachels's words) "the central concern" of human ethics, how can one account for the facts of history that make it so clear that both humans and nonhumans alike have been harmed? Historical evidence seems to contradict a major aspect of the exceptionalist tradition, namely, the claim that all humans are important.

The answer is, of course, that the exceptionalist tradition is a factor when the focus of discussion is politics and its ideals. At this level, discussants often talk in general terms along the lines of "humans and animals," such that humans are placed in a group to be elevated and all other living beings are grouped together for subordination purposes. When one begins looking at politics from the vantage point of on-the-ground realities, however, politics is much messier than it is through the lens of a tidy dualism like "humans and animals."

Something similar happens with law and public policy—these topics are conceived in highly idealistic and generalized terms that contrast humans as a group with all other living beings as another group. When the practicalities of specific situations are the focus, however, it is easy to recognize that not all humans are beneficiaries of particular laws or specific public policies. Most often, something far less noble is transpiring, with one group of humans benefiting and others being disadvantaged. This less-than-noble manner in which politics, law, and public policy work in situations of human conflict has prompted many serious critiques (see chapter 4). It has also prompted humorous but dark observations like Will Rogers's comment, "people who love sausage and respect the law should never watch either one being made."

Because Animal Studies works to illuminate the real world, it assesses not only the how, why, and when of claims about nonhuman animals in political settings generally (the exceptionalist tradition), but also how human-to-human politics impacts nonhuman animals in different situations (see chapter 11). The exceptionalist tradition is a very powerful and very harsh form of human-centeredness that is invoked when human-versus-nonhuman issues arise. In this setting, discussions tend to contrast humans with nonhumans generally. So talk is of humans' dignity and superior intelligence.

But when the issue pits the interests of one human group against another, the exceptionalist tradition is no longer the focus, and instead political power controls the outcome. The dignity of each and every human commands far less attention because the primary focus becomes which human subgroup has sufficient power to protect its own privileges at the expense of the other human group. Nonhuman animals, of course, have long since been forgotten because they are, under the exceptionalist tradition, dismissed as inferior. This contrast helps one see that the exceptionalist tradition has the features of an ideology used in dogmatic ways to justify harms to nonhuman animals even as it remains silent and hollow about one human subgroup dominating other human subgroups.

In one sense, then, Animal Studies ranges widely enough to deal with the marginalization of any animals in politics, engaging the broad dismissals of nonhumans even as it also examines the marginalization of humans. The latter is very important in Animal Studies because it can be integrally involved in the marginalization of nonhuman animals as well—there is, for example, much overlap between harms to nonhumans in slaughterhouses and harms to the human workers in such industries,[54] just as there are noteworthy environmental and social costs for local communities that host some agribusinesses.[55] Animal Studies is ideally situated to comment on the connections between problems faced by marginalized humans and a range of harms done to nonhumans, just as it is ideally suited to suggest that protections for such humans and nonhumans can be dealt with fully and well in political circles if there is the political will to do so.

The Definition Trap and Questions about the Future

Lasswell's 1936 definition of politics—*Who Gets What, When, How*—raises many questions, some of which focus on why and how societies co-opt scientists and produce scientific practices that are aimed at ensuring human power and privilege over the natural world. A crucial first step in animal studies as it approaches political issues, then, is to see that, in the matter of nonhuman animals, politics has long controlled and can shape the practice of science in various ways to further humans' subordination of all other living beings. Few people doubt that Lasswell meant to include only humans within the group that politics benefits. Similarly, when people are asked to identify the "public" in the phrase "public policy," very few hesitate as they answer—humans are the "public," the beginning and end, the raison d'être of public policy.

Animal Studies naturally foregrounds the ways in which politics and public policy in many familiar societies (such as the large, industrialized nations of the early twenty-first century) deal with nonhuman animals. It addresses, for example, which nonhumans legislators, judges, or the public more generally include within anticruelty protections. It looks as well at which animals are protected on the basis of environmental concerns, and how the same sort of animal might be protected when thought of as a companion animal but treated very differently when used in a laboratory or eaten. For example, horses can be treated in startlingly different ways depending on which category policymakers use for them. The same is true for dogs, cats, pigs, and a number of other nonhuman animals. Animal Studies looks at these and many other patterns and anomalies evident in both official and unofficial policies of a society. It also attends to changing patterns over time and future possibilities. This is one reason that any Animal Studies program benefits greatly from studying a wide range of human societies, for by no means have all human societies followed the trend evident in today's industrialized societies of reducing any and all nonhuman animals to mere property to be owned by humans. Through comparative studies of different cultures, Animal Studies can say much about public policy options in this area and the salient fact that every society features some individuals with qualms about owning other living beings. In such matters, Animal Studies uses future-focused questions to remind us that the question of how individual humans and their cultures might now and in the future interact with other living beings remains open. In fact, what can happen at the human-nonhuman intersection in the future is, of course, *radically* open in the powerful, etymologically based sense of "getting to the root (Latin, *radix*) of the matter." That some cultures and modern nations have outlawed ownership of certain nonhuman animals reveals that humans can, whenever we wish, establish a variety of ways of relating to other-than-human animals and the more-than-human world. Admitting this important fact related to humans' future makes a crucial point about politics—in an altogether real way, the choices made in politics project an imagined future, both short term and long term, onto the present citizens of society, coming generations, other-than-human animals, and the more-than-human world generally.[56]

The question of what individual citizens can do *now* is, then, wide open in some important senses. But this question can seem to be closed—for psychological reasons, humans are often attached to their inherited and present practices (see chapter 8). Yet the ethical realities of humans are such that choices about the future are possible in both consumer-based and non-consumer-based societies. It is possible today to imagine forms of politics and science that do something other than ensure humans' power over other living beings. Each individual can, in the role of voter, consumer, or active citizen play an integral part in choosing to continue to dominate or, alternatively, make choices that shape a world in which humans live alongside a great variety of other animals.

Animal Studies, in fact, is capable of describing ways of living with other animals that avoid harmful practices that are legacies of cultures that are dismissive of other-than-human animals. Many ancient religious traditions protected real animals by placing the obligation to avoid harms to them (and humans, of course) among the most cherished notions of human identity and the moral life (chapter 2). The peoples of India, for example, have coexisted with cows in urban environments in ways that often astonish outsiders. Further, familiarity with small-scale cultures provides hundreds of visions and ethical approaches that are far more protective of nonhuman animals than those found in modern politics or public policy. In many ways, the ferment introduced in this book reflects that same possibility. Finally, it must be added that science as practiced today in a number of instances provides the means to live with other-than-human communities, as evidenced by certain efforts in fields such as conservation biology, restoration ecology, reintroduction biology, and ecological economics.[57]

Present-Day Walls

Since, on the whole, contemporary political realities in many societies feature much resistance to inquiries about other animals, let alone serious discussion of the moral dimensions found at the human-nonhuman intersection, Animal Studies has a formidable task in exploring the many different ways that contemporary humans choose to dominate or otherwise marginalize a great variety of nonhuman individuals and communities. This phenomenon is complex and features many factors—one root of the limits of present-day politics can be glimpsed in the opening lines of a 1913 book by the deeply respected Indian polymath Rabindranath Tagore:

> The civilisation of ancient Greece was nurtured within city walls. In fact, all the modern civilisations have their cradles of brick and mortar. These walls leave their mark deep in the minds of men. They set up a principle of "divide and rule" in our mental outlook, which begets in us a habit of securing all our conquests by fortifying them and separating them from one another. We divide nation and nation, knowledge and knowledge, man and nature. It breeds in us a strong suspicion of whatever is beyond the barriers we have built, and everything has to fight hard for its entrance into our recognition.[58]

Tagore contrasted such a walled-off, city-based life with the forest-based life that his own Indian civilization experienced in early phases:

> The west seems to take a pride in thinking that it is subduing nature; as if we are living in a hostile world where we have to wrest everything we want from an unwilling and alien arrangement of things. This sentiment is the product of the city-wall habit and training of mind. For in the city life man naturally directs the concentrated light of his mental vision upon his own life and works, and this creates an artificial dissociation between himself and the Universal Nature within whose bosom he lies.

Even though the ferment on animal issues described throughout this book has created many urban and nonurban circles in which people today discuss nonhuman animals (for example, those who wish to discuss wildlife can easily find a forum attended by many citizens who have a passion to conserve "nature" and "the environment"), modern political institutions remain decisively shaped and dominated by a form of walled-off politics that is thoroughly and unrelentingly controlled by the exceptionalist tradition. This preoccupation with human privilege and power dominates education, science, religion, and much else, such that both political and nonpolitical participants are trained not to notice the relevance of nonhuman animals to much of human life. One venue where this preoccupation continues to thrive, effectively maintaining the walls that protect the exceptionalist tradition, is the realm of contemporary public policy studies.

Public Policy

Public policy circles epitomize much about the human-centered preoccupations of politics. Such driving values contrast with the idealized image of science as a beautiful and truth-seeking enterprise. There is, however, the important parallel between the narrowness of politics and policy as practiced and the shortcomings of science as practiced. Public policy also falls far short of the ideology of the exceptionalist tradition. When some disfavored human group is marginalized, it may be government action that directly excludes and/or harms the marginalized humans, but sometimes it is government inaction through lack of enforcement of a policy that on its face seems to protect all citizens of a particular jurisdiction.

When lack of enforcement becomes the norm and thereby causes some humans to be disadvantaged, it is reasonable to suggest that the real policy of a government—what can be thought of the operative or de facto policy—is to allow such harms. Such situations prevail surprisingly often and at both local and national levels. Consider a 1965 finding by an American investigatory commission (the President's Science Advisory Committee report, *Restoring the Quality of Our Environment*) convened in the wake of Rachel Carson's 1962 book *Silent Spring*. The commission concluded bluntly that a major factor shaping national policy in the United States in the 1960s was not the benefit of the public generally, but instead a far narrower agenda: "The corporation's convenience has been allowed to rule national policy."[59] While commercial enterprise can obviously be a powerful force that creates wealth in a community, large-scale businesses in industrialized societies can, if unchecked, also harm humans in extreme ways. The 1980 volume *A People's History of the United States: 1492–Present* provides numerous examples of nineteenth-century business monopolies harming men, women, and children.

If a society tolerates some humans being harmed even though its leaders claim that each and every human is sacred, the plight of any nonhuman animal in that society is likely to be dire indeed.

Animal Studies and Public Policy

One of the driving forces so evident in the diverse, worldwide developments regarding other-than-human animals is the affection people develop for companion animals. This personal connection has led many citizens to pressure their elected representatives to enact laws and other public policies that protect these nonhuman animals. Each year, literally thousands of legislative proposals are made in countries around the world which, if enacted, would create public policy that provides one form or another of additional protection for some nonhuman animals. While only a fraction of the legislative proposals actually become law, the sheer number being proposed means that hundreds upon hundreds of new laws or revisions of existing laws are enacted and thereby impact public policy on nonhuman animals.

Such developments indicate that animal protection issues have become a factor in various policy discussion circles around the world. That this is a new development in some countries is evident in two comments by MIT environmental studies professor Steven Meyer.[60]

These comments reveal key features about what "public policy" means in powerful circles where the subject matter of Animal Studies is not yet of particular interest. When Dr. Meyer was asked, "What are the best books addressing 'animals and public policy'?" he replied immediately and unequivocally, "There are none." When asked next, "What would you say is public policy on nonhuman animals?" Dr. Meyer replied just as firmly, "There is none—animals do not count in public policy."

The volume of animal protection legislation suggests that a decade later, public policy circles may be starting to change. At national and local levels, not only are more laws being introduced and enacted, but also some animal protection laws are more stringently enforced. It can thus be said that, despite many kinds of opposition and the often-polarized milieu of the politically and geographically diverse animal movement, some segments of the movement are making astonishing progress even though other segments proceed at a slow pace or not at all. The changes have prompted one historian to conclude that the modern version of the animal movement has had "a far greater impact on society than previously suggested."[61]

There remain features of public policy that give educators in Animal Studies and many others (such as animal protectionists and environmentalists) reason to pause, including the risks created by the prevalence of a paradigm for nonhuman animals based primarily on companion animals (maintaining human-centered features connected to domination over these living beings, and failing to learn anything essential about nondomesticated animals). Because much of the legislation now being enacted focuses overwhelmingly on companion animals, the important openings this legislation creates remain within a narrow range.

In some ways Meyer was right, since most nonhuman animals still "do not count in public policy." Animal protection continues to be marginalized in important public policy discussions—in the United States, for example, the major graduate programs where students study public policy are dominated completely by the exceptionalist tradition (as evidenced by curriculum, faculty interests, and publications).

Consider, then, the power of two questions about public policy that Animal Studies prompts. First, given that public policy continues to be dominated by the exceptionalist tradition, how is it that protections for living beings outside the human species might find a place in public policy? One answer, of course, is that ordinary people, completely apart from government-initiated and sanctioned policies, can choose to protect their nonhuman neighbors in a variety of ways. The power of this first question is that it suggests that a society's public policy can be influenced by the daily decisions of private individuals. Further, the question implies that when a sufficient number of citizens take such responsibility, the official public policy of the political realm may no longer be dominated by the narrow-mindedness that typifies the exceptionalist tradition.

With such implications in mind, Animal Studies can ask a second powerful question— what is public policy really, and who makes it? This question goes to the very heart of public policy studies because it asks the scholar or student to get beyond facile assumptions that now dominate public policy circles. As this issue is discussed below, notice how yet again Animal Studies prompts one discipline after another to engage its own foundations.

One of the most common definitions cited in public policy materials is that public policy is whatever a government chooses to do or not do.[62] Animal Studies has major reasons to question the adequacy of such definitions. Similarly, it has reasons to be as frank as possible about the fact that nonhuman animals have either no or scant importance in public policy discussion other than as mere resources in service of a human-centered agenda.

There is power as well in queries prompted by Animal Studies' need for environmentally aware perspectives—does public policy make humans "more effective vandals of the earth"? Insights flow, too, from Animal Studies grappling with the de facto policies implied by lack of enforcement of laws that, nominally, commit a society to anticruelty protections. When one recognizes that lack of enforcement can effectively undo a legislature's enactment of a law, one is led to ask a further question—can government officials also undo deep cultural commitments like anticruelty sentiments that have been part of social consensus for centuries? Modern laws and other contemporary public policies have exempted large corporations, but not individuals, from anticruelty laws passed centuries ago (see chapter 7).

By virtue of its interdisciplinary resources, then, Animal Studies easily enriches the notion of public policy to encompass a wide range of social acts that go far beyond legislation, ballot initiatives, judicial decisions, or administrative decisions. It examines values rooted in cultural heritage and society-wide ethical values, such as concerns about cruelty, from which government-based public policy ideally draws its energy and legitimacy.

Getting beyond Traditional Public Policy

The principal approach to public policy—hence its description as "traditional" or "mainline"—is to focus narrowly on government issues using conceptions drawn from economics. Major textbooks regarding public policy analysis nonetheless regularly observe that there is no consensus on many issues at the heart of traditional public policy analysis—one observer suggests, for example, "reaching a consensus on the precise definition of public policy has proved impossible."[63]

Many people are dissatisfied with the highly theoretical approaches found in public policy textbooks—some criticize the traditional approach as overly reliant on questionable assumptions, such as the claim that humans can be understood primarily in terms of their economic decisions. Some critics notice the extremely heavy reliance on cost-benefit analysis in mainline public policy circles—in *Priceless: On Knowing the Price of Everything and the Value of Nothing* the authors quote one critic who describes the "unbelievable alienation, reductionist thinking, social ruthlessness and the arrogant ignorance of many conventional 'economists' concerning the nature of the world we live in."[64] Similar concerns appear in Deborah Stone's best-selling textbook on public policy:

> The field of policy analysis is dominated by economics and its model of society as a market. A market, as conceived in microeconomics, is a collection of atomized individuals who have no community life. They have independent preferences and their relationships consist entirely of trading with one another to maximize their individual wellbeing. Like many social scientists, I don't find the market model a convincing description of the world I know or, for that matter, any world I would want to live in. I wanted a kind of analysis that starts with a model of community, where individuals live in a web of associations, dependencies and loyalties, and where they envision

and fight for a public interest as well as their individual interests. This kind of analysis could not take individual preferences as "given," as most economists do, but would instead have to account for where people get their images of the world and how those images shape their desires and their visions.[65]

Animal Studies needs for multiple reasons to push those who assume that public policy can be well understood through traditional, mainline approaches of market models, cost-benefit analysis, and other economics-inspired concepts. First, such an approach is worse than narrow and myopic, for it advances the ideology of the exceptionalist tradition and provides no room for alternative approaches. Additionally, such approaches ignore many other inputs, such as cultural values, social consensus, and scientific findings about other animals. As discussed in chapters 4 and 7, while the larger field of law often epitomizes the exceptionalist tradition, legal education in many countries today is home to a growing discussion about the use of law to create a range of protection for certain nonhuman animals. This developing "animal law" reflects important contemporary social realities—for example, more households in a number of industrialized countries have companion animals than have children (chapters 4, 7).[66] A very high percentage of citizens in these countries want greater protections for their companion animals because they deem these nonhuman animals to be treasured "family members."[67]

Such facts are not yet, however, particularly influential in many traditional, mainline policy discussions around the world even though they reflect, to use Stone's words, "a web of associations, dependencies and loyalties" that goes beyond the species line to at least some nonhuman family members and, arguably, some other nonhuman animals as well. If discussion in traditional public policy circles remains unresponsive to such popular issues, it is likely that other, less popular but nonetheless key issues are ignored as well.

One of the reasons that legal education reflects concern for nonhuman animals is that this type of education uses a different notion of public policy that is far more capable of recognizing such human valuing beyond the species line. Law students have long been trained to use a broad notion of public policy that is much more flexible than the traditional, mainline notion of public policy. The Supreme Court of Missouri in 1946 described how the American legal system uses "public policy" as a synonym for "public good" or "public morals":

> The term "public policy," being of such vague and uncertain meaning, and of such variable quantity, has frequently been said not to be susceptible of exact or precise definition; and some courts have said that no exact or precise definition has ever been given or can be found. Nevertheless, with respect to the administration of the law, the courts have frequently quoted and often approved of the statement that public policy is that principle of the law which holds that "no one" can lawfully do that which has a tendency to be injurious to the public or against the public good; ... and also has been defined as "the public good."[68]

This notion of public policy as the public good offers sufficient flexibility to account for not only the "web of associations, dependencies and loyalties" but also, most clearly relevant to Animal Studies, the changes in social values that are taking place because so many house-holds now include nonhuman family members.

Further, as social values evolve and take to heart environmental issues, concerns for social justice, scientific findings, and commitments to compassion, this law-based notion of public policy can offer even more openings than can the traditional, mainline notion of public policy so hamstrung by the narrow-hearted assumptions of the exceptionalist tradition. The Supreme Court of Missouri explained this flexibility:

> [Quoting *Corpus Juris*, a major legal treatise] "One of the best definitions [of public policy] perhaps is that of Justice Story, which applied the term to that which 'conflicts with the morals of the time, and contravenes any established interest of society.'" ...An excellent definition is also found in *Black's Law Dictionary*, where it is said: "certain classes of acts are said to be 'against public policy,' when the law refuses to enforce or recognize them, on the ground that they have a mischievous tendency, so as to be injurious to the interests of the State, apart from illegality or immorality."[69]

Animal Studies has before it the task of opening up public policy discussions of all kinds—those reliant on economics and the exceptionalist tradition, and those anchored in this law-based sense of the public good—beyond the species line. Arguably, it is today "against public policy" in the sense of "conflicting with the morals of the time" when compassion is ignored, species are extinguished, science focuses only on profits, or human interests alone control social policy.

Equating public policy with a broad sense of public morals is more than a common-sense move based on the similarity of the terms "public policy" and "public morals." It has distinctly democratic and populist features. For example, the scholar who went on to found the modern field known as policy studies observed in 1971 that any truly public policy must be grounded in the multiple parts of society. At the beginning of his most important work, *Pre-View of Policy Sciences*, in which he proposed that the study of policy be an interdisciplinary matter, Lasswell reflects how policy is grounded outside of government: "A commonplace of experience is that the decisions nominally made by governments often register determinations that are made outside government—whether in a bishop's palace, a club of industrialists, or a trades-union headquarters. More generally, in many sectors of human life the norms of conduct are formulated and made effective outside the machinery of legislation, administration, and adjudication."[70]

Mainline public policy, as discussed in government circles and as taught in graduate-level programs other than law schools, is noticeably narrower than Lasswell's vision. Accordingly, protests in the spirit of Stone's longing for a "model of community, where individuals live in a web of associations, dependencies and loyalties, and where they envision and fight for a public interest as well as their individual interests" or Ackerman and Heinzerling's lament about those who "know the price of everything and the value of nothing" have been common.

Given that the impoverished assumptions of the exceptionalist tradition continue to dominate those government circles that formulate, enforce, and interpret laws and public policies, it is hard to suggest that such protests have had anything more than a very limited impact on public policy circles. The commercial realm also reflects little impact. But educational domains harbor developments that challenge the exceptionalist tradition.

Education has a complex battery of approaches known variously as "interpretive," "narrative," "deliberative," or "discursive" approaches to policy analysis. These approaches, which are very intellectualized and theoretical, offer one feature that makes them relevant to Animal Studies—they embody broad ways for thinking about and assessing what is really happening in and behind public policy pronouncements. In this regard, they offer help in thinking about how and why public policy impacts other-than-human animals. These approaches can, for example, illuminate how language choices work, although most such analyses are squarely focused on exclusively human problems. But the methods and tools of these approaches can help scholars and students interested in Animal Studies identify how words, ideas, interpretations, and stories impact the more-than-human world. They can help identify when simple dualisms like "humans and animals" shut down debate, or when the companion animal paradigm is subtly assumed to help one understand every kind of nonhuman animal.

The relevant insight of these approaches, then, is that how people talk about a particular situation's underlying set of problems can dramatically affect what those people see as options for solving these problems. These different approaches to public policy open up thinking about how official public policy is shaped by the exceptionalist tradition, as well as when and how government-enacted policies marginalize others who may be humans or nonhumans. In effect, these additional perspectives on public policy provide greater frankness about the importance of identifying who is really being protected and why some official government enactments are not enforced. For example, lack of enforcement of anticruelty laws at the local level has been a recurring problem. A particularly telling example of a systemwide failure in oversight requirements for slaughterhouses (thus causing meat inspectors not to carry out specific inspection tasks mandated by US federal law) was described by the chairman of the National Joint Council of Food Inspection Locals (the federal meat inspectors' union) during an interview published in 1997. Note in the following how both humans and nonhumans stand to suffer because of this failure.

> I knew that the Humane Slaughter Act regulations gave inspectors the authority to stop the line when they saw violations. But I also knew that they did not authorize inspectors to visit the plant's slaughter area hourly, daily, weekly, or ever, for that matter. "So how often does someone go down to the slaughter area and look?" I asked.
>
> "And leave his station?" Carney [chair of the federal meat inspectors' union] replied. "If an inspector did that, he'd be subject to disciplinary action for abandoning his inspection duties. Unless he stopped the line first, which would get him into even more trouble. Inspectors are tied to the line."
>
> "So what's the procedure for checking humane slaughter?" I asked.
>
> "There isn't one," he answered.
>
> "Hold on. You're telling me that inspectors have the authority to stop the line when they see humane violations, but basically, they're never allowed to see them?"

"That's right," he said. "Inspectors are required to enforce humane regulations on paper only. Very seldom do they ever go into that area and actually enforce humane handling and slaughter. They can't. They're not allowed to."

"Besides," he continued, "our inspectors are already overwhelmed with their meat inspection duties and the agency has never addressed the responsibility of humane slaughter. . . . Inspectors are often disciplined for sticking to regulations and stopping production for a contamination problem—meat safety—which has a much higher priority than animal suffering."[71]

Although these alternative ways of analyzing public policy are at times highly philosophical and theory based, they help anyone see what is at stake in public policy debates (in the above case, corporate profits at the expense of both the nonhumans being killed and the human consumers who buy the end product). They can also help everyone see why many people are frustrated when mainline, traditional, walled-off public policy ignores consumer protections or goes forward without anyone challenging the economics-based calculus that allows food animals to be treated as mere property rather than living beings who are part of humans' larger community.

Continued Dominance of Traditional Public Policy

Even in the face of alternative perspectives and challenges, the traditional approaches that foreground economics-based factors and marginalize certain cultural norms (like equality for humans and anticruelty traditions regarding nonhuman animals owned by large businesses) remain enormously influential. They predominate, of course, in government circles and business, but also in the professions and classrooms within the educational establishment. Government decision makers and many other people continue to assume that it is obvious (some would think it the only rational choice) that the "public" in "public policy" is and must be the human species alone. Many also assume that a cost-benefit analysis is the clear way to figure out which people-centered policy option might be chosen by government officials and then enforced. Not surprisingly, then, traditional public policy discussions remain a bastion of "humans and animals" thinking, with the result that the exclusions of the exceptionalist tradition are not even noticed. The results are predictable—policy people, when they do talk of animals, tend to talk only about endangered or pest species or the production parameters of industrialized uses of food animals.

Ferment Issues

There are, however, some important openings. The companion animal paradigm has been mentioned already, and other inroads have opened up as well because so many people are concerned about risks to life on Earth created by the exceptionalist tradition. Environmental circles now include pioneers who push traditional fields like economics to recognize that ecological systems must be healthy to support humans. Such proposals, while often anthropocentric, have important potential to foster vigorous communities of other-than-human animals. For example, discussion of "natural capital"—an idea by which business, community, and government decision makers consider the value of "ecosystem services" such as clean air and water, flood protection, irrigation, hydropower production, drinking supply, crop pollination, and climate stabilization—is decidedly anthropocentric in its motivation. But such a concept is a step toward a multispecies world in which other-than-human animals can still have a chance to thrive in their natural communities. This concept only slightly moderates the approach that threatens destruction of the very ecosystems upon which any economy must be based. But it provides an opening because it is a corrective to both traditional economics and mainline public policy that have long been virtually autistic about the natural world.

A related approach that focuses on specific animals suggests that massive ecological imbalances develop when all of the top nonhuman predators and herbivores are eliminated. Wolves, large cats, herbivores like bison, sharks, and great whales can change ecosystem dynamics dramatically. When these other-than-human animals are removed, ecosystems suffer from changes in soil, water quality, and vegetation, and there are increases in infectious diseases and invasive species.

There is also much ferment regarding the use of food and research animals as mere resources. Although the exceptionalist dimensions of many critiques of these uses are impossible to ignore, there are nonetheless powerful voices addressing grave problems and calling for changes. Major institutional voices within the last decade have addressed a number of serious problems related to food animals. The 2008 report published jointly by the Pew Charitable Trusts and the Johns Hopkins School of Public Health concluded, "By most measures, confined animal production systems in common use today fall short of current ethical and societal standards." The United Nations' Food and Agriculture Organization report *Livestock's Long Shadow* concluded in 2006 that livestock production facilities are "probably the largest sectoral source of water pollution" and are major causes of an array of serious environmental and human health problems, including emergence of antibiotic resistance, erosion, pollution of lakes and rivers, dead zones in coastal areas, and degradation of coral reefs. The report's focus is, to be sure, human interests, as can be seen in the way its conclusion was framed: "the concentrated animal waste and associated possible contaminants from [intensive factory farming] systems pose a substantial environmental problem for air quality, surface and subsurface water quality, and the health of workers, neighboring residents, and the general public."[72]

The 2008 report from Pew and Johns Hopkins was even more blunt about the economic harms that factory farming causes to human communities. Noting that "the costs to rural America have been significant," the report describes harms suffered by communities that go beyond the loss of family-owned farms and reduced civic participation rates:

> Although many rural communities embraced industrial farming as a source of much-needed economic development, the results have often been the reverse. Communities with greater concentrations of industrial farming operations have experienced higher levels of unemployment and increased poverty.... Associated social concerns—from elevated crime and teen pregnancy rates to increased numbers of itinerant laborers—are problematic in many communities and place greater demands on public services.[73]

The bottom line of both of these reports is a human-centered conclusion—the economics-driven phenomenon of factory farming creates serious environmental risks for humans even as it is at the same time not helpful economically to local communities. While the focus is not on nonhuman animals, these reports do make clear that it is both humans and nonhumans who are suffering.

Research animals have also received much attention—including laws or administrative fiats that ban the use of nonhuman great apes in experiments. More broadly, the philosopher Bernard Rollin cites federal regulations governing laboratory animals enacted in the United States during the 1980s as reflecting a change in social consensus on the moral issues involved in using nonhuman animals as experimental subjects.[74] This development, which is not abolitionist but nonetheless reflects an admission that moral considerations do extend to other-than-human animals, was enshrined in law despite opposition from the medical research establishment and a lack of support by laboratory animal veterinary associations.

Another policy change reflected in legislation is the breadth of the movement to ban circuses. Such bans have been put into place at both national and local levels (hundreds of cities around the world ban circuses and a number of countries ban the use of wild animals in circuses, with Bolivia banning even domesticated animals).

Zoos, however, remain powerful and popular organizations in many countries. Attendance worldwide amounts to hundreds of millions. Although attendance is down in a number of countries over the last half-century, in some countries, such as Japan and England, attendance remains robust—for example, it has often been claimed that annual attendance at the 130-plus accredited zoos in the United States exceeds the number of paying attendees at all of the professional football, baseball, hockey, and basketball games combined.[75]

While zoos today often use such numbers to make education-based arguments and cite their contributions to conservation to justify their continued existence, Animal Studies offers a wide array of information and critical thinking skills by which students can assess whether such arguments are powerful or mere rationalizations of a traditional form of human domination over other-than-human animals.

Companion Animals Reprise—Policy Realities

One of the reasons that public policy proposals focused on companion animal issues have gained some traction is that owned companion animals are so often deemed family members. This focus brings votes and popular support. While Animal Studies has many other tasks to accomplish, it can make evident that this limited area is very complex, as is virtually every topic area dealing with other-than-human animals. An astonishing number of issues arise regarding companion animals, including how to deal with homeless and feral animals, the use of human-friendly animals as therapy animals (for human health), and the special place of honor accorded owned horses, cats, and some dog breeds but not others. An emerging pet trade that markets nonnative or "exotic" nonhuman animals creates severe problems, as does the fact that some societies like to eat the very animals that other societies treasure as companion animals, sacred animals, or some other valued category.

Companion animals have played a major role in developing the discussion known generally as "the human-animal bond," which has played an important role in the recent history of veterinary medicine. The number of nonprofit organizations pursuing companion animal issues is important (because of both their cooperation and their failure to work together). Extremely difficult issues arise with the topic of cloning of companion animals, rental businesses involving companion animals, hoarding of animals, or domestic violence issues sometimes known as "interlocking oppressions" or "the link" (see chapter 11) whereby both humans and companion animals are at great risk of harm.

Thus, even though companion animals represent the more-than-human world in only minor ways, this category will no doubt continue to play a large role in Animal Studies. The attention lavished on many of these animals and the proposals brought to protect them and their owners in legal systems reflect that, given popular support, protection for some animals beyond the species line easily, even naturally, fits into public policy discussions.

Policy beyond Companion Animals

There is a significant reason that Animal Studies pays particularly close attention to the free-living animals traditionally called wildlife. As a nonhuman animals that are less subject to direct domination by humans, they offer important perspectives on other-than-human life. Today, of course, most people recognize that virtually all wildlife communities and migrating groups are impacted in multiple ways by human presence, pollution, and habitat loss. Nonetheless, the category of wildlife, because it involves free-living individuals and communities, offers an even more diverse range of issues that need humans' imagination and humility.

Animal Studies also must pay attention to this category, for while it calls to mind the animals' natural features, the category of wildlife is constructed. It is, after all, a catchall grouping; further, our understanding of what is "natural" and "wild" is also deeply impacted by cultural presuppositions. This can be seen, for example, in the variability of legal definitions of what counts as a wild animal. Finally, other animals' ubiquity is an important reality that impacts both specific and general issues across the entire spectrum of Animal Studies.

Getting free-living nonhuman communities and individuals into discussions of public policy—however one defines it—is essential if humans are to learn how to think and talk knowledgeably about other-than-human living beings. The very act of learning how to find and then become responsibly informed about and even protect many different kinds of animals "out there" in the more-than-human world creates a series of challenges that go to the very heart of Animal Studies.

As If Other Animals Matter: Beyond Science and Politics

If one asks whether science and politics, through public policy, have the capacity to go forward "as if other animals matter," the answer is obviously yes. Science naturally reaches for other animals' realities and clearly has the capacity to deal with the many kinds of other animals that exist "out there." Politics and public policy also can address how we might deploy our remarkable human abilities to care beyond the species line. These important acknowledgments apply not only to science and politics but to the additional fields covered in the following chapters. Indeed, if Animal Studies is to realize its potential, it must engage many other fields essential to fostering careful thinking about other animals in something other than a blatantly human-centered key.

4

Early Twenty-First-Century Animal Studies
Three Cutting Edges

This chapter introduces three active, influential, and highly developed forms of contemporary thought to underscore key features of today's Animal Studies. These three—law, philosophy, and Critical Studies—are now accepted approaches in academic circles, and later chapters explore additional details of how these fields reflect the pervasive presence of nonhuman animals in human life (chapters 6 and 7). While these three fields are early stars in the Animal Studies firmament, the academic world includes a bewildering array of other fields in the natural sciences, the social sciences, and the humanities that engage some segment or another of the more-than-human world and its nonhuman citizens.

First, each of these three fields blossomed dramatically in the last decades of the twentieth century—in each field, one can see harbingers of much else that is coming. Second, these three areas sustain academic endeavors that model a commitment to the important role of critical thinking skills as we engage our inherited perspectives, ideas, and ethics about other living beings. Third, these fields feature approaches that prompt "interdisciplinary humilities," that is, frank recognition of our need to learn from a variety of disciplines as we do Animal Studies. These humilities expand each field's subject matter relative to its traditional forms of education. While these fields thus reveal how individual disciplines develop Animal Studies themes, they cover, even in combination, only a fraction of the issues about other-than-human animals raised in modern societies. They also provide opportunities to notice how the ferment about humans' connection to other-than-human animals often arises at the margins of modern industrialized societies, such as local nonprofits trying to save some of the many animals being harmed or in little-known academic fields raising issues that the mainline academic world ignores. As with all successful social movements, though, as time passes advocacy for the cause can be found increasingly close to the center of a number of mainline institutions. Animal studies has moved beyond very humble beginnings in philosophy and law to become an easily recognized issue receiving support from mainline institutions such as law schools, professional organizations, national media, and a business sector that sells, literally, tens of billions of dollars of merchandise to consumers each year.

A First Cutting Edge: Law and Other Animals

Although law is a fundamentally conservative field, today in many industrialized societies it is a major vehicle by which change is brokered. Thus even though governments control people through law—recall that law and public policy are a projection of an "imagined future" upon subsequent citizens' reality (chapter 3)—law has also provided a means by which societies can move away from traditional biases, prejudices, and harms. It is this tradition creating change, rather than controlling the status quo, that has allowed lawyers, legislators, and educators to tap into the ferment on animal issues. Of particular importance in this regard are the following developments in legal education that offer a paradigmatic example of Animal Studies.

A Story of Student Demand

The emergence of a field called animal law within legal education over the last few decades illustrates fundamental features of Animal Studies strikingly well. In 2000, Harvard Law School offered its first animal law course, called Animal Rights and taught by a leading proponent of specific legal rights for certain nonhuman animals.[1] The Harvard course was the direct result of petitions for such a course that were signed year after year by scores of students at this high-profile law school.

From 1977, when the first such course in the United States was taught at Seton Hall Law School, to 2000, fewer than a dozen such courses were offered at US law schools. But when Harvard Law School announced its course, the American legal education establishment took notice. So did the media, and courses in animal law multiplied rapidly in law schools around the world—within the following decade, the number of such courses increased tenfold in the United States.

Such an increase would be significant in any educational field, of course, but this development was particularly significant for a variety of reasons. First, the demand for new courses was driven almost solely by students. Second, the development took place at a level of education (law schools) characterized by an entrenched tradition of open discussion that is among the most developed in modern education circles. Often referred to as the Socratic method, this approach has long been used in law schools because training students to be advocates requires the freedom to make arguments and discuss their implications without fear. Third, law is the paradigmatic public policymaking tool, with lawyers functioning effectively as policy mavens in the sense made popular by Malcolm Gladwell in his 2000 bestseller *The Tipping Point*—individual lawyers, law-based educators, and legal activists diligently gather information about what the legal system can do for nonhuman animals and then circulate their proposals widely and effectively, thereby creating new trends.

The developing student demand is anchored in "a deeply personal dimension of connection with nonhuman animal individuals themselves" (chapter 1). There are almost 200 accredited law schools in the United States. In a single decade the number of these graduate-level professional schools offering an animal law course shifted from fewer than a dozen to more than two-thirds (including virtually every one of the fifteen most prestigious schools). This shift is inherently interesting, but the inauguration of such courses due to student petitions is most telling—students want education that is relevant, and students find open discussion of industrialized societies' present treatment and future possibilities with other animals to help them consider what sort of changes would foster the kind of compassion-intensive society in which they would like to live.

Another harbinger is that since a large majority of law schools in the United States, Australia, and some other countries now offer this kind of course, each year around the world several thousand law students are taking them. Given lawyers' proficiency at being heard in public policy circles (compared, especially, to veterinarians and scientific researchers, whose graduate-level education and professional practices in no way expose them to public policy matters or mechanisms or lead them to engage extensively in public policy discussions), the emergence of so many animal law courses means that future policy discussions touching upon animal issues will continue to be lively.

Further, law schools' deep commitment to the set of critical thinking skills that come with the Socratic method dovetail well with the place of such skills in Animal Studies more generally. Thereby, the commitment within Animal Studies to careful reflection is deepened, as is the practice of critical thinking in legal education more generally. It should also be noted that the combination of critical thinking skills with personal interests of the kind that law students bring to animal issues is particularly powerful.

Benefits of Open Discussion

The development of animal law in the midst of an educational tradition featuring the deepest of commitments to open discussion, as well as expertise in political and public policy discussions, has allowed animal law to emerge as a leader in Animal Studies in several senses. The existence of animal law courses produces much discussion among students. It also produces media attention. Perhaps most importantly, it prompts proposals for change via different policymaking mechanisms—legislation, administrative regulation, and court-based decisions about animal-related issues. The net result is the development of even more interest in other fields that provide information and perspectives on other-than-human animals, all of which further enriches Animal Studies generally. When students seek forms of education that they can invest with personal commitment, optimal circumstances exist for teaching important skills like rigor, critical thinking about the role of values, and recognition of the power of diversity and tolerance.

Further, skills of many other kinds are fostered when students (or anyone) look beyond the species line. Doing so helps, for example, with the process of each of us taking responsibility for assessing the implications of the ideas and values we have inherited. Such skills can foster a deeper sense of both the riches of our sciences and the humanities and, as importantly, their limits as well. Animal law, for example, offers perspectives that help us appreciate the idea that our sciences remain "terribly silent about everything close to our hearts, everything that really counts" (chapter 3). It takes perspectives found in law—as well as those nurtured and exercised in some of the humanities, arts, faith communities, and social movements that promote justice in daily life—to help us see how we might take advantage of what Schrödinger called "factual information" put by science into "magnificently coherent order."

These features of law itself and of legal education more particularly suggest a robust future for animal law. They also reveal how the subfield of animal law is already a developed form of Animal Studies because it readily supports careful, wide-ranging, future-oriented education and thinking in general. Many legal traditions' reliance on evidence-based approaches also opens the possibility of learning new facts and, when appropriate, *unlearning* debilitating biases on which traditional exclusions have been based. These developments also provide an example of the ferment surrounding Animal Studies, as well as hints of more ferment to come.

Law is one of the most respected, even privileged ways of talking in modern societies. Nonlawyers often accord an automatic respect to discussions framed in terms of the ideas and vocabulary that dominate legal thinking. While respect for the tradition of legal discourse can be deep and informed, such respect can be so automatic as to be unthinking—law has often been the means by which harms to both human and other-than-human animals have continued (chapter 7). Given that some precincts of law are extremely unfriendly to change, automatically respecting law-based analyses can be a mindless, rather than mindful, approach.

The privileged status of the legal tradition is such that social movement advocates aggressively employ legal systems to gain advantage for their cause. Ideas such as the rule of law, specific legal rights, and other law-based protections create opportunities for social change and protection from harm. The prevalence of open debate within legal education allows new ideas to surface regularly, thereby modeling the value of open debate in any field. Such openness is particularly useful as humans engage our inherited perspectives, ideas, and ethics about other living beings.

Interdisciplinary Challenges

Because law is powerful on its own and also regulates so many domains of human life, it begs features that foster interdisciplinary humilities. Without good input from other fields, for example, laws governing humans' interactions with other-than-human animals are likely to be inadequate. They will be adequate only if they are informed by input from sociologists knowledgeable about many different features of the human-nonhuman intersection, ethicists who frankly assess all relevant issues, and natural scientists from a range of fields that provide evidence-based information about the animal themselves.

But like so many fields in Animal Studies, animal law is only now emerging into a vigorous interdisciplinary phase. Law draws input from other fields in a variety of ways, of course, but, as one of society's most privileged discourses, it can slight even prestigious science-based work. Thus legal systems have often been ambivalent about science, using specific science-based findings only reluctantly and at times in ways that perpetuate rank injustices.[2]

Simply said, the legal tradition has long relied on its own conceptual resources, not on outside sources. Today, however, natural science findings are remarkably important—*Rattling the Cage* and *In Defense of Dolphins* (chapter 3) both show how advocates of greater protection for certain nonhuman animals utilize detailed scientific research from, respectively, primatology and marine mammal studies. Wise in *Rattling the Cage*, for example, contrasts the poignant story of Jerom, a chimpanzee confined to a sterile environment because he is an experimental subject in a biomedical research project, with scores of scientific studies about chimpanzees' and bonobos' personalities, intelligence, communication, and social needs. White's *In Defense of Dolphins* advocates that dolphins be granted "personhood" based on scientific findings about these animals' self-consciousness, intelligence, free will, and abilities to form deep social bonds.[3]

Indeed, the power of science regarding other animals is such that animal law may help develop legal traditions' now somewhat ambivalent attitude toward science. This is important because legal systems for a variety of reasons often demonstrate great reluctance to challenge human-centered values. This tendency is no doubt related to the overwhelming influence of tradition in law, but other factors are the respect and deference accorded key enterprises anchored by the exceptionalist tradition, like economics.

First-Wave Animal Law

The newly emerged interest in the use of the legal system to reduce serious harms to other-than-human animals has led to an important social movement that can be seen as the opening stage or wave of animal law. The leaders of today's animal law movement clearly think of themselves as leaders in not just law, but also in animal protection and ethical (values-based) issues more broadly. This is surely a fair claim, because first-wave animal law ventures beyond concepts of law already established within the exceptionalist tradition.[4] This initial wave has, for example, prompted discussion of new kinds of protections for certain nonhuman animals such as owned dogs and cats, chimpanzees, gorillas, orangutans, bonobos, dolphins, whales, and elephants.

Those developing this first wave of animal law have, naturally enough, attempted such work through traditional legal ideas and methods. This explains why the use of property law ideas and specific legal rights (both of which undergird the whole system) remain dominant in first-wave animal law, just as it accounts for the continued use of traditional methods like litigation-based approaches, legislative lobbying to pass new legislation and amendments, an emphasis on work that can produce income for lawyers, and development of casebooks for teaching. First-wave animal law thus has worked within the existing legal system, using traditional legal reasoning patterns to create litigation and legislative challenges to existing harms to humans' favored groups of nonhumans.

First-wave animal law has featured noteworthy variety along a continuum running from one pole called "welfare" to another pole called "rights." Some advocates have foregrounded questions about improving conditions of nonhuman animals within existing human uses (like laboratories and farms). Others have advocated creation of very specific legal rights for the most cognitively sophisticated nonhuman animals (chimpanzees, for example).[5]

In its most aggressive work, first-wave animal law has been preoccupied with two categories of nonhuman animals—those we dominate and live with (companion animals) and those who are our closest evolutionary cousins. This preoccupation with companion animals has been an important populist move. Such a focus is evident in many places, for example, legislation or challenges to civil wrongs, such as negligence, that lawyers call torts. Companion animals have also figured prominently in organizing groups of lawyers (such as the American Bar Association's Animal Law Committee). While attention to primates has a certain popular appeal, since such animals have received much media attention through the work of Jane Goodall and others, it also has a strong scientific basis because of discoveries about these animals' remarkable intelligence, personalities, and social complexity.

Making companion animal issues the dominant paradigm for animal law parallels the problem of covert human-centeredness in Animal Studies. Human-centeredness may also be detected in the widespread tendency to focus on primates, because these animals have cognitive abilities like those of humans. Any heavy preoccupation with concepts and categories constructed "by humans for human purposes" risks a surreptitious affirmation that humans' abilities and interests are the measure of what really should matter to any intelligent, moral being.

Despite these risks, first-wave animal law has produced important opportunities. It has been instrumental in mobilizing activism seeking changes within government policy circles, such as courts, legislative bodies, and administrative agencies. First-wave animal law has also promoted discussions of important protections for research animals, and more recently amelioration of the harsh living conditions of food animals.

It is common in first-wave animal law for discussions about companion animals, research animals, or food animals to reflect some acceptance of human domination over other living beings. This is particularly so when a weak sense of welfare prevails, that is, where the word "welfare" signals that only minor concessions to the well-being of nonhuman animals, such as larger cages or providing toys that "enrich" a sterile environment, will be made even as the overall situation of human domination continues. Such a use of "welfare" falls far short of its original meaning, which relates to the more substantial issue of quality of life from the captive animal's point of view.

Further, while emphasizing cognitively sophisticated nonhuman animals has risks, it has some benefits as well—first-wave animal law has often made points relevant to wildlife by raising issues of captive primates, dolphins, whales, and elephants. Such points extend to all individuals of these kinds, the wild populations as well as the captive. Some forms of first-wave animal law focus on wild animals as individuals, not just at the species level as is typical of traditional law, politics, and education. Some forms, however, continue to center primarily on extinction risks or population welfare rather than the harms to and sentience of individual animals.

Pointing out these limits is not in any way meant to deny the achievements of first-wave animal law, which has, in fact, prompted what promises to be a revolution in the scope and tenor of law. Some harms have been eliminated, and others ameliorated, which are signal achievements in a human-centered environment. But this is true for only a tiny fraction of the nonhumans that our species harms. Thus, in spite of developments in animal law and the emergence of other fields of Animal Studies, the vast majority of harms remain in place, and in some cases are worsening. But genuinely important gains have been achieved even though the exceptionalist tradition continues to dominate legal systems generally. Law continues to place humans alone in the treasured category "legal persons with rights," with the result being all nonhuman living beings remain in the categories that make them mere "legal things" that can be owned by humans in one way or another.

It is important to note that some animal law advocates remain dissatisfied with what has been achieved to date. In particular, the preoccupation with only cognitively impressive (to us) nonhumans or those animals who can become family members deeply troubles some animal protectionists. It risks, as noted above, yet another affirmation of humans as the definitive measure of the value of life—if the upshot is that those living beings that are not like us count less, both animal law and Animal Studies will proceed in an impoverished way. Many ethicists, scientists, and animal protectionists are deeply uncomfortable with this sort of cognitive hierarchy as either the leading edge of Animal Studies or the measuring stick by which we prioritize our ethical choices and create legal protections beyond the species line.[6] A number of different insights drive this discomfort. One is the vision at the heart of the most famous quote in the Western world's animal protection movement, Bentham's penetrating observation, "The question is not, Can they reason? nor Can they talk? but, Can they suffer?"[7] Another insight is the spirit of the Axial Age sages' recognition that our own possibilities as moral, intelligent creatures are rooted in our abilities to care about a wide range of nonhuman others, not merely those who are like us or otherwise please us.

Such criticisms must not obscure, however, one more noteworthy success: the work of first-wave animal law through its combination of personal meaning and novel approaches in the tradition-oriented realm of law models well how those who pursue Animal Studies often develop approaches that fuse together activism, creativity, and science-based approaches. Animal law's achievement in this regard has yet to produce culture-wide effects, for legal and public policy venues are not the circles of our society that create and sustain our social values (chapter 3). As suggested below, it is possible that second-wave animal law will concentrate not merely on public laws, public debates, and government-based actions, but also on the broader and deeper set of values we signal with words like "cultural norms," "social ethics," and "public mores." These are the deep, sustaining foundation of human life in our communities. These values must be addressed if any field of Animal Studies is to encounter the core problems driving the radical subordination of all nonhuman lives in today's legal systems.

Second-Wave Animal Law

The future of a robust animal law field requires more and different approaches that amount to a second wave of animal law. This second wave will be far more interdisciplinary, characterized by increased recognition of the fact that law needs to work hand-in-hand with many other disciplines to discover, explore, and understand nonhumans' realities. Second-wave animal law will also nurture other disciplines' capacity for caring about others beyond the species line, just as the best of our human-centered law nurtures caring about other humans. This will lead to an imaginative, humble, ethics-sensitive engagement with other living beings that produces a qualitatively different stage of animal law in which important perspectives of nonlawyers play a role.

Note that second-wave animal law builds directly on the first wave's success in foregrounding questions about legal rights. In crucial ways, then, the first wave has opened doors and minds (which is historically and psychologically important). But such openings have their limits, for even the powerful tool of creating specific legal rights can only do so much. In fact, even in law circles this tool is but one among many others in the legal toolbox. Invoking "rights for animals" is part of the larger moral revolution, and while specific legal rights for some animals, such as chimpanzees, can work well in some contexts, there are other tools that can also work as well or better. For example, banning ownership of chimpanzees altogether, while not necessarily involving specific legal rights, can be extremely practical in its application and effects, thus providing fundamental, effective protections for chimpanzees. Further, since many nonhuman animals do not have the kinds of abilities that would support individual-based legal rights, foregrounding specific legal rights for individuals risks prioritizing a strategy that works for relatively few nonhuman animals.

Wildlife remains a category not fully addressed by first-wave animal law, no doubt in part because animal law gained a foothold in our imagination through its appeal to those who enjoy companion animals. The vast majority of people involved in both the worldwide animal protection movement and today's animal law developments are acutely aware, however, that there are many other animals "out there." To be productive, animal law needs to engage the earth's more-than-companion animals with creativity and in light of their actual realities in their own contexts. Such an engagement will necessitate entry into new realms that go far beyond the concepts that now control how law, lawyers, judges, and elected lawmakers think of nonhuman animals. It will also require careful reexamination of legal concepts that undergird the exceptionalist tradition. Foremost among these is the key notion that humans alone are "legal persons with rights" and all other living beings are designated "legal things." This dualism promotes law's valuing of only those nonhuman animals who can provide an economic benefit to the legal owner.

Similarly, animal law needs to move well beyond the limited approach that dominates environmental law. Under this approach, a nonhuman animal has legal significance only if its species is deemed threatened by extinction. Animal law has much capacity to challenge such one-dimensional approaches that are based solely on species-level realities. In particular, the second wave can expand animal law by creating richer, more interdisciplinary approaches that create further opportunities important to the integrity of animal law and future possibilities in Animal Studies more generally.

A Second Cutting Edge: Philosophy and Other Animals

The concern for other-than-human animals among philosophers has a number of salient features in tension with each other. Philosophical reflection about other animals has occurred since the very dawn of the discipline, suggesting in the spirit of Lévi-Strauss that other animals are indeed "good to think." Nonetheless, philosophers' commitment to think as carefully as possible about nonhuman animals has sometimes been worse than dormant—it has, in fact, been absent for prolonged periods of time. Further, those who dared to raise animal-related questions at times found themselves at odds with those philosophers deemed by the exceptionalist tradition to be the hallmark of careful human thinking.

Today, however, philosophers again robustly engage a battery of issues that arise as humans try to think as carefully as possible about the great variety of living beings outside our species. The work of many philosophers has been, and continues to be, a prominent feature of worldwide efforts to extend protection to some nonhuman animals. The two philosophers most often cited in contemporary animal protection literature are the Australian

Peter Singer and the American Tom Regan, both of whom in the 1970s and 1980s produced seminal literature in the social movement known variously as animal liberation, animal protection, or animal rights.[8]

Singer and Regan represent two different approaches within philosophy. Singer produced the widely read *Animal Liberation* in the mid-1970s, which focused on blatant harms to farmed and research animals. Singer's book clearly energized many individuals to revive, in many industrialized countries, the ancient but marginalized tradition of animal protection. Singer approaches ethics on the basis of utilitarian calculations in the tradition of Jeremy Bentham, whose question "Can they suffer?" in 1789 put nonhuman animals on the agenda of mainline Western philosophy.

While utilitarian calculations assess whether an act is right or moral by attempting to measure whether it produces more good than bad (with, admittedly, much guesswork), Regan's work is anchored in a detailed argument in favor of a rights-based approach to animal protection.[9] By invoking the rights tradition, Regan denied that harms could be inflicted upon certain nonhuman macroanimals simply because human benefits from the intentional harms were deemed so great.

The seeds planted in the 1970s and 1980s by Singer, Regan, and other philosophers focusing on liberation and rights have sprouted in many different ways. The worldwide animal protection movement today features a great variety of efforts aimed at abolition or amelioration of the harms done intentionally to the macroanimals used for research, food, companionship, entertainment, and so on, as well as wildlife.

In 1980 the Society for the Study of Ethics and Animals was created by the American Philosophical Association. This group was among the first academic groups to offer organized discussion at professional meetings of issues going beyond resource-focused use of other-than-human animals. This group has opened up many opportunities for academic philosophers to discuss a wide range of ethical and other philosophical issues raised by humans' encounters with other-than-human animals. At such meetings, the important tradition of careful thinking, which philosophers have helped develop in a variety of ways, allowed discussants to range widely in topics not traditionally covered in philosophy courses. In this manner, this group modeled for others how a full engagement of issues arising from other animals' realities prompts critical thinking.

Asking about Knowledge Claims

It is in the very nature of the philosophical enterprise to ask questions about the quality of knowledge claims. Early Greek philosophy was born through such inquiries, and the seminal thinkers of the Western philosophical tradition have again and again addressed this issue. Sometimes philosophy has been seen as the leading edge of this inquiry, as when John Locke referred to philosophers as "under-labourers": "It is ambition enough to be employed as an under-labourer in clearing the ground a little, and removing some of the rubbish that lies in the way to knowledge."[10]

This description includes overtones of humility, although it can be read as suggesting a unique role for philosophers regarding human thinking. While thinkers in other fields also have been known to take the view that their own approach to human knowledge is the best, careful thinking of the kind that philosophers aspire to is by no means their exclusive province. It is a hallmark not only of the best work in the natural and social science traditions but also in history, ethics (both religious and secular), cultural studies, psychology, and other fields.

One of the principal ways philosophy contributes to animal studies is through the greatly expanded and very diverse work of academic philosophers on animal issues. One of the best-known modern philosophers, Martha Nussbaum, included nonhuman animal issues as one of three topics in her 2006 book *Frontiers of Justice: Disability, Nationality, Species Membership*: "When I say that the mistreatment of animals is unjust, I mean to say not only that it is wrong of us to treat them that way, but also that they have a right, a moral entitlement, not to be treated in that way. It is unfair to *them*."[11]

The philosopher Daniel Dennett has noted problems in thinking about other animals' consciousness by many people who aspire to be careful, moral thinkers:

A curious asymmetry can be observed. We do not require absolute, Cartesian certainty that our fellow human beings are conscious—what we require is what is aptly called moral certainty. Can we not have the same moral certainty about the experiences of animals? I have not yet seen an argument by a philosopher to the effect that we cannot, with the aid of science, establish facts about animal minds with the same degree of moral certainty that satisfies us in the case of our own species.[12]

Many other philosophers have noticed similar anomalies. Rachels, for example, has observed the following dilemma about attempts to use nonhumans in research: "In order to defend the usefulness of research [researchers] must emphasize the similarities between the animals and the humans, but in order to defend it ethically, they must emphasize the differences."[13] The veterinary ethicist Bernard Rollin, having pondered this dilemma, suggested, "From a strictly philosophical point of view, I think that we must draw a startling conclusion: If a certain sort of research on human beings is considered to be immoral, a prima facie case exists for saying that such research is immoral when conducted on animals."[14]

The Reluctant Establishment

Despite the fact that respected twentieth- and twenty-first-century philosophers have engaged a variety of issues regarding other-than-human animals, by no means have philosophy departments accepted such topics into the curriculum offered general philosophy students. This may be because most of those now teaching philosophy courses were trained, as students, to ignore nonhuman animal issues. Further, much training in philosophy departments for the last century has focused on highly technical versions of the discipline that hold little relevance to daily life. The widely respected philosopher Kwame Anthony Appiah in his 2008 *Experiments in Ethics* describes how academic philosophers beginning in the nineteenth century moved away from the field's long-standing engagement with the complex, difficult-to-resolve problems we put under the umbrella "human nature," turning instead to highly technical, conceptual analysis of moral terms and language more generally.

In 2010, when Appiah pushed readers of the *Washington Post* to ask, "What will future generations condemn us for?," he answered his own question by suggesting that at least four messy problems of human nature now need attention—our prison system, industrialized meat production, the institutionalized and isolated elderly, and the environment.[15] Issues of central importance to Animal Studies appear twice in this short list. Concern for food animals has dramatically expanded in the last decade, and the broad topic of the environment involves myriad animal issues (through, for example, habitat protection).

Appiah's list reflects not only his concern that philosophers return to daily life issues— it also reflects the ferment discussed in this book. Like law and public policy, philosophy is both informed and energized by trends started outside formal education and other academic bailiwicks. As such concerns surface and mature, philosophers and other critical thinkers play an important role in clarifying issues, for as the pioneering philosopher of biology Ernst Mayr once observed, "Our understanding of the world is achieved more effectively by conceptual improvements than by the discovery of new facts, even though the two are not mutually exclusive."[16] The ferment on other-than-human animal issues continues, then, to be prompted by a combination of grassroots ideas and energies from critical thinkers practicing in a variety of areas. This combination today is stirring philosophers as fully as it moves active citizens, people of faith, educators, artists, and others whose lives connect to other-than-human animals in our shared, more-than-human world.

To use Locke's term, then, there are many "under-labourers" in a variety of domains thinking in many different ways. The energies drawn from a wide range of sciences, creative fiction and other arts, religion, formal and informal education, and many academic disciplines offer novel ways of comprehending the world around us. Indeed, the innovations that count as "conceptual improvements" (Mayr) and "removing some of the rubbish that lies in the way to knowledge" (Locke) include many ideas and perspectives from nonphilosophers.

Animal studies, by virtue of its interdisciplinary reach, can provide philosophers and others a helpful perspective on what is happening not only in philosophical circles today but also in many other domains. Clearly, neither philosophers nor other intellectuals are in charge of how people are supposed to think about other-than-human animals. But philosophers do have a major task, as do all critical thinkers and students in Animal Studies: listening carefully to many different people who encounter nonhuman animals. If one listens in this way to a wide variety of people about why nonhuman animals are significant, then one begins to comprehend that any culture—and its social movements—will feature a great variety of explanations why other-than-human lives are important to individual humans.

What makes contemporary philosophy an important leading edge in Animal Studies is that this forceful, highly intellectualized tradition has deep commitments to both open-minded exploration and key humilities that emerge when a field is hospitable to interdisciplinary inquiries. Philosophers' great commitment to and skill in critical thinking are among the most valuable contributions to creating a hospitable environment for a robust form of Animal Studies. This is one of the reasons that philosophy has, as a field, welcomed efforts like those of Lévi-Strauss, Derrida, Singer, Regan, Nussbaum, Dennett, Rachels, Rollin, Mayr, and so many others. These thinkers have enhanced our abilities to see more clearly the impact of human choices on the nonhuman citizens of the more-than-human world.

Philosophy itself, then, is a cutting edge of Animal Studies because it models so well the range and depth of an enhanced capacity to nurture inquiries of many kinds—such as ethics—that are relevant to daily life. Chapter 7 explores further why this nurturing of the existential dimensions of humans' encounter with other-than-human animals in daily life is crucially important in Animal Studies.

A Third Cutting Edge: Critical Studies and Other Animals

We turn here to a set of highly intellectualized, mostly academic inquiries known as Critical Studies that have an important presence in contemporary Animal Studies. The word "critical" as used in this field, which has important links to cultural studies and critical theory (see chapter 7), signals that advocates of this approach are concerned to point out oppressions and challenge the assumptions and value structures that anchor them. While the principal focus of Critical Studies has been human-on-human dominations, the very process of identifying such oppressions and their injustices creates skills needed to learn about and acknowledge harms to other-than-humans as well. Further, coupling identification of human-on-human domination with advocacy for change creates a special skill set that transports well to domains where human-on-nonhuman oppressions occur. Once such oppressions are noticed, they can be analyzed carefully and then challenged directly as needed.

Critical Studies draws power from its employment of many basic tools of critical thinking. For example, important perspectives on human thinking and valuing are generated when one uncovers assumptions and practices that promote different forms of "conditioned ethical blindness" permitting individuals not only to tolerate but even to justify oppression in their midst (see chapter 10). Thus Critical Studies repeatedly invokes humans' special analytical and ethical abilities, which in combination reveal that this kind of inquiry has implications that are far more than intellectual and conceptual—they are social, political, legal, educational, and more.

It is the combination of descriptive analysis coupled with ethical sensitivity that has allowed Critical Studies to emerge along with law and philosophy as one of the more developed forms of contemporary Animal Studies. Integration of ethical questions into inquiries about oppression facilitates recognition of the day-to-day aspects of practices and problems that motivate many to pursue Animal Studies.

As advocates of Critical Studies analyze particular oppressions and then call for disavowal of the inherited claims and practices that hold such oppressions in place, they provide one opportunity after another to identify and then engage specific oppressions. Further, by challenging received values and traditional exclusions, Critical Studies calls upon many different disciplines. Thereby, Critical Studies contributes analyses that speak to the commitments and tasks at the very heart of Animal Studies. For example, by consistently questioning assumptions, Critical Studies provides habits of mind that challenge the ideological features of the exceptionalist tradition. Crossing disciplinary lines regularly, advocates of Critical Studies point out how claims that "all humans are important" are often mere smoke screens for much narrower agendas, like class advantage, political power, religious domination, or some other form of exclusion.

In effect, the provocative critiques of human practices that define Critical Studies share a fundamental spirit with Animal Studies as an enterprise that necessarily focuses on exclusions prompted by many humans' lifestyles and actions. The exclusions that have gained the attention of Critical Studies are, admittedly, more often human-involved, but the prominent willingness of Critical Studies advocates, for example, to examine inherited perspectives about other animals, to grapple with new ideas on their own merits, and to talk openly about the central role of inclusive ethics strengthens humans' ability to examine the quality of claims about other living beings.

By its very nature, then, Critical Studies has a breadth that leads to interdisciplinary humilities. Like philosophy, to which it has important historical debts, Critical Studies relentlessly examines knowledge claims through one lens after another, thereby creating discussion that needs viewpoints enriched by studies of history, religion, social sciences, psychology, literature, and cultural studies. Such work reveals that there is much to learn from inquiries that are willing to be "radical" in the etymologically based sense of "going to the root."

Critical Studies offers powerful challenges to complacency about traditional forms of education. Scrupulous examination of how traditional dismissals of other-than-human animals are anchored, especially because so many claims can be attributed to a refusal to investigate, is particularly important when prevailing practices are justified on exclusively human-centered grounds (such as claims that only humans have emotions or are intelligent). Opening minds to both evidence and the power of questions can help students see the quality of inherited, human-centered justifications. In such a situation, conditions exist that enhance the possibility that Animal Studies can be pursued freely.

Challenges for Critical Studies

Because Critical Studies is deeply invested in "thinking about thinking," Animal Studies can push Critical Studies advocates to be as clear and accessible as possible in their writing and claims. Critical Studies is characteristically pursued in a highly academic and intellectualized style, such that newcomers often are frustrated by the complexities of the concepts, vocabulary, and scholarship. The writing may feature virtuosic displays of jargon, and in the modern academy many people dismiss Critical Studies as nothing more than a barrage of high-sounding words (see chapter 7).

Animal Studies can push Critical Studies advocates to see where their analysis is dominated by intellectual pyrotechnics and pretensions—not unlike challenges to scientists who practice and speak in human-biased ways. Animal Studies also challenges ethicists whose work fails to problematize the exceptionalist tradition and philosophers whose focus is so unduly technical that it is existentially irrelevant or inaccessible.

With patience, however, one can recognize that Critical Studies offers opportunities to engage not only oppressions, but also the challenges and paradoxes such oppressions create for those who insist that humans are to be understood as intelligent, caring moral beings. The field's challenges to received views foreshadow more ferment on both human and nonhuman animal issues. Critical Studies also invites deeper and more informed analyses of the overlap between, on the one hand, human-on-human domination and, on the other hand, human-on-nonhuman domination.

Generalizations across These Cutting Edges

The fields of law, philosophy, and Critical Studies represent vibrant and developing efforts, and in the manner that distant thunder heralds a summer storm, these fields have caught the awareness of many people. As examples of early but still developing subfields, these enterprises model Animal Studies' commitment to produce insights that are educational, philosophical, and values-revealing. Further, each of these inquiries invites us to work with interdisciplinary approaches and the humility that a multidisciplinary approach requires. In addition, each of these fields regularly engages the question of ethics in connection with living beings beyond the species line. If together these three fields can open up human minds and hearts, no doubt a combination of many other fields can also do this in even richer, broader, and more compelling ways, prompting everyone to examine past developments, present realities, and future possibilities at the human-nonhuman intersection in as honest and sensitive a manner as possible.

Animals in the Creative Arts

"Animals have taken over art, and art wonders why." This headline appeared in June 2000 above a two-page feature article in the New York Times.[1] The movement of nonhuman animals into the center of an arts tradition has happened before—in Paleolithic times, interest in other-than-human animals was such that, as one art historian observed, "Never perhaps in the whole history of animal art, even in China, has the animal appeared so magnified, so sublimated, without ever losing its reality or naturalness, than in Paleolithic art."[2] Another art historian, speaking of the "awe of animals" characteristic of primal peoples, suggests, "The evidence for these facts is so overwhelming that it is recognized across academic boundaries from anthropology to art history to religious studies to the sciences."[3] Our tradition of portraying other-than-human animals in creative fashion is so long-standing and widespread that yet another observer suggests, "In all of art since the cave paintings, it is probable that animals are represented more often than any other class of things in nature."[4] Arts traditions of various kinds, then, reflect that many humans from time immemorial have been deeply interested in and respectful of some other-than-human lives.

Human-Centeredness in the Arts

Such observations are not meant to deny that many discussions of the arts feature an overwhelmingly human-centered bias. The arts are, after all, a pinnacle of human creativity and can easily be pursued by focusing only on humans.

Human-centeredness in the arts can come in forms that are more sinister than simply ignoring other-than-human animals. A small amount of art has presented the public with an opportunity to harm nonhuman animals, with the intent to shock. Some artists, for example, have intentionally designed exhibits at which visitors had the opportunity to kill living animals—one widely discussed example was the 2000 incident at the Trapholt Art Museum in Kolding, Denmark, in which goldfish were put into blenders that could be switched on by members of the public.

But even though there have been, and surely will continue to be, arts traditions that share something of the dismissive spirit of economics-driven industries like factory farming, the arts have often reminded people that they live in a more-than-human world. Further, historians of art have become increasingly sophisticated in exploring the patterns of humans' recurring fascination with other-than-human animals in arts traditions.[5] As a number of art historians have observed, even when nonhuman animals are dismissed by a society's institutions, educators, and leaders, they often remain in the margins of artistic endeavors.[6] The presence of nonhuman animals in the human arts makes it unlikely, then, that those inquiring about arts and other animals will be presented only with views that conform to the exceptionalist tradition.

Creativity beyond Human-Centeredness: Artists as Teachers and Pioneers

Many different kinds of art have benefited from and opened special vistas on nonhuman animals for Animal Studies. Artistic endeavors give us multiple, flexible forms of expression that are not tied to the limits and biases of words used primarily for analytical description. Through animal-involved symbols, which may or may not have anything to do with the individuals of the biological species whose generalized image is invoked, much can be accomplished on matters where empirical inquiries are difficult for one reason or another. Scientific approaches have great power but in no way exhaust our world. Patient observation, as important as it is in learning about the world, reveals often only minor details or surface issues.

Arts can help with at least some matters "close to our hearts…that really count." They can alert us in unique ways to intangibles such as the relationship between figures and ideas. Individual arts, some of which are addressed below, can provide pioneering perspectives, vistas on unique experiences, and one-to-one connections of the sort one is often invited to consider by poets, painters, and photographers. Further, because the arts are capable of focusing on encounters with other animals as individuals, they have special abilities to communicate beyond the highly generalized concepts through which language, science, and the humanities work. Individualized artworks can facilitate this process in creative or eclectic ways that prompt new, fertile consideration of the realities of other-than-human individuals. They provide novel perspectives that can prompt humans to focus afresh on subjects that have been marginalized. In effect, any focus on marginalized subjects demands that we think outside the box, that is, think outside inherited paradigms, such as human-centeredness. In this, the arts can prompt a kind of self-awareness through critical thinking.

In these respects, individual artists through their work can pioneer new awarenesses within human communities. Those who break new ground, as it were, in human thinking inevitably face challenges from those invested in established claims to knowledge—we are familiar with the scientific pioneers (Newton, Darwin, Einstein) who prompted new ways of looking at the world in which we live. Artists, too, can push us to see the world anew—said another way, they can challenge what it is acceptable to claim to know.

Education as a human endeavor has its conservative features—recall that it is "rarely a place of daring." To be credible in established circles, one may have to forgo true innovation of the kinds that threaten established ways of thinking and seeing. While any of the arts can be co-opted in this way (think of the "successful" painter whose goal is to please wealthy patrons in order to get lucrative commissions), there are many artists whose passion is "art for art's sake" and therefore go it alone. Art that makes unseen connections or fosters awareness of features of human possibility that are presently unnoticed or undervalued can be unpopular and controversial. But our social values are often in need of pioneers who are able to look at what has been previously taught (chapter 8). When this is done, the community has greater prospects of putting aside those ideas and claims that are inadequate to the task at hand, such as attempting to learn about the actual realities of other animals.

Animal Studies is acutely in need of such pioneers because of the human-centered social values that prevail in so many realms. Both realism and humility about the place of humans in a more-than-human world are needed as humans inquire about connection and other animals' realities. For this task, art can offer perspectives, free up imaginations, invite hearts, tantalize minds, and reveal communities and connections—thereby making change possible.

Animal Studies Teaching Artists and Art Historians

It can be argued, too, that those who pursue arts-related versions of Animal Studies will be able to add new perspectives on art as well. They will be able to help individual artists explore the frontiers of human understanding of other-than-human animals. Further, Animal Studies is full of resources that can help historians inclined to frame the history of a particular art through a story drawn from the exceptionalist tradition. Given the long, varied, and fecund history of artists teaching the world about other-than-human animals, failing to notice how individual arts can do this is to risk a failure to tell the whole story.

Finally, while scholars have only just begun to plumb the complex roles that our creative arts play in social movements generally, it is already clear that the worldwide animal protection movement has often used various forms of literature, painting, sculpture, music, and dance to call attention to problems or prompt awareness of the need for fundamental change. In this sense, many different arts can contribute to helping humans focus on different challenges that arise at the human-nonhuman interface.

Before There Were Words

Ancient rock art found around the world predates the earliest written human records by thousands and thousands of years. As visual art, rock paintings and carvings take advantage of humans' dominant sense (vision is similarly dominant in virtually all primates). Today many people are familiar with the markings and paintings of Paleolithic humans at the Lascaux and Chauvet caves in France. Both feature paintings that are evocative and surprisingly realistic as they picture many different kinds of animals.

It is intriguing to wonder what such markings might tell us regarding either the humans in the communities from which the artists came, or the realities of the nonhuman animals. One scholar describes Chauvet Cave as "decorated with stunning images of animals by a few Late Paleolithic humans around 30,000 BC. The cave walls are filled with complex scenes—confronting rhinoceroses, snarling lions, herds of animals drawn as if rapidly moving through the cave—420 animal images in all (and only six human images)."[7]

Since the art motifs vary greatly, featuring carnivores or herbivores, the same scholar frankly observes, "it is largely unknown why the art motive changed from carnivores to herbivores.... At most cave sites there is no direct relationship between the species depicted on the walls and ceilings and the bones found scattered about, indicating that the artists did not usually draw the species eaten by the group."[8] We are so removed from the time and worldview of the artists that such conclusions are, as scholars regularly acknowledge, tentative guesses. But one factor does connect us to these paintings—the animals depicted are recognizable and we can count them and assess which animals appeared most often. "The species most often depicted in cave paintings were horse (30 per cent of all drawings)..., bison and aurochs (another 30 per cent), deer, ibex and mammoth (another 30 per cent), and bears, felines and rhinos (10 per cent)."[9]

The detailed drawings of these other-than-human animals beg speculation of several kinds. Certainly the Chauvet artists had a keen sense of observation and substantial knowledge of animal behavior, as shown in their depictions of animal anatomy, animal-on-animal confrontations, and accurate illustrations of the "social" behaviors of certain species, emphasizing for example the difference between gregarious species such as mammoths, lions, and bison and solitary species such as bears and panthers.[10]

The details may have been noticed from up close, for some speculate that our remote ancestors' familiarity with the depicted animals was possible because they could get much closer than we can to such animals—"it is only because of modern weapons that hunted animals have learned to run from a distance."[11]

From an Animal Studies standpoint, though, it is important to acknowledge that such reconstructions rely on guesswork that, even when seemingly obvious to us, can be completely wrong. The remoteness of the events about which we speculate can be astonishing—for example, evidence suggests human association with wolves dates back more than 100,000 years. This association was most likely a hunting partnership, for there is evidence that 120,000 years ago Paleolithic humans built shelters in caves with the skull of a wolf intentionally placed at the entrance.[12]

We can ask as well about the human side of ancient human-nonhuman interactions—do we, for example, know anything at all about the individuals in the hunter-gatherer societies who are the likely creators and observers of these images?

On the Question of Intentions

The British scholar Kenneth Clark in Animals and Men said, about how nonhuman animals impacted some early human artists, "Personally I believe that the animals in the cave paintings are records of admiration. 'This is what we want to be like,' they say, in unmistakable accents; 'these are the most admirable of our kinsmen.'"[13] In one sense, because we know nothing at all specific about the artists whose work we see, we can speak freely of them and their intentions. But certitude of any kind about their intentions eludes us, such that claims about the meaning of such ancient art reveals far more about the claimant's personal sense of the possibilities of human fascination with living beings.

One of the tasks of art-focused Animal Studies is to call out forthrightly the quality of different claims attributing significance and meaning, if any, to such images. Some questions beg an answer—what was the relationship of the images in ancient caves to the actual animals whose form is part of the image we now view? We recognize that many images are anatomically correct such that we can identify the particular species depicted. But beyond this, we often must admit a profound ignorance, although the awe that these paintings inspire within us may prompt insights that help us in our search for other animals' realities.

Perhaps we can inquire about issues that might have arisen during the transformation of a hunting-gathering lifestyle to an agriculture-based and pastoral lifestyle. The implications and consequences of domestication, agriculture, and urbanization on humans' view of the earth's other animals are central issues in Animal Studies, of course, and there is much scholarly speculation about such issues.[14]

A fundamental question that takes the inquirer in a completely different direction is, What options might have been available to a human in one of the ancient caves where art is found today? Could this individual really understand the lives of the nonhuman animals that shared that particular ecological niche? How do the options for Paleolithic individuals compare to the options now available to us?

Animal Studies can pose such questions from several different perspectives: of ancient humans in any number of different societies, of citizens of industrialized societies who have had the benefit of a thoroughly modern education, of small-scale societies familiar with their own local flora and fauna, and from a great variety of comparative vantage points. All of these questions have more power than their answers—for example, if someone asks about the meaning of a cave painting, one can wonder what such a question means. Does art ever have meaning in the sense that sentences in a book have meaning? Can a work of art have its own meaning of a kind that words fail to convey? Does art have significance when it helps someone discover the meaningfulness of enjoyably engaging an image? Might a picture, a photograph, or some other portrayal using an animal image simply invite one to become aware of another animal's individuality, or its integration in a social grouping, or its contentment with a trusted friend, or some other uniqueness? Might many works of art be like music that is enjoyed for its sheer beauty and not at all for sentence-like meaning?

The task of responding intelligently to these questions pushes anyone doing Animal Studies to engage various critical thinking tasks needed to understand what is at stake when someone answers such questions. Further, as so often happens in the human-centered environments where Animal Studies can now be attempted, such questions touch on some of the most debated topics in the study of creative arts, including those that have nothing to do with other-than-human animals. Art-focused Animal Studies, then, introduces issues arising solely within the rich complexities of human art. These issues are completely apart from the complexities that arise when humans try to convey something meaningful or artistically beautiful about nonhuman animals.

Such challenges reveal that the arts are a rich source of human exploration of the world. They may or may not be verbal, but they can facilitate awareness and connection in astonishing ways. Thereby, humans' artistic achievements play remarkably important roles in our lives.

Music, Dance, and Other Animals

The origins of music are integrally tied to human fascination with nonhuman animals. As with dance, the perceived realities of other animals' lives greatly impacted early music. Animal imitation was a highly practiced art through both voice and instrument, and early musical instruments were made from animal parts and often carved into animal shapes.[15] Not surprisingly, then, one finds animal-related themes throughout music traditions. Even classical music from the human-centered Western tradition, for example, includes compositions that mimic animal sounds or use animal images in titles or as story pieces.

Dance as an expressive art also has origins in imitation of nonhuman animals. One scholar in 1964 suggested, "dancing was originally nothing more than a completion of the animal disguise by the appropriate movements and gestures."[16] Another respected scholar explains that the movements of some other-than-human animals "capture human imagination and inspire imitation." Some dances have "verisimilitude" because they mimic the rhythms of a nonhuman animal or "perfectly reproduce the behavior and antics" of some particular animal.[17]

For many reasons, both dance and music have communicative possibilities that are unique and appealing to humans, capable of conveying much that words by specialists, anthropologists, or even poets cannot convey. As one scholar suggests, "In dance man can lose himself."[18]

Animal Studies asks which realities of other-than-human animals intrude on our imagination—it may be the highly choreographed courting dances of a certain animal that intrigue, or the stalking behavior or communication patterns of other animals that inspire, even compel, the artist. In some human cultures, it is claimed that the nonhuman animals invoked through dance are related to the dancers' clan. Sometimes the animal being imitated is the principal food and resource animal for the tribe. Sometimes a story is "told from the point of view of the animals... [which] says something about the mutual respect and reciprocity expected between hunters and animals."[19]

Animal Studies is also capable of assessing how anyone—whether dancer, observer, or scholarly analyst—evaluates a dance event. For example, Lonsdale couched one of his observations in the male-oriented language choices that prevailed in 1980s scholarship: "Man is the supreme dancer."[20] Such a conclusion is an example of the great variety of human-centerednesses that prevail as we engage the more-than-human world. As our recent history has revealed regarding the challenges to language that is racist, sexist, homophobic, and ethnic-centered, language-based habits can be softened or changed altogether—a fact that Animal Studies can make pertinent to how we speak of the more-than-human world. Note how such a perspective allows one to see the strident overtones in the following passage as it begins with connection but moves quickly to human-centered themes: "Man is an animal. Paradoxically, the animal dancer exhibits infinite superiority over the beast while at the same time humbling himself before the animal model, his god."[21]

Whether this claim captures the attitude of "the animal dancer," as well as what meaning might be attributed to the phrase "the animal model, his god," is a topic that Animal Studies does well to explore, for there is much at stake in phrases like "infinite superiority" and "humbling." Similarly, the association of certain nonhuman animals with dancing deities, as in the divine figure of Shiva in India or when shamans dance in animal form in order to influence other realms,[22] can be explored with creative questions from multiple disciplines.

In general, Animal Studies needs many of its constituent disciplines to explore what is at issue when gods and fantastic figures, such as therioanthropes like the "goat-man," are involved in dances. It similarly needs to accommodate scholars who suggest that concentrating solely on fantastic figures rather than actual animals is squarely within the domain of Animal Studies. The study of dance reflects well the complexity of humans' attitudes toward other animals. One scholar suggests that an "extreme degree of familiarity with animals" (meaning, of course, nonhuman animals) exists alongside "hostility towards them."[23]

Dancers or scholars sometimes invest performances with highly symbolic overtones. Apart from such claims, there may be features in the dance itself, or its origins, connected to other-than-human animals in telling ways. Such connections can also be found in myths that a culture tells about the origin of dance, or in symbolic overtones that a modern audience finds in the art.[24]

Today, Animal Studies has resources to explore these issues because scholarly work on dance and its relationship with the more-than-human world has been developed for decades. In addition, though, Animal Studies pushes beyond previous scholarship by asking a rich array of additional questions. Notice, for example, how Animal Studies can return the focus to the other animals by asking, Do (nonhuman) animals dance?[25] Similar questions can be posed about music making—the point of such questions about whether any nonhuman animals create works of art is to move away from the relentless human-centeredness of modern scholarship.

Since we can plausibly ask questions about other animals' creativity, consider established art critics who, when presented with work done by an undisclosed artist (who was, in fact, a nonhuman animal), judged the work to have great artistic merit. Of course, one must also mull over that there are critics who question whether chimpanzees, elephants, dolphins, and other nonhuman animals who have been advanced as artists are, in fact, doing work that qualifies as art. Animal Studies has a wide array of critical thinking skills at its disposal to engage both positive and negative claims on these paradigm-breaking questions.

Deeper, Wider Questions about Symbols

Complicated issues arise as Animal Studies balances the central place of other animals' realities with humans' long-standing and vibrant preoccupation with animal-connected symbols. In the past, some scholars have taken the study of symbols employing other animal images of one kind or another to be the major thrust of Animal Studies. Entire books on this subject have been written, however, that do not explore in any way questions about actual biological individuals. Such an approach need not signal that other-than-human animals' realities have been completely marginalized, for the study of animal-connected symbols is itself a revealing subject. But the scholarship performed during previous centuries has in fact constantly allowed human-centered inquiries, such as symbolic value, to push any inquiry about other animals' realities to the margins. Said another way, animal-connected symbols, but not the biological animals themselves, have often been treated as the natural and full range of any human's concern about "animals."

A few basic divisions help distinguish some of the fundamentally different issues at stake when humans' complex use of symbols is foregrounded as a principal endeavor in Animal Studies. Symbols are, of course, human tools. When they involve images of nonhuman animals (through either words or shapes or sounds), such symbols can be closely or remotely based on a nonhuman animal's realities. They can be based on but one part of the animal, or on a legend mentioning some caricature of the animal. It is not at all uncommon for them to be based on outright factual error, that is, a nonreality associated with the animal because of ignorance or apathy. Symbols can, then, be so fundamentally unrelated to other animals that they have, as it were, a life of their own. Real animals, however, can themselves become a symbol, as in the following example.

> And he rocked, constantly, tugging on chains that bound his legs to the slightly raised platform on which he stood.... This bull was never let out of the pavilion.... So for decades now, he had been here on his raised dais, rocking, straining, surging back and forth with unfathomable power.... Surging, swaying, pulling this way and that, forever and a day—the heaven-sent king of elephants, born of clouds and rain, colored like the sacred lotus, a captured god but now an obsolete one, something out of a distant time and kingdom, his purpose all but forgotten.... Alone in his dark, golden-spired pavilion. Forever alone. Colossal. And very likely insane. That was the message in those eyes: madness.[26]

The elephant subjected to this obviously attenuated life is held captive because Buddhist captors intend, ironically, to honor him. Given a name—Pra Barom Nakkot—and chosen because the local humans deemed him to have the distinctive, auspicious features of a white elephant, Pra Barom Nakkot was merely lighter in color than most elephants. He possessed a number of the seven features (his gait, carriage, and overall shape) traditionally associated with sacred elephants.[27] The preoccupation with sacred elephants is based on the legend of flying white elephants found across Asia. Pra Barom Nakkot was, because of his special markings, "seen as [one of the] descendants of the original winged elephants that roamed the cloudscapes above Earth and as avatars of the Buddha."[28]

While some forms of Animal Studies conceivably might focus on Pra Barom Nakkot's symbolic significance only, failure to mention the fact that his captivity deprived him of any chance whatsoever to pursue the development of his own interests (that is, to live as a unique individual in an elephant community) is an extremely one-dimensional form of Animal Studies. It is, in effect, merely human-centeredness accomplished under the guise of talking about animal symbols—said another way, this form of Animal Studies is not at all about Pra Barom Nakkot, but far more about human animals.

Critical thinking requires that Animal Studies call out the factors that limit any account to a single dimension. Such an approach would identify why failing to call out the obvious human-centered features of Pra Barom Nakkot's subordination hides a values-driven agenda. In fact, omitting any revealing portion of the whole story of why Pra Barom Nakkot is "alone in his dark, golden-spired pavilion" does a disservice to the portions of the story that are told. The whole story is rich, because it includes the fact that this individual elephant was denied the possibility of interacting in the complex social network that characterizes all elephants' lives. Denied the normal social network through which he would have learned to deal with the natural world and communicate with other individual elephants, Pra Barom Nakkot, as a captive from an early age, was given limited training by humans. All of this made his subordination all the easier.

In effect, any form of Animal Studies that focuses solely on the symbolic aspects of his predicament, while ignoring the obvious harms to this biological individual, is so truncated as to be inevitably impoverished.[29]

Scholars of the Arts and Imagined Animals

The above comments are not meant to suggest that work on symbols of nonhuman animals is unimportant. Such work is, in fact, crucial. As one of the most insightful mid-twentieth-century thinkers about nonhuman animals suggested, "Everyone lives in a mythic world, however ignorant of it they may be."[30] Animal symbols work in almost countless ways, and wonderful contributions about human interactions with animals have been made by those who work regularly with symbolic features of human communication, meaning, ethics, and knowledge.

There are, nonetheless, animal-related symbols that say little, if anything, about the other-than-human animals they depict. These can be very trivial, such as sports team logos (the Dolphins, the Tigers), but they can also be rich and diverse, such as totems or clan-affiliated symbols that speak of both connection and even possible ancestry. Extensive scholarship has also been directed to studies of fantasy-based animal images, as well as to cryptozoology, which is the study of, literally, "hidden animals" whose existence has never been proven by physical evidence but are referenced in myths, lore, or anecdotal reports.

Animal Studies adds nuance to all of these issues by calling out when work on animal-related symbols is merely human-centeredness displacing a search for truth, and when such work is pursued because it illuminates animal life as a complex phenomenon in our lives. Animal Studies is, thus, perfectly capable of engaging the plain fact that some animal-connected symbols have nothing whatsoever to do with living nonhuman animals themselves. Similarly, Animal Studies can identify symbols that harm other-than-human animals, or those that create sympathy and compassion. Historical work makes clear that whole eras have been dominated by symbols in ways that are alien to modern humans: "References to nature in ... the Gospels, have been persistently understood from the perspective of modern urban people, themselves wholly alienated from nature, for whom literary references to nature can only be symbols or picturesque illustrations of a human world unrelated to nature."[31]

Thus, Animal Studies needs to be vigilant about the fact that even if it is easy for moderns to slip into dismissive, negative views of symbols, fairness and critical thinking mandate that we view such symbols in their own historical context. One facet of the Western cultural tradition known as the Renaissance has, in fact, been characterized as "the emblematic tradition" because of the prevalence of symbolic accounts. During this extended period, narrative forms of all kinds were characteristically dominated by many animal-related symbols. One researcher describing the work of Renaissance scholars on the natural history of animals suggested that ordinary citizens had an "emblematic world view," and thus perceived not what twenty-first-century citizens see on the basis of our science-saturated preoccupation with actual realities, but instead "a world where animals are just one aspect of an intricate language of metaphor, symbols, and emblems." This emblematic worldview was not just religious, for, the researcher comments, it was "the single most important factor in determining the content and scope of Renaissance natural history." Thus, Renaissance natural histories of, say, peacocks would characteristically list many associations, but would say very little about real peacocks themselves: "if what you seek is a collection of true statements about the peacock, or an anatomical description, or the peacock's place in a taxonomic scheme based on physical characteristics, then you are bound to be disappointed."[32]

Today, Animal Studies offers abundant, multidisciplinary approaches that go beyond not only human-centered concerns but also modern preconceptions of ancient viewpoints, misconceptions of other cultures, and the shortcomings of impoverished, one-dimensional historical accounts. Through a commitment to open and inclusive scholarship, Animal Studies thereby models a willingness to assess where on a continuum any symbolic reference sits—at one end are symbols that are in some way connected to the biological individuals, while at the other end are symbols that are used primarily as affirmations of humans' special ability with symbols. In making such distinctions, Animal Studies has the capacity to prompt broad, healthy inquiries consonant with the basic goals of education, science, and ethics.

Literature: The Power of Words on Their Own

The legendary capabilities of literature are captured by a prominent critic's suggestion that literature "is the human activity that takes the fullest and most precise account of variousness, possibility, complexity and difficulty."[33] Of course, as suggested above, writing that calls in some way upon images of nonhuman animals may not be about biological animals in any meaningful way. As one reviewer of a new novel using lion-related themes suggested in 2007, "so burdened are lions with symbolism that it's surprising they manage to stagger even a few paces, let alone spring at their prey."[34]

Yet both poetry and prose offer many examples of the extraordinary power of words unadorned by pictures or figures of any kind. Such mere words can convey much about the human intersection with other-than-human animals.

Unacknowledged Legislators of the World

Percy Bysshe Shelley suggested in 1819, "Poets are the unacknowledged legislators of the world."[35] Another famous poet claimed, "Poetry increases our feeling for reality."[36] A professor of philosophy adds that poetry allows us "to focus on that which we normally pass over in our everyday activity: the world." As he explains Wallace Stevens's poetry, this philosopher argues that poetry has "a range of observation, power of expression and attention to language that eclipses any other medium."[37]

These claims introduce the possibility that certain language arts have transporting power that helps some humans engage reality in special ways. If one wishes to evaluate whether poetry helps human engagement with other-than-human animals, reading the opening lines of the poem "The Summer Day" by the contemporary poet Mary Oliver as it focuses on a grasshopper raises that possibility. A second issue is whether poetry helps connect readers to other-than-human lives in ways different from or merely supplemental to the connections and perspectives possible by virtue of personal experience, the promptings of science, or other forms of critical thinking.

> Who made the world?
> Who made the swan, and the black bear?
> Who made the grasshopper?
> This grasshopper, I mean—
> the one who has flung herself out of the grass,
> the one who is eating sugar out of my hand,
> who is moving her jaws back and forth instead of up and down—
> who is gazing around with her enormous and complicated eyes.[38]

By the medium of words sequenced in a series of lines, Oliver moves awareness from the generalized ("the swan," "the black bear," and "the grasshopper") to a particular grasshopper. Using critical thinking, we easily recognize that each grasshopper is an individual. We also recognize that each person meets different individual grasshoppers, not merely "the grasshopper." We are, it is surely true, nonetheless used to the kind of blunt generalizations of the opening lines—the swan, the black bear, the grasshopper. These are, however, inadequate to experience, for they homogenize, respectively, all swans, all black bears, all grasshoppers into vagueness.

It is simpler to think and talk in such superficial ways, and doing so is licensed by the tradition of asserting that all members of the same species share what some call an "essence." But this assumption is, in reality, only a laziness—what is real to us is the actual individual grasshoppers that we really do encounter, if we but notice, as this poem's opening lines ask, "This grasshopper, I mean."

The transition from the generalizing frame of mind to the specific, local, and individual—"This grasshopper, I mean"—actually tracks our experience. Noticing this moves us beyond the habit of seeing only categories to the far more existentially meaningful encounters with individuals that actually take place in our daily lives. The poet finishes this poem with a line that addresses each reader as an individual: "what is it you plan to do with your one wild and precious life?"

Different emphases appear in the opening lines of another poem, "The Swan":

> Did you too see it, drifting, all night, on the black river?
> Did you see it in the morning, rising into the silvery air—
> An armful of white blossoms,
> A perfect commotion of silk and linen as it leaned

into the bondage of its wings; a snowbank, a bank of lilies,
Biting the air with its black beak?[39]

Is the "you" here an acquaintance, or any and every reader, or perhaps any being that has eyes to see what the words picture? Whatever one's answer, these lines present through familiar words a very distinctive picture of a single bird. The poem closes with questions that give this bird an almost universal quality:

And did you feel it, in your heart, how it pertained to everything?
And have you too finally figured out what beauty is for?
And have you changed your life?

Opening with images of a single living being, but ending with questions that touch upon feeling, connection, and meaning in the reader's own life, these lines permit questions of meaning, connection, spirituality, philosophy, ethics, and the animals' actual realities to emerge and mingle. Such writing fits Trilling's claim that literature is a "human activity that takes...account of variousness, possibility, complexity and difficulty."

Senses of connectedness akin to those invoked by these lines of this popular American poet can be found in culture after culture, era after era, and in art form after art form even though the form of expression varies considerably. The fact that Animal Studies is driven by personal connectedness anchored in meeting different other-than-human animals is one reason that art forms of different kinds have a presence in courses and publications in the field.

The Power of Prose

Examples abound of prose energized by a power and vision of the kind that invest poetry and visual arts with transporting energy. Below is a series of passages written by a Pulitzer prize–winning author, Scott Momaday, seven years before Peter Singer's seminal *Animal Liberation* and seventeen years before Regan's "case for animal rights." They reveal how literature can, like poetry and the visual arts, be extremely sensitive to the diversity and realities of the myriad creatures who share the larger earth community with humans. Even though these passages are artificially removed from their artistic context (a bit like pulling a plant out of the soil to show someone how alive it is), they reveal nicely how living beings can be said to inhabit the author's awareness.

There is a kind of life that is peculiar to the land in summer—a wariness, a seasonal equation of well-being and alertness. Road runners take on the shape of motion itself, urgent and angular, or else they are like the gnarled, uncovered roots of ancient, stunted trees, some ordinary ruse of the land itself, immovable and forever there. And quail, at evening, just failing to suggest the waddle of too much weight, take cover with scarcely any talent for alarm, and spread their wings to the ground; and if then they are made to take flight, the imminence of no danger on earth can be more apparent; they explode away like a shot, and there is nothing but the dying whistle and streak of their going. Frequently in the sun there are pairs of white and russet hawks soaring to the hunt.[40]

Later in the same passage (which is pages long), Momaday mentions more animals:

In the highest heat of the day, rattlesnakes lie outstretched upon the dunes, as if the sun had wound them out and lain upon them like a line of fire, or, knowing of some vibrant presence on the air, they writhe away in the agony of time. And of their own accord they go at sundown into the earth, hopelessly, as if to some unimaginable reckoning in the underworld. Coyotes have the gift of being seldom seen; they keep to the edge of vision and beyond, loping in and out of cover on the plains and highlands. And at night, when the whole world belongs to them, they parley at the river with the dogs, their higher, sharper voices full of authority and rebuke. They are an old council of clowns, and they are listened to.

The language here is intentionally creative in ways that open the reader to other animals' mystery and complexity. More animals are then brought into the reader's imagination:

Higher, among the hills and mesas and sandstone cliffs, there are foxes and bobcats and mountain lions. Now and then, when the weather turns and food is scarce in the mountains, bear and deer wander down into the canyons.

As the passage adds yet more animals, a connection is made in this panoramic view to the local town and "man's imagination":

Great golden eagles nest among the highest outcrops of rock on the mountain peaks. They are sacred, and one of them, a huge female, old and burnished, is kept alive in a cage in the town. Even so, deprived of the sky, the eagle soars in man's imagination; there is divine malice in the wild eyes, an unmerciful intent. The eagle ranges far and wide over the land, farther than any other creature, and all things there are related simply by having existence in the perfect vision of a bird.

　　These—and the innumerable meaner creatures, the lizard and the frog, the insect and the worm—have tenure in the land. The other, latecoming things—the beasts of burden and of trade; the horse and the sheep, the dog and the cat—these have an alien and inferior aspect, a poverty of vision and instinct, by which they are estranged from the wild land, and made tentative. They are born and die upon the land, but then they are gone away from it as if they had never been.

The power in Momaday's passage clearly is sustained by his roots in American Indian culture, for in many indigenous cultures it is no accident that other-than-human animals appear so fully in stories of all kinds. Thomas Berry in *The Dream of the Earth* (1988) explicitly recognized that some human cultures had long featured what he terms "human intimacy with the earth": "Fortunately we have in the native peoples of the North American continent what must surely be considered in the immediacy of its experience, its emotional sensitivities, and its modes of expression, one of the most integral traditions of human intimacy with the earth, with the entire range of natural phenomena, and with the many living beings which constitute the life community."[41]

Momaday's book provides an example of how fecund literature inspired by an indigenous culture can be. This can be significant for those who come from a culture so stifled by the exceptionalist tradition that its members are not encouraged in any way to take seriously our abilities to notice and be in relationship with nonhuman animals.

There are, of course, countless other examples of literature taking an especially creative role regarding human animals trying to imagine the realities of nonhuman animals. Anna Sewell's classic 1877 novel *Black Beauty,* which was from the very beginning a best-selling book (50 million-plus copies sold to date), is told from the horse's point of view. While it was not the first story told in this manner, Sewell's book spawned many other accounts that attempted to create a voice for nonhuman animals. This is a good strategy from which one can, to use Sewell's own words about her motivations, "induce kindness, sympathy, and an understanding treatment" of the animal being discussed.

A more recent example of fiction that has sold widely (1 million-plus copies) is Daniel Quinn's *Ishmael.* The unnamed gorilla who is the leading figure in the book summarizes nicely how malleable humans can be depending on the story they are told. "There's nothing fundamentally wrong with people. Given a story to enact that puts them in accord with the world, they will live in accord with the world. But given a story to enact that puts them at odds with the world, as yours does, they will live at odds with the world."[42]

Children' Books, Children's Invitations

The issue of children and animals appears in this book in chapter after chapter, in part because children so easily and naturally connect with other animals. Another reason is that the topic of children is a paradigmatic way in which we raise issues about the future—chapter 11, for example, addresses how children's learning about other living beings is shaped informally, which of course will impact how future generations deal with the human-nonhuman intersection. Since Animal Studies makes consideration of future possibilities a central task, it stands to reason that children will be part of many different inquiries it pursues.

Children's natural interest in other animals has long created demand for literature. Walking into the children's area of a bookstore, one will immediately notice the overwhelming presence of animal images. The abilities of children to relate to other animals are quite suggestive of native human abilities in this regard—this is one way in which a focus on children teaches adults.

Another lesson that children can teach is suggested in the traditional maxim memorialized by Gerard Manley Hopkins's poem "The Child Is Father to the Man." Adults can learn not only by looking at children's books but also by asking why and how modern "education" curtails growth of children's native interest in other animals (chapter 11). There are interesting parallels in the conception of adulthood as "putting away" this fascination so evident in childhood and Western civilization's disparagement of indigenous peoples as "savage" because they have reverence for nonhuman animals.

With such questions in mind, it is worth walking again into the children's area of a modern bookstore. Animal Studies will help one see the astonishing inconsistencies of immersing children in animal images to help them learn but later encouraging them to put away this native concern if they wish to enter the all-important world of exceptionalist adulthood.

Honoring Our Fundamental Flexibilities

The arts reveal well how flexible our minds are, how expressive our language can be, and how flexible and necessary our nonword arts can be. This chapter has given only a few examples of the many and altogether diverse human creative arts meant to prompt recognition of the fundamental flexibility of nonword expression, thinking, feeling, and communicating. So many more—photography, films, mixed media and so on—also provide diverse means of deepening awareness of the significance of living in a multispecies world. What creates the most astonishing flexibility, however, is that these modes, which are often thought of as nonanalytical, are extremely expressive of dimensions that mere words cannot convey. The upshot is that these creative arts are important dimensions of human life in many areas, not just matters involving other-than-human animals and the more-than-human world. But it should be clear that in animal-related matters, these nonverbal approaches provide us with insights that not only supplement, but even surpass, what our word-based, analytical approaches achieve.

To be sure, humans' analytical skills are at times extraordinarily impressive—to us. Yet they are obviously limited. In daily life, humans think in a great many ways that are relevant to engagement with other living beings (whether human or not). For millennia, humans have recognized that intelligence comes in many forms—many cultures have, for example, insisted that many nonhuman animals have some of the forms of intelligence that humans characteristically have, as well as other forms.

Howard Gardner's *Frames of Mind: The Theory of Multiple Intelligences* (1985) points out different kinds of intelligence possessed by humans.[43] Gardner identified seven different kinds of human intelligence—musical, logical-mathematical, linguistic, spatial, bodily, intrapersonal, and interpersonal. In 1997, Gardner added an eighth entry (naturalistic intelligence) to the list. This sort of intelligence has to do with nature, nurturing, and relating information to one's natural surroundings.

The general idea of multiple intelligences is now widely accepted although the specifics are not generally agreed. In exploring whether other animals have forms of intelligence that are like—or unlike—any of the multiple forms of human intelligence, some people become uncomfortable. Yet for many other people it is obvious upon exploration of the world that some other-than-human animals have unique kinds of intelligence that human animals do not possess, and that some possess some of the kinds of intelligence we have but in different ways and degrees.

Such questions have their roots in common sense—we are, after all, obviously animals even if our language habits like "humans and animals" constantly gloss over what "everyone knows." Questions about intelligence also follow from the careful, honest acknowledgments mandated by critical thinking. Experience shows that recognition of multiple kinds of intelligence contributes to our understanding of each other and our nonhuman neighbors, thus enriching each human's awareness and deepening the multiple intelligences already at work.

These observations imply that analytical, word-based intelligence may often serve us best when it recognizes itself as a channel through which other kinds of intelligence, thinking, and feeling are shared. Indeed, mere analysis without inputs from other forms of human life is empty, just as other forms of human life can, when pursued without the benefits of critical thinking, be blind. Discussing the importance of literature, Nussbaum comments in this vein: "It is all too easy to see another person as just a body—which we might then think we can use for our ends, bad or good. It is an achievement to see a soul in that body, and this achievement is supported by poetry and the arts, which ask us to wonder about the inner world of that shape we see—and, too, to wonder about ourselves and our own depths."[44]

Animal Studies is capable, because of its interdisciplinary commitments, of affirming how often human cultures have found "a soul in that body" when engaging their nonhuman animal neighbors. The only thing that will make this surprising to some readers is a misleading idea that circulates in both education-dominated and education-wanting segments of Western culture—that nonhuman animals do not have souls.

The Christian community offers a number of different claims about whether nonhuman animals have souls. The official theology of the Catholic tradition, for example, follows Thomas Aquinas, who passed along Aristotle's position that soul is indeed widespread in living beings. While some Catholic and non-Catholic Christians assume the narrow position that only humans have souls, in ordinary life many adherents of the diverse Christian tradition simply ignore such denials of a soul for nonhumans.[45]

Islam, the world's next most populous religious tradition, asserts that nonhumans do indeed have souls, as do the traditions that originated in the Indian subcontinent. Indigenous peoples, too, characteristically assume that nonhuman animals have souls. The simple fact is that, across time and place, the vast majority of humans have readily affirmed that soul-like ideas apply as fully to nonhuman animals as they do to humans. Frankness about this salient fact is one of the contributions that Animal Studies can make to the general understanding of how humans as a group have thought about other-than-human animals.

This not-so-well-known fact that the majority of humans think some other-than-human animals have souls also helps one see certain additional features of Nussbaum's observation. It is, as she says, "all too easy to see" another individual (including, for example, an animal outside one's own species) "as just a body—which we might then think we can use for our ends, bad or good." She notes, however, "the achievement" of "see[ing] a soul in that body" is supported "by poetry and the arts, which ask us to wonder about the inner world of that shape we see." Whether one chooses to employ the notion of soul regarding any nonhuman animals, Animal Studies prompts sufficient attention to both the world around us and to our own realities, choices, and ethical abilities to acknowledge that humans are not the only fascinating animals in the larger community of life. For all these reasons, it is not surprising that in 2000 the New York Times observed that "animals have taken over art."

Animals in Philosophy

"Philosophy" and "philosopher" derive from two Greek words—*philein*, which means "to love," "to be fond of," or "to tend to," and *sophia*, knowledge or wisdom. The alleged perspicacity of these lovers of wisdom is the subject of any number of stories and fables. There are, for example, many versions of the story that William Temple, an Englishman who became the archbishop of Canterbury in 1942, asked his father, "If philosophers are so smart, why don't they rule the world?" His father, who at the time was the current archbishop of Canterbury, replied, "They do—500 years after they're dead!"

The possibility that our community and institutions might be ruled by the claims of dead philosophers can be viewed in a number of ways—for some, it will be a consoling vision because it suggests that our lives and values are guided by careful thought, continuity, and even wisdom. But for others, this idea is frightening—the sixteenth-century philosopher Montaigne, quoting the first-century-BCE philosopher Cicero, suggested, "Nothing so absurd can be said that it has not been said by some philosopher."[1]

The accuracy of this oft-repeated generalization aside, there is one broad claim about philosophers that is clearly accurate—many of the humans we put in this revered category have had very strong views about nonhuman animals even when they knew relatively little about the vast majority of the nonhumans in their own locale, let alone the wider world. In fact, it is possible to use some of the most influential philosophers in the Western tradition to suggest that philosophy as a whole has inclined to negative views of other-than-human animals.

The influential German philosopher Kant never traveled more than eighty miles from his hometown of Konigsberg, which means he himself never had any exposure to those nonhuman individuals and societies that are the most complicated outside our own species— he spent no time with elephants or whales or dolphins or bonobos in the contexts where their evident abilities play out most fully. Nonetheless, Kant denied that individual humans as moral beings owed any direct duties to any nonhuman animals—such beings are, in Kant's words, "not self-conscious and [thus] are there merely as a means to an end. That end is man.... Our duties to animals are indirect duties toward humanity." In another lecture addressing "duties toward others," Kant framed this startling claim in a slightly differently way that makes the exceptionalist features of this human-centeredness even more apparent: "our duties to animals are duties only with reference to ourselves."[2]

Kant's claims, which, on their face, appear to dismiss any sort of rich engagement with animals outside our own species, are by no means the most famous denials of human duties to other living beings. The extraordinarily influential philosopher René Descartes made even more thorough dismissals regarding living beings outside our own species. Descartes asserted famously that all other animals—but not humans, of course—"act naturally and mechanically like a clock which tells the time better than our own judgment does." Such a view might seem to be a compliment, for it might be taken to suggest that other animals somehow are more accurate at the challenging task of timekeeping. Further, if coupled with Descartes's admission that some nonhuman animals have body parts like us, one might suspect that Descartes in fact was somewhat open to, perhaps even held somewhat positive views of, some other-than-human animals.

To the contrary, however, Descartes argued again and again that any and all animals outside our species have no mind and are devoid of reason.[3] Although Descartes traveled much more than Kant, it is pertinent to any assessment of the quality of his thinking and, in particular, the breadth of his dismissals of the earth's nonhuman animals that Descartes was not, in fact, at all well traveled. He did not seek out other animals in the context of their lives, and although he is often said to have used a method by which he "doubted everything," he showed no inclination to speculate that his own ignorance about other-than-human animals might limit his ability to generalize about them. Accordingly, Descartes simply passed along the prevailing views he inherited from the culture into which he was born.

It is a historical accident that the intellectual traditions in which Kant and Descartes were nurtured developed in environments separated from virtually all of the Earth's nonhuman individuals who are large-brained, long-lived, socially complex animals. For example, all of the nonhuman great apes were absent from Europe and the Middle East; nonterrestrial whales and dolphins lived only offshore, an environment which, even if near a human coastal settlement, remained inaccessible to humans. The few elephants known to Descartes's and Kant's teachers and cultural forebears were captives, respected for their intelligence but ultimately detached entirely from their families and natural environments and often insane by virtue of a lifetime of isolation and instrumental use.

These circumstances provide grounds for wondering whether either Descartes or Kant—or others in the same cultural milieu—could think carefully about the realities of such animals, let alone speak of them in informed ways. Attitudes, perspectives, and comments about other animals developed and passed along by humans in such circumstances are, on the whole, likely to be framed on the basis of misinformation or even total ignorance.

If one applies some of the simplest canons of critical thinking to Descartes's broad dismissal, as well as his own declared method of sweeping away all that he had learned, one finds Descartes's approach not only wanting but intellectually inconsistent. His approach is clearly beset by limited exposure to the vast majority of life outside the human species, and not at all characterized by a humility that prompts one to look with an open mind. As importantly, Descartes's approach is inconsistent because, despite his famous claim to employ complete, root-discovering doubt in order to pursue the truth and even the possibility of knowledge of the external world, he clearly radically failed to do this on the issue of nonhuman animals. Thus while Descartes nominally committed himself in his writings to "reject[ing] as absolutely false everything as to which I could imagine the least ground of doubt, in order to see if afterwards there remained anything in my belief that was entirely certain,"[4] his method of radical doubt was never deployed in connection with the views he inherited regarding nonhuman animals. Said another way, Descartes assumed the propriety of the exceptionalist tradition, and this in turn caused him to fail in his attempt to deploy systematic or hyperbolic doubt.

It is surprisingly obvious to anyone who takes a single course in Animal Studies today that Descartes chose not to take other living beings seriously. Some features of his ignorance, then, were arguably self-inflicted. Descartes by his own admission was, in fact, ignorant of much of the real world: "the little which I have learned hitherto is almost nothing in comparison with that of which I am ignorant."[5] Despite being so ignorant of other animals, and despite his proclaimed commitment to doubt all, Descartes was not in the least restrained when opining about other animals' importance or actual realities.

Implicitly, Descartes's failure to study other animals suggests that he felt such work unimportant. Further, from the vantage point of developing critical thinking skills, Descartes failed to examine at all the mental habits and patterns of speaking about other animals that prevailed in his seventeenth-century milieu. He uncritically accepted the validity of sorting the world out along the lines of "humans and animals" and was thereby easily deluded into concluding that his certainties about some animals must be true of them all. This way of thinking still prevails today in many circles despite our obvious awareness, based on everyday experiences, that all nonhuman animals share no common trait other than being nonhuman.

The ironies are, thus, that while Descartes did identify a number of helpful principles or rules when seeking knowledge (such as clearly recognizing one's subject, dividing the subject fully and fairly, proceeding from simple to complex in the divisions made, and being comprehensive), he ignored these rules when exploring the more-than-human world. The result is that his thinking remained mired in the exceptionalist assumptions that produced his astonishing myopia.

It is this one-dimensionality that Descartes bequeathed to posterity. While Cartesian approaches remain powerful today, critical thinking suggests in almost countless ways that we are by no means bound by this heritage. In fact, the combination of blind acceptance of existing prejudices regarding the realities of other animals with an unrestrained willingness to talk about all other animals flies in the face of common sense, not to mention Descartes's notorious pretense of having examined knowledge carefully and systematically. Descartes's approach was one-dimensional, and by virtue of its combination of narrowness and arrogance exemplifies one aspect of much "urban thought": "Ethological investigation, once it was vigorously set on foot in this century, has shown that Western urban thought was (not surprisingly) often even more ill-informed than local superstition on many such questions [about the realities of other animals], and that it had consistently attributed to animals a vastly less complex set of thoughts and feelings, and a much smaller range of power, than they actually possessed."[6]

Descartes's Aggression

Descartes aggressively went after those who disagreed with him, stating, "There is no prejudice to which we all are more accustomed from our earliest years than the belief that dumb animals think."[7] Critical thinking and Animal Studies in tandem prompt one challenge after another to Descartes's reasoning here—there is, in fact, at least one prejudice to which we are certainly more accustomed than the alleged prejudice that some nonhuman animals think. This is the common assumption that we can talk about all other animals easily by virtue of the simple generalizations and dismissals we have inherited. Animal Studies, through its critical thinking commitments, is ideally suited to revealing the different ways that inherited forms of speaking and thinking, to which we are far more accustomed than the notion that other animals think, are in fact misleading and poor ways of thinking and speaking. This is so whether they are measured by common sense, science, ethics, or simple canons of careful thinking.

Identifying the Influence of Self-Inflicted Ignorance

Descartes's views were influential in his own times—there are gruesome descriptions of experiments on dogs carried out by followers of Descartes at Port-Royal-des-Champs:

> They administered beatings to dogs with perfect indifference, and made fun of those who pitied the creatures as if they had felt pain. They said that the animals were clocks; that the cries they emitted when struck, were only the noise of a little spring which had been touched, but that the whole body was without feeling. They nailed poor animals up on boards by their four paws to vivisect them and see the circulation of the blood which was a great subject of conversation.[8]

Of particular relevance to contemporary Animal Studies is that Descartes's worldview continues to be highly influential—Descartes's dualism separating humans from all other animals is, for example, often said to support the claim that there are no moral problems when science-based researchers use nonhuman animals as experimental subjects in modern research facilities.

Given modern science's ideals and given what modern science has discovered about other animals, Cartesian-like approaches are clearly at odds with so much that humans now know about other-than-human animals. In the words of Paul Shepard, Descartes's views on other animals are a kind of "species solipsism," even "a wildly perverse view."[9] What makes such critiques significant, of course, is that dualistic views focused on humans alone continue to be advanced by both individual humans and collectives such as industries, governments, and mainline institutions in religious traditions and education.

Descartes's generalizations about other-than-human animals, then, are not those of the humble explorer. Rather, they are driven by an unexamined ideology overdetermined by tradition and underdetermined by actual realities. Through foregrounding critical thinking and interdisciplinary approaches, Animal Studies has the potential to produce views that are more responsibly anchored in fact-based explorations and informed by more careful and responsible generalizations.

Descartes's dualism can be sustained, Animal Studies shows repeatedly, only by a failure to inquire, a refusal to be reflective about one's own thinking, and a refusal to unlearn inherited biases. Such refusals kept Descartes's otherwise brilliant mind inattentive to earth's wide range of nonhuman animals. Animal Studies, in contrast, attempts to maintain processes and values that help call out obvious biases even as they remain committed to discovering the realities of other living beings to the extent that is possible for our capable but limited human minds.

On Putting the Horse before Descartes

Bernard Rollin's 2011 memoir *Putting the Horse before Descartes* calls for correcting Descartes's influence on thinking about nonhuman animals.[10] Rollin's singular successes in effecting social change in public policy in a number of different countries give him a powerful voice in addressing Descartes's views on nonhuman animals as naive, unabashedly species- and self-serving, and, thus, deeply troubling.

As or more significant, though, is that Descartes's views of other animals also make a radical break from a potentially positive idea that preceded his writing. Descartes denied outright the long-standing recognition within Western culture that some nonhuman animals are sentient, possessed of feelings and emotions, and subject to pain and suffering. Such claims are, of course, plausible in light of readily observable behaviors in, for example, familiar domesticated animals or wildlife. Descartes's view that other animals are mere automata with no minds breaks away from virtually all of his predecessors who had, in one sense, put the horse before the cart in matters of sentience.

But even though a tradition that other animals are clearly sentient existed before Descartes was born (1596), one major figure after another in the Western intellectual tradition had dismissed other-than-human animals for more than a millennium. These figures may have known the difference between the horse and the cart, but they nonetheless found ways to assert human superiority over all other animals.

This dismissal was not handed down from time immemorial, however, for, as the historian Richard Sorabji points out, in the centuries before and after Jesus was born, the Greek world had featured an astonishingly vibrant debate about whether other animals had speech, intelligence, and many other cognition-, emotion-, and intelligence-conferring abilities. Sorabji observes, however, that one particularly dismissive school of Greek thought (the influential group we know as the Stoics) influenced certain Christian thinkers in ways that produced the views that now prevail: "the Stoic view of animals, with its stress on their irrationality, became embedded in Western, Latin-speaking Christianity above all through Augustine. Western Christianity concentrated on one-half, the anti-animal half, of the much more evenly balanced ancient debate."[11]

So, Sorabji surmises, from the time of Augustine, who died in 428 CE, until Descartes began his work, there prevailed in many influential circles a profound dismissal of nonhuman animals. However, it was mixed in one sense—it dismissed other animals but not in the radical manner of Descartes. Augustine, for example, whose influence rivals or exceeds that of both Descartes and Kant, observed that other animals were far more than machines or automatons because they had important, recognizable levels of perception—for example, Augustine suggested birds sing because they enjoy their own songs.[12] In his highly influential treatise *De Trinitate*, he suggested, "beasts perceive as living, not only themselves, but also each other, and one another, and us as well."[13]

Thus while on the whole it can be said that the official views of mainline authorities in the Western tradition's best-known theology, philosophy, and education promoted the subordination of other animals, the Western cultural tradition was also home to, like so many other human cultures, individuals who engaged other animals seriously in one way or another. Francis of Assisi, who died in 1226, is a well-known example of an individual who openly talked of compassion for other animals. Animal Studies works to assess such diverse threads in Western culture; it also seeks to assess the historical role of mainline philosophers like the Stoics and Augustine in shaping views of nonhuman animals. It also engages how Descartes has influenced now-prevailing views of other animals (such as the views that support experimental uses of nonhuman animals). As importantly, Animal Studies also features many skills relevant to assessing whether ordinary people at the grassroots level of different societies put the horse before the cart, as it were, in their daily lives. It is important to distinguish what might have been said and done in the day-to-day world of ordinary citizens from the claims made by well-known historical figures in influential circles that actively dismissed nonhuman animals as inferior or otherwise subordinated them to humans.

Familiar Territory in the Western Philosophical Tradition

Positive views in the Western cultural tradition seem only rarely to have blunted radical subordination of nonhuman animals to human interests. Aristotle, for example, held a worldview that subordinated all other animals to humans even though he certainly thought nonhuman animals far more than machines. His writings include a simple, overwhelmingly influential analogy that to this day causes some to opine that Animal Studies is inferior to "the study of man":

> We may infer that, after the birth of animals, plants exist for their sake, and that the other animals exist for the sake of man, the tame for use and food, the wild, if at all, at least the greater part of them, for food, and for the provision of clothing and various instruments. Now if nature makes nothing incomplete, and nothing in vain, the inference must be that she has made all animals for the sake of man.[14]

Although many formidable figures in the history of Western thought, such as Cicero,[15] have passed along this analogy, Animal Studies is also heir to many different challenges to such a claim. For example, many people in both ancient and modern times have seen such a claim as transparently foolish. Many ancient commentators observed that many nonhuman animals seemed in no way to be designed for humans. Most whales swam too fast to be caught until steam-powered vessels were invented in the nineteenth century, and "vermin" such as scorpions had no redeeming *pro-human* value of any kind. Modern commentators have been particularly dismissive of Aristotle's reasoning—the historian of ideas Glacken, who authored the pioneering study of ecological views *Traces on the Rhodian Shore*, saw Aristotle's comment as "disappointingly crude."[16]

From the standpoint of critical thinking, the analogy involves several grave problems—there is a category mistake or equivocation, since the category "plant" excludes any animals, while the category "animals" does not exclude humans. Relying on the second claim "all animals are for all humans" on the basis of the validity of the prior claim "all plants are for all animals" is a questionable move, then, because the comparisons are not logically equivalent. In short, the second claim relies on an artificial removal of humans from their natural category.

The Continuing Tradition of Dismissal

The analogy was Aristotle's attempt, of course, to justify his culture's existing practice of subordinating all nonhuman animals to human interests. A great many after-the-fact rationalizations purport to explain why inherited traditions by which humans harm other living beings present no moral problems. After Descartes, justifications of humans' right to harm other living beings became ever more shrill and one-dimensional,[17] and their importance in secular realms is reflected by the fact that they continue to prevail in the production of food, the use of nonhumans in research designed overwhelmingly to protect humans' interests, and more.

These rationalizations have become so virulent that Midgley referred to the Western cultural position as "absolute dismissal," which describes well modern agribusiness's attitude toward the moral rights of nonhuman animals.[18] Such a dismissal is, in fact, the same as the worldview of the economist Walras that "man alone is a person; minerals, plants and animals are things" (see chapter 2). It is the mainstay of the legal dualism "humans are legal persons, while other living beings are mere legal things without rights" and it is the public policy that supports businesses and research institutions that use other animals as though they were completely inanimate resources or trash.

Dismissals are found within certain religious institutions that are sustained by the same roots—a 1994 Catholic pronouncement regarding other animals provides an example: "Animals, like plants and inanimate things, are by nature destined for the common good of past, present and future humanity."[19] Such a claim shows that Descartes's dismissal had deep roots even though he suggested his views were anchored in careful, doubt-inspired thinking. In reality, Descartes's dismissal is one strain of thinking within the Western cultural tradition and is consistent with the ancient theological claim expressed so tersely by the nineteenth-century theologian Rickaby (chapter 2): "Brutes are as things in our regard: so far as they are useful for us, they exist for us, not for themselves."

Such a strain of thinking may eclipse other, more empirically based strains, but it does not entirely erase them. Such a nuanced, multifaceted account is part of the larger story that needs to be told.

Important Reflective Tasks

Some scholars have suggested that Descartes himself, as opposed to his followers, did not deny consciousness, pain, pleasure, and even joy in all nonhuman animals.[20] While Descartes's views seem on their face to be an unequivocal, seemingly ideological dismissal of all nonhumans, Descartes owned a dog and named him Monsieur Grat (Mr. Scratch). Such actions mildly suggest that Descartes's radical dismissal of all other animals as mere machines could simply have been rhetorical overstatement.

Critical thinking has the capacity to juxtapose verbal claims with actual choices and actions, for, as Gandhi said, "The act will speak unerringly."[21] Given that specific actions in daily life often speak unerringly about what one truly believes, Animal Studies can engage the facts of daily life of everyone, including dismissive-minded thinkers like Descartes, when in search of the full story of humans' interactions with other living beings.

Aristotle's work, for example, can be read with breadth and depth on the nonhuman animal issue. Aristotle made comments that seem to dismiss nonhuman animals, such as his claim that "animals have but little of connected experience."[22] This claim is, we know, inaccurate, for the connectedness of many animals' experience is evident to those who live with an average dog. What is odd about this generalization is that Aristotle, who was commonly referred to in the Middle Ages as "the Philosopher" and who is, by some accounts, the most influential philosopher ever, produced extensive treatises on nonhuman animals that were for more than a thousand years deemed the pinnacle of human systematic knowledge of other living beings. Thus, this thinker's contributions to our way of thinking are as great as, perhaps even greater than, those of any other figure in history.[23]

Aristotle was, like any other human, deeply impacted by, even a potential prisoner of, his birth culture. There are legendary examples of what seem to be insensitivity, including his dismissive views of women and his belief that some humans are by nature intended to be slaves. These negative views are far more often talked about than his occasional one-dimensionality on nonhuman animals. Yet today Aristotle's generalizations about other animals seem, in light of recent empirically verifiable information, overwhelmingly ideological and even fundamentalist. Those familiar with large-brained social animals such as nonhuman great apes (orangutans, chimpanzees, gorillas, and bonobos), dolphins and whales, elephants, and others, as well as dogs and cats, puzzle at a number of Aristotle's claims. The complexities of Aristotle's shortcomings can be seen in Durant's assessment that "Aristotle makes as many mistakes as possible for a man who is founding the science of biology."[24]

Critical thinking helps place Aristotle's passage in its full context. We can ask, did Aristotle actually make the analogy claimed in his name? Yes, he did. Was his language on this occasion meant to apply to all animals, just to those known at the time, or to some other grouping of other living beings? This is open to debate. The point is simply this—Aristotle thought and wrote against a particular backdrop of cultural assumptions that he inherited, as well as political and social realities amid which he lived. He was a pioneer in many areas, and his work on other-than-human animals was, in its time and for over a thousand years after, an astonishing achievement even if, by contemporary standards, it is troubling at any number of points.

Kant may also be read in ways that go beyond the one-dimensionality of his claim that nonhuman animals are "not self-conscious and are there merely as a means to an end. That end is man." While this claim seems to promote the exceptionalist tradition, the Harvard philosopher Christine Korsgaard has worked for a decade to assess if one can read Kant in ways that are kinder and more generous toward other-than-human animals.[25] In this work, Korsgaard stands with many other prominent philosophers today whose work, though not thought of as part of the animal rights movement, clearly reflects on humans' thinking about and acting beyond the species line. Prominent public intellectuals in many countries now regularly address animal protection issues as part of social values—see, for example, the questions raised by the psychologist Steven Pinker about how work on behalf of animal rights plays out with regard to human rights.[26]

Another important, reflective task for critical thinking is to assess some less obvious features of claims about our fellow living beings. In order to lend moral authority to "the cause," sometimes quotes on animal issues are attributed to famous people even when they have not spoken to the animal protection cause. For example, while one may read that Abraham Lincoln said, "I am in favor of animal rights as well as human rights. That is the way of a whole human being," there is no evidence that he said or wrote it. Scholarship and critical thinking in combination can help identify who said what regarding the rich and ever-so-diverse human-nonhuman intersection.

Philosophers as Social Animals

An English philosopher observed that, with regard to wolves, what had been "popular with philosophers" was not a fact-based view of the actual animals themselves, but instead a misleading "folk-figure" (chapter 2). Because philosophers often used this caricature "uncriticized, as a contrast to illuminate the nature of man," it is worth investigating whether philosophers often have spoken freely precisely because they knew little or nothing about most other-than-human animals. Some philosophers might be described as having violated their own standards of careful thinking as they followed the crowd and made what amount to uninformed and misleading claims about nonhuman animals.

Philosophy at its best, however—which admittedly is not a simple standard to achieve—has prompted individual philosophers and students to aspire to careful, critical thinking. This involves asking question after question on important, sometimes politically unpopular topics. Such a process aims to elucidate our own minds' resources and processes, what we count as evidence, the way that our assumptions impact our conclusions, how we handle inherited claims, and much more.

The eminent philosopher of psychology Daniel Robinson has suggested, "Philosophy is created when the mind turns from practical matters of avoiding danger and uncertainty to a form of critical inquiry in which its own resources are objectified and subject to critical scrutiny."[27] Although the claim that "the Greeks invented philosophy" is an overstatement since many human cultures have displayed inquiry skills like those of philosophy, this claim can stand for the important insight that seminal Greek thinkers "transformed inquiry from an essentially practical or ritualistic/religious enterprise into a form of abstract and theoretical thought."[28] The Greeks may not have been the first to do this, but they are well known for early development of multiple schools that fostered critical inquiry and created a tradition that has impacted virtually every subsequent philosopher.

A key figure was the legendary Socrates, who, "at least as he is revealed in Plato's dialogues," displays "a commitment to objectify the self and hold it up to scrutiny."[29] Critical inquiry in the tradition of Socrates asks about, and thereby tests, the most fundamental views, claims, convictions, and values in one's world. The Greeks asked such questions about themselves in relation to other humans, and in this process they bequeathed to all humanity a developed tradition of examining how powerfully custom can impact our ideas of the world. The Greeks also modeled that critical inquiry does not yield to established ways of thinking, but can unlearn and relearn. In this enterprise, the thinker is humble about her or his own views and heritage, and yet at the same time recognizes that there is no viable alternative to using the mind's own resources, which are often limited in very important ways.

Reflexive thinking offers help in identifying key problems in popular habits, like human-centeredness. It can also help identify self-inflicted ignorance and the risks of a closed mind. All of these elements play a part in the following example taken from Plato's dialogue "The Statesman." In this passage, one of history's seminal thinkers wonders about the ways in which our common humanity might distort the world of other creatures and reality in general.

A participant in the dialogue responds to the young Socrates's notion that the world can be split into two divisions ("man being one, and all brutes making up the other"):

> Suppose...that some wise and understanding creature, such as a crane is reputed to be, were, in imitation of you, to make a similar division, and set up cranes against all other animals to their own special glorification, at the same time jumbling together all the others, including man, under the appellation of brutes,—here would be the sort of error which we must try to avoid.[30]

The speaker entices Socrates to examine how any thinking being's point of view might be impacted by factors that are often not noticed. Plato's point has extra interest for the field of Animal Studies because it illustrates nicely how humans can be taught about their own mental habits with a story about another animal supposed to be intelligent. So if cranes divide the world, we can see that the cranes err by leaving valuable humans out. The story invites us to notice that this strategy is precisely our strategy—and critical thinking needs to call this out even if doing so is unpopular.

Critical Thinking and Other Animals' Realities

Philosophers can contribute to the obviously important question of how humans, somewhat able to understand each other (through sharing language and perception capacities), can work more effectively at understanding specific nonhuman individuals within their communities. For example, it begs question after question to claim that the nonhuman individuals brought into human society are fully representative of all nonhuman individuals' abilities.

If Plato's cranes reasoned this way (for example, if they claimed that humans held captive in crane society exhibited the full range of human behavior), humans would easily notice the shortcomings in such reasoning—as Plato suggested, "here would be the sort of error which we must try to avoid." The thrust of Plato's crane story, then, is that we can avoid errors by liberally employing humility, critical thinking, and an open mind.

It is instructive and humbling to notice that the academic field we call philosophy has at times wandered away from critical thinking, occasionally degenerating into the exclusive domain of elite classes. Historically, some humans were excluded from circles where philosophical exploration was developing, just as some people were excluded from the elite groups allowed to recite certain religious traditions' sacred texts. Women, for example, in many societies were excluded from philosophical circles and access to revealed scriptures, just as they have been barred from education and public debate in cultures around the world.

Because of these exclusions and other types of bias against "outsiders," philosophical discussion has often gone forward without full input from a broad spectrum of the human species. The restriction of philosophical discussion to a privileged class or group violates key insights that drive critical thinking, such as the importance of asking questions about received values (including rules as to who is allowed to participate). Closed circles often lack open-mindedness, or a willingness to ask questions that illuminate the very process of critical inquiry. The upshot is that philosophy at some points has been decidedly one-dimensional, favoring one class's ideas and thereby sanctioning harms, oppressions, and much worse for the excluded groups. Often, those called philosophers have foregrounded merely one set of stilted, agenda-laden questions even as they have ignored other questions that might problematize the privileges and exclusions that the dominant class enjoys.

The result has been unresponsive, even oppressive thinking done as philosophy. Like Plato's hypothetical cranes, certain human groups counted only members of their own group as elevated. When such exclusions prevail, the vaunted "love of wisdom" can become an ugly affair.

Critical thinking skills have, fortunately, a way of resurfacing in human minds. In the fashion of Socrates's and others' commitment "to objectify the self and hold it up to scrutiny," critical thinking about exclusions and other injustices has broken through in circle after circle. Sometimes this has been a practical affair, addressing pressing needs of the community such as violence or harms to certain human groups. Sometimes it has been a response to the need for better means of surviving, or the need for an intelligible world, or for enabling education that is responsive to the real world. The Greeks used philosophy to reflect on basic questions that have, in the real world, very practical consequences—including how we can govern ourselves, how we can assess honestly what we reliably know, and how individuals or societies as a whole should act in day-to-day affairs or crises.

On the whole, however, breakthroughs in critical thinking have often been only piecemeal or partial. There have only rarely been, for example, periods in which the breakthrough impacted every aspect of political debate, or in which every exclusion was closely, sensitively, and honestly examined.

The purpose of this recitation of philosophers' foibles is to foster recognition that philosophy can stumble when answering straightforward questions about other animals' realities. Indeed, in a manner eerily like Plato's hypothetical cranes, philosophy has been human-centered so often that its practitioners have failed again and again to see their own exclusions. And, of course, one issue that has characteristically been excluded from the force of critical thinking has been humans' domination over and one-dimensional thinking about the more-than-human world.

So it is precisely in the context of philosophers' many attempts to solve problems in ways that protect their own culture's superiority, or an elite group's privileges, or even the interests of "the whole human race" that philosophers' abilities to help in Animal Studies must be placed. The field is perfectly capable of asking questions about humans' relationship possibilities with both other-than-human animals and the more-than-human world, but this has only occasionally been done well. The love of wisdom needs to prevail over the love of one's own inherited background, one's own political group or sex, or one's own species. In other words, philosophy, to be worthy of the name "love of wisdom," needs to entail more than finding ways to justify our own species' interests and privileges.

Failure of philosophers to consistently practice this form of philosophy has had consequences, for many people have been led by philosophers in the exceptionalist tradition to the impression that careful reflection readily justifies a human-centered world. Philosophers have, then, been one of the interest groups that have often advanced reasons for ignoring the interests of nonhuman animals. Thereby, many philosophers have helped justify denials, subordination, dismissals, and a general failure to acknowledge possibilities of complex individuals and communities beyond the species line.

Yet humans' philosophical capabilities can sustain those who ask powerful questions about inherited views regarding humans' alleged superiority. One of the touchstones of a full engagement with the world is a responsible, critical engagement with the question of other animals' realities. The emergence and reemergence of such inquiries have by no means originated primarily from those called philosophers. Rather, such questions come from across the continuum of humans, often being prompted by day-to-day lives as much as by pure speculation.

Animal Studies as a Form of Philosophical Frankness

The bottom line is that the best human philosophical traditions promote critical thinking skills that prompt frankness about selfishness, narrow-mindedness, fantasy, and bias. Animal Studies can be seen as one form of such frankness. The special virtues of philosophy itself, however, include at least four features that are crucial to Animal Studies.

First, the history of philosophical reflection reveals that many philosophers have been stimulated by the presence of other animals, and have in response commented on the place of humans relative to nonhumans. Importantly, this trend is evident in many philosophical traditions, not merely the mainline philosophical tradition of the culture in which a history of philosophy course is taught.

Second, some philosophers have not only ignored the more-than-human world but have also worked hard against any conclusion that moral issues go beyond the species line. As already noted, some revered philosophers have, despite being uninformed, perpetuated prejudices. Like any human thinker who relies uncritically on biases and other culturally approved generalizations, philosophers can produce inadequate analyses of problems and say transparently false and foolish things about the living beings beyond the species line.

But the history of philosophy also shows that, through frank evaluations of contemporary practices, some philosophers have identified certain practices that are extremely problematic. The many animal-friendly philosophers identified in chapter 4 stand in this tradition of challenging inherited ways of making coherence, prevailing thinking patterns, dominant worldviews, historical claims, and culturally significant myths. While the great popularity of the exceptionalist tradition has at times jeopardized those brave enough to call out the weakness of such human-centered thinking, it is nonetheless possible to see the formative role and power of questions asked by some philosophers regarding other-than-human animals.

A third point is that some philosophers have added much to our awareness of the limits of human thinking about other living beings. It was noted above that Augustine suggested that "beasts perceive as living, not only themselves, but also each other, and one another, and us as well." While such a conclusion starts out as a guess, of course (given that the realities of nonhuman "beasts" are so elusive), this kind of thinking can reach the status of educated guess based on close, careful, and repeated observations of other animals.

The limitations in our knowledge have been used against other animals in ways that are astonishingly revealing of a debilitating human-centeredness within modern science traditions. The Harvard scientist Donald Griffin described the reluctance of scientists to confirm consciousness in other-than-human animals as a double standard. Griffin suggested that claims about other-than-human lives were being retarded by a "paralytic perfectionism" that creates an "insidious barrier to scientific investigation." Note how Griffin's observations are similar to the "curious asymmetry" noted by Dennett:

> [The] tendency to demand absolute certainty before accepting any evidence about mental experiences of animals reflects a sort of double standard.... The antagonism of many scientists to suggestions that animals may have conscious experiences is so intense that it suggests a deeper, philosophical aversion that can reasonably be termed "mentophobia." The taboo against scientific consideration of private, conscious, mental experiences is more prevalent when nonhuman animals are concerned.... This mentophobic taboo has become an obstacle to scientific progress.[31]

Nuancing Anthropomorphism

Nonscientific and science-based speculation about thinking and feeling in other-than-human animals is often called "anthropomorphism" when it attributes human characteristics to other living beings, imaginary beings, inanimate objects, ecosystems, and even the universe. Philosophers, scientists, animal activists, and others interested in Animal Studies have reflected on this problem a great deal.[32] Griffin asserted, for example, that the "customary" charge of anthropomorphism had not been thought through as carefully as possible, and thus is "a serious error [that] suffers from circular reasoning.... Consciousness is assumed in advance to be uniquely human, and any suggestion to the contrary is then dismissed as anthropomorphic...merely reiterating a judgment that consciousness is uniquely human."[33]

Griffin noted that J. A. Fisher concluded, "The idea that anthropomorphism names a widespread fallacy in common sense thinking about animals is largely a myth...and the use of the term as a critical cudgel ought to be given up. It cannot stand for what it is supposed to."[34]

The primatologist Frans de Waal uses what amounts to a critical anthropomorphism, that is, a careful, realistic use of terms we use for ourselves.[35] He challenges those who refuse any such exploration as in "anthropodenial," which he defines as "a blindness to the human-like characteristics of other animals, or the animal-like characteristics of ourselves": "Those who are in anthropodenial try to build a brick wall to separate humans from the rest of the animal kingdom. They carry on the tradition of René Descartes, who declared that while humans possessed souls, animals were mere automatons."[36]

Griffin and de Waal have been among the most courageous in exploring other animals' unfathomably mysterious realities. A respected Catholic leader has seen such mystery as a positive rather than a negative:

> ...the world of brute animals. Can any thing be more marvelous or startling...than that we should have a race of beings about us whom we do but see, and as little know their state, or can describe their interests, or their destiny, as we can tell of the inhabitants of the sun and moon?...We have more real knowledge about the Angels than about the brutes. They have apparently passions, habits, and a certain accountableness, but all is mystery about them.[37]

Just as various scientists, theologians, and philosophers challenged the exceptionalist tradition, many people outside these academic realms relied on their familiarity with the day-to-day events of the natural world to challenge views that subordinated all other animals. Calling upon experience and evidence, John Muir commented, "I have never yet happened upon a trace of evidence that seemed to show that any one animal was ever made for another as much as it was made for itself."[38]

A final point is that philosophers continue to play a major role in the modern animal protection movement. In the wake of the formative roles of Singer and Regan, many other contemporary philosophers have supplied important insights into the possibilities of human individuals and societies acting responsibly toward nonhuman animals. Further, in 1980 the Society of Study of Ethics and Animals was the first of many professional organizations focused on nonhuman animal issues. Similar organizations now exist within the American Bar Association, American Academy of Religion, American Psychological Association, American Sociological Association, and American Association of Geographers.

Philosophy and the Basic Tasks of Animal Studies

Forthrightness about the complexity of the human-nonhuman story in the history of philosophy is one of the tools needed to accomplish the four basic tasks of Animal Studies.

Through commitments to the clearest and fairest forms of thinking, philosophy can make basic contributions to the task of telling the entire story of human interactions with other-than-human animals. Philosophy's penetrating insights into the range and limits of human abilities to know specifics about the world we coinhabit with so many other forms of life put philosophers in a good position to contribute to Animal Studies as it pursues the task of learning other animals' realities even as we identify human limitations in pursuit of this goal.

Philosophy also has an important role in helping individuals, communities of humans, and our species as a whole identify possible futures on a shared earth. Philosophy's frankness is especially pertinent to the ethical dimensions of humans' impact on and domination of other lives. While many humans justify virtually any action in pursuit of human goals such as knowledge, control, and privilege, many philosophers have wondered about the morality of human choices for other living beings. Philosophy has much power to parse our choices. Equally, philosophers can help identify the roles humans can take in nested communities composed of different levels. Moving beyond family at the closest level, humans encounter national, religious, and cultural groups at a more general level. Philosophers have a role in assessing how humans relate to even more general communities, such as our whole species, our fellow primates, all mammals, the visible or macroanimals, all living beings, ecosystems, the earth, and even more.

Owning Up to a Certain Arrogance

Some philosophers have reflected a particular kind of arrogance. This is the tendency of each academic field to claim that its own way of thinking and reasoning is the definitive form of human thinking and reasoning. Theologians have shown a similar arrogance in claiming that their own field produces the truest and best of human knowing because theology is—or at least in medieval times was said to be—the queen of the sciences, that is, the most important of human endeavors and thus a field against which all other fields were measured and found less important. Economists, too, have often displayed a smug superiority and self-importance, as have certain scientists who claim that science is the only true knowledge and thus the measure of all other human claims. Business leaders, too, sometimes make the same faulty assumption.

Love of Wisdom around the World: Many Traditions

Just as the Greeks developed systematic inquiry, other human cultures have offered insights about human strengths and limits, especially as they pertain to both the realities of specific nonhuman animals and humans' role in the more-than-human world. A philosopher of an indigenous people describes other animals as "the Natural World people."[39]

Small-scale societies have, in one sense, significant advantages over urban dwellers when it comes to understanding other animals—their commitment to tribe and place provides a combination of community, opportunity for encounter, and connection, and these elements "work toward relatedness, especially in knowing animals."[40] Many people from small-scale societies willingly recognize that other tribes and cultures also can have an intense, orienting connection to their particular place and local world. Thus these reflective thinkers often refrain from exclusivist interpretations of religious experience:

> The question that the so-called world religions have not satisfactorily resolved is whether or not perspectives on nonhuman animals that are grounded in religious experience drawn from one place and time can be distilled from their original cultural context and become, with regard to the moral standing of any or all nonhuman living beings, abstract principles and general rules applicable to all peoples in different places and at different times. The persistent emergence of diverse perspectives about other animals, many of which are grounded in the most basic of religious commitments to ethical concerns for others (be they human or otherwise), suggests that cultural context, time, and place are the major elements of claims about other animals and that such claims' content, and particularly their dismissal of other animals as inferior beings about whom ethical agents need not care, is by and large illusory.[41]

Romanticization as Disrespect

Such generalizing has its limits and even perils, to be sure. A major risk lurks, for example, in any romanticization of the views of indigenous peoples. Critical thinking requires that specific viewpoints be described with accuracy and realism.

This has often not been the case. Recall the encyclopedia entry about "the savage" in "the lower stages of culture" (chapter 2), which represents how some figures in our intellectual and scientific history have made extremely negative, dismissive statements about indigenous peoples. A different sort of problem is the romanticization of indigenous peoples, the classic example of which was the Noble Savage myth, that had far less to do with the indigenous peoples than with the mythmakers' wish for a version of their own society free of problems and human greed.

In their own way, such romanticizations may even be more dismissive of the actual realities of small-scale societies than are more explicitly dismissive views. The latter might get some details correct but then portray them as negative. Romanticizations, on the other hand, purport to honor but nonetheless use the indigenous people's views out of context and thereby cause the actual, place-sensitive views and realities to be left aside. Some animal protectionists tend to cite the animal-friendly views of certain indigenous groups. This can be intended as respect, to be sure, but if it involves distortion of any kind, such uses violate a simple principle of mutual or reciprocal respect that demands that the views of any people be well described.

Romanticizing a particular people's views for one's own purposes is ironic—since the actual views of the indigenous people are eclipsed, the very people one purports to honor are made to disappear. In the modern world, indigenous cultures, which represent human cultural abilities as fully as any of the major remaining cultural traditions, have been greatly harmed and reduced. Today the remaining small-scale societies are under siege as both igno-rance and apathy prevail—one observer wrote in 2009, "More groups of indigenous peoples have likely been destroyed during our age than in any other comparable time period."[42]

A related risk comes with broad generalizations about whole traditions or cultures. Scholars of different traditions warn of inevitable distortions when outsiders reduce complex cultural or religious viewpoints to mere generalizations. While comparative work has obvi-ous risks, there are also potential benefits because the great diversity of resources helps us see some simple facts—some traditions have been open-minded while others have aggressively asserted their own superiority even as they have failed to recognize any moral implications in their exclusions and subordination of fellow humans (chapters 7, 9).

Two Central, Linked Issues

At times some philosophers operate, as suggested above, in narrow confines on the question of other-than-human animals. Appiah critiqued academic philosophers' embrace of highly tech-nical, conceptual analysis and language more generally that leaves unattended philosophy's historically important engagement with the difficult issues of "human nature" (chapter 4). One finds today entire philosophy departments far more interested in logical flaws or repug-nant implications of some minor issue even as extraordinarily harsh realities in the local or wider world are entirely ignored. As one philosopher has suggested, "Philosophers are not as interested in weakness of will as in irrationality, since they are trained to deal with ideas, not character traits."[43]

Even though philosophers have not deployed these skills particularly well, modern phi-losophers are responding often and diversely to the riches of the human-nonhuman intersec-tion. Thus, philosophy will continue to have a central role in Animal Studies in two major areas—epistemology (exploration of what it is that humans can and do know) and ethics. Although these two fields are by no means all or even most of philosophical, reflective think-ing, they are of central importance to Animal Studies.

In an interesting way, these fields are, for practical reasons, tied together. Many philoso-phers have been so exclusively focused on humans that they have ignored altogether the reali-ties of other animals. It has become traditional to pursue the exceptionalist tradition's agenda in both epistemology and ethics. In ethics, of course, this is an issue because of the harms to other-than-human animals wrought as an exclusivist, human-centered agenda dominates, reshapes, and destroys so many domains in our more-than-human world.

But today both of these traditional philosophical problems are being richly recon-sidered. It is as if critical thinking has finally arrived in philosophy. In fact, a combination of these fields is needed to illuminate the nature of the categories into which nonhuman animals are today characteristically sorted. Epistemology is needed, for it pursues the basis of human claims, which pertains to the categories by which we sort out—or divide up—this world. Ethics is needed for analyses and evaluations of our actions, which inevitably impact many others (of both human and nonhuman kinds).

So a combination of epistemological inquiry and ethical reflection is needed to prompt careful reflection about how humans learn about and treat animals (again, it matters not whether the impacted beings are human or nonhuman). If one does not pay attention to the details of other lives, one's ethical reflections about those beings will be both blind and empty. Thus, both ethical and epistemological inquiries are important to recognizing the kinds of difficulties that will be encountered when studying and making claims about other animals and, similarly, in seeking out and analyzing the great variety in cultural views of other-than-human animals. Finally, both epistemology and ethics are essential to any examination of claims about humans' superior status, especially because most philosophers have found such a claim all too congenial to a positive evaluation of their own self-worth. In fact, philosophers have often been, as the status accorded Descartes reveals, principal proponents of dismissive, subordinating views of lives outside the human species.

Animal Studies, which reflects in many ways that studying other-than-human animals can in fact be good for the philosophical soul, helps all of us do philosophy in an interspecies key. Animal Studies pulls philosophy beyond the humanities, just as it pulls the practice of science out of the stupor caused by the exceptionalist tradition.

Comparative Studies
Legal Systems, Religions, and Cultures

Twenty-first-century education has prospects of being richly comparative for a variety of reasons. Communications of many sorts effortlessly cross national, continental, and cultural lines, making it easy to discuss issues from multiple vantage points. In formal education, the emergence of cross-disciplinary approaches similarly has fostered multifaceted conversations. The interested public, scholars, and students from different cultures have today before them a breadth of information and perspective that allows them to engage problems of political and economic justice for the diverse groups that comprise modern nation-states. Solutions to many different problems are now readily seen as impacted by the democratization of information through communication facilitators such as social media. The upshot is that governments and many different international, national, and grassroots constituencies can interact in a great variety of ways as these problems are addressed.[1]

The emergence of extensive, inexpensive sharing of the stunning breadth and depth of information available has facilitated identification and examination of the dynamics, problems, and possibilities of a multispecies world, particularly when more and more people are given access to the insights and humilities of different comparative approaches. Both learning and unlearning based on cultural differences and interdisciplinary achievements are increasingly possible, which clearly enhances everyone's prospects of understanding local social and ecological realities and, thereby, the global environment as a whole.

The result is that today it is far easier to see how stories differ from one place to another, how variously nations and cultures frame their own history and that of others, and how each society features peculiar choices, even idiosyncrasies, in matters of governance and justification of internal regulation. Similarly, one easily encounters the world's surprisingly different religious traditions—some are overtly polytheistic, others are monotheistic, and still others are not theistic in any ultimate sense. These traditions have ancient roots and when each is examined at the local level, the tradition is easily recognized as internally diverse to an astonishing degree—for example, the number of different Christian denominations is often estimated in the tens of thousands.[2]

Comparisons across cultures also have the particular virtue of illuminating the ways human thinking has been dramatically shaped in the past by the cultural narratives in which individuals were nurtured and came to adulthood. These narratives are very diverse, and the particular version in which one is raised invariably impacts identity and expectations. As discussed throughout this volume, cultural views can distort features of the world we live amid, but they obviously have an equally powerful capacity to enable one to identify life's challenges, clarify their extraordinary complexities, and then live with and even welcome them. In all these ways, cultural views have the power to inform us about, or blunt any individual's sensibilities toward, the ways other-than-human living beings and the more-than-human world are harmed by our choices.

Comparative analysis of humans' lifeways,[3] then, can also set our imagination loose to create ways of living that allow us and other living beings to thrive in a shared world. Of course, comparative work regarding humans' many differences has long been a challenge precisely because human communities feature great diversities and differences on virtually any feature of human possibility. If comparative work is well done, however, it can enhance each person's ability to reflect on his or her own cultural heritage and place in the larger community of life, just as it clearly enables one to reflect on one's neighbors of all kinds. Thereby, comparative work aids us as we go about thinking carefully about our own thinking and that of others.

In Animal Studies, comparative work increases students' awareness of the diversity of nonhuman life, and it offers many opportunities to see the strengths and limits of human abilities so evident in many claims about nonhuman animals. When there is no deep commitment to comparative work regarding humans' inevitable intersection with other-than-human animals, both formal and informal education are at great risk. Among the most debilitating of the risks is the possibility that education about the nonhuman inhabitants of our shared world will remain human-centered in ways that continue to produce harms to other living beings and dysfunctions for humans in our local and global communities.

In the most basic way, then, comparative work helps us see that we can choose among possible belief systems, governance arrangements, consumer styles, and ethical justifications for actions. Then, on the basis of careful thinking, we can take full responsibility for how we choose to act in the real world. To illustrate this possibility, we turn in this chapter to three types of comparative study that already have a significant place in modern Animal Studies: law, comparative study of religious traditions, and cultural studies. The goal here is to engage how these fascinating but complex enterprises offer insights into central issues and tasks of Animal Studies.

Comparative Study of Legal Systems

The most common approach to studying law is to engage it within a single jurisdiction, that is, a specific community or nation. Another approach, which has some similarities to interdisciplinary work, is to compare a number of legal systems as each works out law in its own jurisdiction. A third, more truly interdisciplinary approach calls upon disciplines other than law to help one understand how law works.[4] Such approaches are valuable because they offer a variety of creative insights into the many different ways that law and legal systems have been used—this powerful vehicle can, of course, create freedoms for humans and thus help negotiate and resolve important social tensions. But, history readily reveals, legal systems are often used to maintain oppression or political advantage for one group of humans over others and, in a similar manner, to dominate life beyond the species line.

Traditional Comparisons of Legal Systems

Comparative legal studies offer insights into the similarities and differences that two or more legal systems display in structure, histories, and styles of judicial and legislative decision making. A good example is the field of comparative constitutional law, which examines those foundational compacts or agreements by which a group of citizens agree about how their political and legal systems will be constituted and administered. Some national constitutions provide for protection of humans only, while others recognize protection of other animals and nature as fundamental duties of the state and individual citizens.[5]

Comparative work of this kind is an ancient enterprise—Solon in the sixth century BCE consulted the laws of many different communities as he drafted the laws of Athens, and then two centuries later Aristotle reviewed more than 150 constitutions of both Greek and non-Greek cities as he pursued in his *Politics* the characteristically Greek question, how do we best govern ourselves? Comparative work conducted in this spirit has addressed many different aspects of legal systems. Because those pursuing comparative law before the nineteenth century aspired to develop a precise discipline, they called comparative law "an independent science." This supposedly more precise approach was "devoted to discovering the principles of just law, that is to say law conforming to the will of God, to nature and to reason" but in practice involved "little concern for...the law as it applied in fact." Instead, "the principal study [in universities]...was the search for just rules that would be applicable in all countries."[6]

In the later nineteenth century, modern comparative law began to pay attention to actual laws in practice at the local level. The complexities of this phenomenon, and especially the need for insights from diverse sources, can be seen in the following observation:

> Comparative law scholars have shown...that judges may write as if legal judgments were exercises in deductive logical reasoning, but in fact there is a great deal of ambiguity and uncertainty in the law, and therefore room for discretionary decision-making by judges. Because of the open texture of language, general rules can never determine their own application; thus, adhering to the "official" theory of mechanical judging has only the effect of disguising the need for the exercise of discretion.[7]

This form of comparative study is extremely helpful when one studies laws' impacts on other-than-human animals, for it permits an overview of what is happening around the world in the form of both legislation and court-based contests. Comparative work of this kind reveals that today one can find each year, literally, thousands of legislative and litigation-based examples proposing the use of one legal tool or another to protect living beings beyond the species line. Many such proposals languish, of course, but hundreds and hundreds are acted upon in towns, states, provinces, nation-states, and even international bodies. Some of these proposals may be introduced by a legislator to curry favor with a particular voter constituency, and others are merely idealistic stabs at change that have no realistic chance of current enactment. Yet the thousands of legal challenges taking place each year include many examples of what can and does work. In this way, interested political bodies all over the world (that is, other towns, states, provinces, and nation-states that are looking for feasible ideas) are made aware of one possibility after another.

Interdisciplinary Comparative Work

Many people seek to understand the impacts of different laws and legal systems more generally beyond legal circles and institutions. Multiple disciplines such as history, political science, sociology, social psychology, ethics, and even religion can be brought to bear in assessing such impacts. Calling upon nonlaw disciplines helps one inventory, analyze, and compare the work that legal systems do around the world. Animal Studies looks at law in a robustly interdisciplinary manner to identify different societies' interactions with other-than-human animals.

Since there are many different kinds of legal systems, and since individual legal systems are greatly impacted by local culture and political realities, each legal system needs to be explored specifically in order for Animal Studies to develop an informed and nuanced perspective on the effects, good and bad, of legal systems as a whole on the more-than-human world. Islamic law, for example, has always provided a "right of thirst" for wildlife, and in this it can be contrasted with modern secular law which is, by and large, silent on such issues (sometimes anticruelty statutes require individuals to provide water, but many corporate practices cannot be challenged in industrialized societies that have passed laws exempting "common practices" of businesses from anticruelty legislation).[8]

Such work requires not only a review of many aspects of the particular cultures one is studying, but also information about the identity and realities of the nonhuman animals impacted by a particular legal system. By historical accident, some legal systems have emerged in cultures that had few, if any, of the more socially complex nonhuman mammals who so obviously display personality, intelligence, family connections, and even community and culture.

Comparative work also illuminates the highly specific tools available in legal systems, such as legislation, litigation, administration, and enforcement. Each of these distinctive tools or parts of the overall legal process works with a variety of nonlegal social processes by which any society regulates its individuals, including peer pressure, education, morals from religious communities and other local groups, and other traditions of many kinds. While those studying law often put a great deal of emphasis on government-based processes as the crucial step in social control (which makes sense since government channels are the way in which much power moves in modern societies), many forms of social control well beyond government action and inaction have important roles in the way any individual or group interacts with ubiquitous life beyond the species line.

Looking solely at the internal workings of law, a legal system can be thought of as a box holding many different kinds of tools. From an external vantage point, it is easy to see that these different tools must be put to use in society as a whole. How such tools work in this larger social context can be seen well only when one employs the insights of many disciplines.

Looking in this way at law in its broadest context permits one, in turn, to see that law often has had (and, of course, still has) oppression-laden features. Further, if one sees clearly that legal systems and their many tools have regularly been used to disadvantage, even harm, specific groups of humans, then one easily recognizes, too, that legal systems play a central role in the harms done to countless nonhumans. Relatedly, interdisciplinary work also helps one see that in an astonishing number of historical and contemporary situations, oppression of humans has been interwoven in complex ways with oppression of various nonhumans.

In all these ways, interdisciplinary work makes clear that law is deeply impacted by social values even as it often shapes those values. Subject to social change and also a force in creating social change, law as an eminently human enterprise is, in fact, capable of being very flexible if people choose to make it so. This is the meaning of Henry David Thoreau's oft-cited observation, "The law will never make men free; it is men who have got to make the law free."[9] Activists of all kinds recognize this political reality, refusing to accept, in the words of William Lloyd Garrison, a leading antislavery abolitionist in the United States, the "prevalent heresy...that what the law allows is right, and what it disallows is wrong." Garrison argued further, "All public reform comes about through the 'interrogation' of the law by morality."[10]

From such vantage points interrogating existing law, one can best understand the phenomenon broadly referred to today as animal law. This concerted effort is the legal side of a much larger, worldwide social movement that is attempting to open up legal systems and other important institutions to the insights driving the ancient tradition of concern for justice and compassion beyond the species line.

Agenda-Driven Comparisons: Critical Legal Studies and Theory

Animal Studies needs to pay attention to various approaches collected under the name "critical legal theory."[11] Going well beyond merely identifying issues, these approaches consistently ask each human to act with regard to certain problems and oppressions.

An early form of critical legal theory is referred to as "legal realism," an approach that was inspired by the Harvard Law School scholars John Chipman Gray and Oliver Wendell Holmes. In its fullest expression between 1920 and 1940, legal realism promoted an empirically based approach as part of its attempt to describe how practicing judges actually decided the cases before them. Early legal realists criticized then-prevailing legal theories that suggested judges always follow legal rules closely by employing legal reasoning patterns that track logical, syllogism-based reasoning. Legal realists observed that judges often, though not always, exercise considerable discretion on whether legal rules control the outcome of a case. Sometimes judges decide a case, for example, on the basis of their own political and moral intuitions about the facts of the case before them. These early legal thinkers were thought of as realists because they believed the prevailing theories of law overstated how judges in fact decide many cases. So these legal realists pressed others to admit that judges' political and moral convictions sometimes play a determining role that overrides strictly legal considerations. The ruling is still the product of official activity, but judges are clearly doing more than following preexisting law. In such cases, the task of a legal realist is empirical, namely, identifying which psychological, sociological, and other factors influence judicial decision making.

Although historically this approach was not used in the following way, the theory itself is relevant to how laws involving various nonhuman animal issues might be interpreted and even bent, as it were, in a particular case. For example, a judge's intuition regarding the importance of compassion, or the need to have an owner carry out a duty of care to an owned animal, might lead the judge to the conviction that justice will only be done in a particular case if the law provides some sort of fundamental protection. A similar result might also be mandated because the judge is aware of newly developed scientific information suggesting that the nonhuman animals involved are suffering in some way not previously imagined.

There are other legal theorists who go much further than this first group of legal realists. For example, other legal philosophers have argued that legal realists dramatically understate how a judge's personal convictions impact decisions. This is the position of critical legal studies, which is, in effect, a more radical or "going to the root" approach that offers a thorough critique of mainstream legal thinking about what judges do when they decide cases. The critical legal studies movement suggests that only rarely do judges apply existing law in a simple, logical manner. Rather than logic, it is ideology (in the narrow sense identified in chapter 1) that most characteristically shapes how judges make decisions. What prevails and thus gives content to law (even in a democracy) is, then, the product of narrow ideologies struggling against each other. Said in another way, people driven by one set of ideas struggle against other people driven by a different set of ideas. Such competition produces "ideological struggles among social factions in which competing conceptions of justice, goodness, and social and political life get compromised, truncated, vitiated, and adjusted."[12]

The end product is, in the eyes of critical legal theorists, a level and degree of inconsistency that impacts the law in both its generalities and specifics, creating at all levels what is sometimes called law's "radical indeterminacy." In the face of such inconsistency, a judge has options for how to decide the case, which of course creates an opening for personal, political, and ethical criteria that impress the judge. This view is at odds with the standard image of judges, which holds that judges interpret, not make, the law.

Critical legal studies, through such interpretations, clearly reveals an important agenda—this form of analysis is deeply invested in affirming that some legal decision making produces, under the sway of narrow ideologies, outcomes that create injustices or foster a lack of compassion. This agenda is valuable precisely because it prompts people to look carefully at what judges and other decision makers are really doing. More specifically, through its criticism of the established order, this approach provides a way to understand why existing law (as found in legislation and previously decided cases) has so often in modern times followed a power-oriented agenda in service of the exceptionalist tradition. The combination of power and extremely narrow forms of human-centeredness is the product, critical legal theorists might suggest, of an ideology that has prevailed in the struggle of competing ideas. Importantly, the result may not be truly human-centered—some laws advance only the interests of an elite class that controls the law-making process but effectively masks this narrow agenda by claiming that the law in question treats all humans as equal.

Approaches within the domain of critical legal studies have the potential to explain why a society can feature deep-rooted values like compassion and caring (for example, as part of a religious tradition) but nonetheless enact laws that ignore compassion for nonhuman animals in favor of, say, profits for businesses. In this regard, such approaches contribute an important perspective on how and why nonhuman animals are so often unprotected within legal systems.

Another form of agenda-driven comparative work in law is known as "law and economics." This approach, which argues for the value of economic analysis in the law, has both a descriptive side and a prescriptive, theory-laden side. As description, law and economics advocates claim that many areas of the common law feature content best understood in terms of a tendency to maximize preferences—said another way, economics-based reasoning allows one to see why important areas of the common law tradition, including property, contracts, torts, and criminal law, have taken their present shape. Such features may or may not appear on the surface of legislation, in official explanations of the law, or in judicial opinions. Even when economics-driven reasoning controls the outcome, then, such reasoning may be covered over by more traditional rhetoric about justice, the public good, and other common ways of talking about law. One influential contemporary commentator suggests in his description of the common law tradition (which dominates in England, Canada, India, the United States, and dozens of other countries around the world) how fully this important tradition focuses on human issues alone: "the common law is best (not perfectly) explained as a system for maximizing the wealth of society."[13]

There is also a prescriptive or values-based theoretical underpinning to the law and economics approach. This view of law recommends that judges and elected officials should make their decisions in ways that maximize wealth, and that this goal can be achieved most fully through free-market mechanisms. The concern for "the wealth of society" manifests what at first will seem a part of the exceptionalist tradition, but it is more strictly a focus on the humans within a society who benefit from economics-related activities (in other words, as long as the overall wealth of a society is increased, the fact that some humans do not benefit at all or are even made worse off does not count).

Animal Studies has the capacity to ferret out the controlling assumptions of such an approach—for example, that humans alone have the abilities that matter, namely, autonomy and the calculating abilities to create markets that satisfy human preferences and maximize what can be quantified as wealth. Another assumption is that outcomes by which wealth is measured are properly centered solely on humans—impacts on nonhuman individuals and communities are not a factor unless humans choose to make them so. This approach, then, is a kindred spirit of the preoccupation with humans found so broadly in mainline Western cultural institutions and thinkers such as Descartes and Kant. This biased set of assumptions also reveals how law and economics approaches are both theory laden and driven by a very specific values-based agenda.

A fourth form of agenda-driven comparison of legal systems has been given the name of "outsider jurisprudence." This approach represents especially well how an agenda-laden approach to law and legal systems in general can be dominated by features that are "radical" in the sense of going to the root of certain oppressions. This feature opens to view the obvious ways in which all law is involved in control and protection processes that result in both inclusion and exclusion. Such features beg a fundamental question, as do all ethical and social control schemes: Who and what will be included, and who and what will be excluded?

Outsider jurisprudence insists that legal systems characteristically have structured their social control in ways that promote the interests of only some privileged humans (such as white males) even as they exclude women and other disfavored humans. The upshot is that the protected group receives key privileges in the form of fundamental protections offered by the law. In this vein, critical theory anchored in feminist concerns argues that patriarchal assumptions dominate present law and are manifested in the ways that property rules play out, how crimes are identified and punished, and how key legal rights are allocated or denied. A similar approach often called critical race theory focuses on the role of race as a decisive factor that sustains race-based oppressions. Under this view of legal systems, decisions by judges, legislators, administrators, and law enforcement sustain traditional claims of racial supremacy by favoring one race over another. The upshot is a powerful dismissal of the excluded group's experiences, concerns, values, and even personal histories, all of which thereby become invisible in important policymaking circles and even in explanatory theories that claim to be open-minded and inclusive. Particularly evident in these approaches is the way a Critical Studies–inspired evaluation of law challenges models of judicial decision making that assume neutral rules will be applied with impartiality.

Animal Studies is enriched by such theorizing about law and legal systems because it illustrates how law not only allows, but even promotes, the harsh treatment of some disfavored group. Domination, exclusion, caricature, and apathy among a society's privileged groups create more than dismissal and a harsh legal system—such problems also give birth to and then sustain self-inflicted ignorance among those that control power. Such ignorance can make it difficult to reform a system or to create openings for protections that other living beings need because it, in effect, keeps those in power from recognizing their complicity in perpetuating serious ongoing harms.

In effect, critical legal studies offers a template to those who wish to evaluate the exclusions and harms that law imposes on other-than-human animals. At the same time, Animal Studies also assesses the positive side of law, namely, the unquestioned potential of legal systems for protecting some nonhumans when those who control the content of law choose willingly to provide fundamental protections (for example, how many affluent dog and cat owners now lobby for better legal protection for owned animals).

Liberation for Some, Oppression for "Others"

The exclusions that law holds in place have drawn many other wide-ranging critiques. A legal system may be controlled by people or politically powerful institutions that proclaim "all humans are equal" even as reality falls far short of this standard. It is so common that, in actual practice, only some humans receive the law's protection that many people have come to expect that a legal system will protect "the rich" but disadvantage "the poor."

Candor about such issues helps everyone recognize that while legal systems provide protections for a privileged group, they also create and anchor exclusions of "others." Such shortcomings have prompted indictments far angrier than the humorous quip that "people who love sausage and respect the law should never watch either one being made." The populist orator Mary Ellen Lease challenged the imbalances of the American legal system at an

1890 political convention in her home state of Kansas when she thundered, "Our laws are the output of a system which clothes rascals in robes and honesty in rags."[14] This critique underscores, to be sure, only the human-on-human problems that lawmaking can create. It is a short step to conclude that, when legal systems developed within an intensely human-centered cultural tradition create problems for human others, they characteristically fail to nourish citizens' capacity to become richly aware of nonhuman others.

Law as Projecting Futures

Rivaling negative views of lawmaking and legal systems, of course, are soaring views of what legal systems can do. This tradition of hyperbole about law assumes that legal systems are for humans alone. The philosopher Cicero once said, "we are all, in the end, slaves of the law that we might be free."[15] Cicero's reasoning pattern is simple enough to follow, for individuals' willing submission to "the law" is a bargain that can create important benefits for those individuals and groups willing to be "slaves to the law."

Such generalizations fail to identify, of course, the impact of any society's laws on those who are not willingly governed. Even if we assume that it is a minor problem that some humans cannot be claimed to consent to be slaves to rules, Cicero's adage glosses over the fact that law as he understands it might be seen to make "slaves" of nonconsenting individuals from all species. While Cicero's slave image is, of course, merely a literary device focusing on law as a social compact, the slavery image has the power to explain some of the harsh realities that legal systems impose on so many nonhumans.[16]

Cover's insight that "law is the projection of an imagined future upon reality" shows that legal systems and other forms of public policy dictate an imagined future for far more than the citizens of a society and its future human inhabitants. Our human laws create a bewildering and morally charged set of effects beyond the species line. Animal Studies must grapple with these as just one part of our many, ever-so-human projections onto the realities of nonhumans. In a great variety of ways, our laws tie other-than-human animals down for solely human purposes, destroy habitat and community, distort and impoverish individual lives, and create many captive circumstances that fit the descriptions "worse than slavery" and "worse than death."

Legal systems can, to be sure, project more than a future dominated by harms—they can, in fact, do many positive things for nonhuman animals, such as create wild habitat protections and anticruelty limitations and other benign conditions for the nonhuman animals in our midst. The harms created by legal systems are, then, by no means the full story, nor a basis for a thorough evaluation of legal systems' possibilities. It is disingenuous to foreground a system's failures without discussing its contributions in other areas. Thus, even when legal systems protect an array of oppressions and harms, they may also provide the tools to remedy specific problems.

Animal Studies will, then, inevitably deal with the fact that law and legal systems are complicated animals with many legs and arms, as well as long, evolving histories. As fully as they are subject to—and often trumped by—complicated political realities that dominate their host society, legal systems have been the means by which human groups have achieved noteworthy triumphs.

Comparative Frankness about Human-Centered Values

Many analysts of law and human achievement find frankness about such matters to be worse than a mere triviality—it is held by many to be a moral absurdity that contradicts the exceptionalist foundations of not only law, but also the prevailing views of ethics, education, religion, politics, and much more. Insensitivity to nonhuman animal issues can prevail because so many influential and even respected moral authorities (elected officials, government agencies, educators, religious leaders) support exclusions justified solely by human-centeredness. Because of such dismissals and apathy, harms to nonhuman animals are very often deemed to have no moral consequence whatsoever.

When this kind of reasoning is subjected to the canons of critical thinking, it is wanting in many respects. To argue that traditions of human-centeredness justify a legal system's failure to be concerned with the interests of other-than-human beings is to argue in a circular fashion. The argument assumes away what is at issue, namely, what is it that justifies an exclusivist form of human-centeredness in the first place?

Impatience with those who state the plain facts of harms to nonhuman animals is common, of course, but such impatience ignores the self-transcending genius of humans' ethical abilities. Such abilities clearly can include, when the human imagination permits it, living within a species-transcending ethic. Realization of this potential, which historical and cultural studies show has occurred in a great variety of places and times, is by no means totally absent in modern industrialized societies—many students are, for example, impressed by this possibility, as evidenced by the demand for animal law courses. Yet in mainline decision-making circles, such breadth of concern reaching beyond the species line remains, at best, a subordinated value and, at worst, a topic that policymakers work hard to ignore.

A comparative, interdisciplinary analysis of cultures can reveal that even though Animal Studies has emerged in an historical era when mainline cultural institutions often feature a pronounced lack of awareness about nonhuman animals, there remain an astonishing variety of ways in which individual humans, some businesses, a number of governments, and many marginalized groups and small-scale societies continue to honor the ancient insight that some nonhuman animals have their own point of view and interests.

Further, a comparative analysis of legal systems can reveal a similar though more limited insight. Prohibition of some forms of cruelty to nonhuman animals can be found in the official legislation of virtually every society. While enforcement of such laws remains very inconsistent, it is nonetheless common for judges to note that the purpose of such legislation is protection of the nonhuman animals themselves. An example can be found in the 1888 case of *Stephens v. Mississippi* dealing with an anticruelty statute: "This statute is for the benefit of animals, as creatures capable of feeling and suffering, and it was intended to protect them from cruelty, without reference to their being property, or to the damages which might thereby be occasioned to their owners."[17]

Despite the common sense of this position (many would agree that such statutes are, as the court said, intended for the nonhuman animals' sake), problems arise. The first is inevitable—what counts as protection is seen from the human vantage point because, as one veterinarian has said, nonhuman animals "are perpetual 'others,' doomed to have their interests represented to humans by other humans."[18] The second problem is more purely a political dilemma—compassion-focused protection is almost always a very low priority in industrial-ized societies and thus is easily trumped by human interests like making money or avoiding inconvenience. In many industrial societies, in fact, laws have diluted and even erased duties of compassion—such as the wide-ranging exemptions granted by many American states to corporations pursuing industrialized agriculture. The exemptions mean that individual citi-zens must still observe long-standing prohibitions on cruelty, but large corporations are no longer subject to these traditions.[19] This is important because the astonishingly large number of food animals involved in modern production processes called "factory farming" make food-on-the-plate the principal way that citizens interact with nonhuman animals.

Reacting to an Absence of Legal Compassion

As a leading edge of Animal Studies, the field of animal law is helpfully understood as a straightforward reaction to industrialized societies' denial of the compassion traditions that sit at the heart of so many human cultural traditions. Exclusionary attitudes and tactics are so dominant in many government and private institutions that a "human and animals" dualism is fairly said to verge on the status of cultural presupposition. Educational institu-tions now teach this dualism much as they once taught racism or sexism as the natural order. Science instructors continue to use the phrase "animals" to mean only nonhuman animals even though they readily acknowledge that humans are clearly primates, mammals, and ver-tebrates. Perhaps most tellingly, in some circles espousing the "humans alone matter" point of view will function as the measure of morality, sanity, rational judgment, and good taste.

Litmus Test: Which Realities of Other Animals Are Noticed?

Since a simple, unavoidable question in Animal Studies is the level of commitment to learn-ing the actual realities of animals outside our species, a natural question is whether those governed by a specific legal system are prompted to consider nonhuman animals as anything more than mere resources. If so, are the realities of other animals taken seriously?

Animal Studies also poses a battery of practical questions. It is relevant, for example, to ask whether what is being noticed about any nonhuman animal is so thoroughly dominated by human preconceptions that it becomes meaningless. Are inquiries about nonhuman animals successful in getting beyond cultural caricatures or other misleading constructions of other animals' realities? Does a prevailing bias for human-centered inquiry continue to plague those who try to answer questions about other animals' realities? These questions open the door to the most fundamental issue, namely, are other animals themselves really noticed?

The Question of Issues and Tasks

Developing perspectives that can be drawn from sophisticated comparative work in law requires careful study and realistic appraisals of the specific provisions and enactments of many different legal systems, including their enforcement realities. Such perspectives permit Animal Studies to address fully its core issues and tasks, including recognition of the differ-ent ways human societies have answered the central, recurring question, Who and what are animals? Equipped with information about the diversity of approaches that human societies have developed, Animal Studies is able to see a broad range of possibilities for legal systems and thereby foster inquiries important to the human community.

As Animal Studies develops an increasingly informed and detailed perspective on animal law developments around the world, it is also easier to see its rationale. For example, on the issue of personal connection and meaning, answers can be anchored in details like grassroots developments and student demand—both of these features reveal the ways in which Animal Studies creates an open, responsive, compassion-based approach that allows humans freedom to interact with other living beings with integrity, in a science-literate manner, and with ethically informed sensitivities.

Law-based comparative work also helps the larger story emerge, providing opportunities to explain how law as one of humans' most powerful tools is utilized around the world. The larger story can also be told more cogently when details of how legal systems project an imagined future can be assessed.

Comparative work, then, can make clear the many ways that law is a central component in how our species chooses one particular future over another. Such work underscores that today's humans must choose whether our species will foreground compassion and coexistence, or continue with the present realities of domination of nonhuman animal communities and even extinction of entire species.

Comparative Study of Religious Traditions

Comparative approaches to the study of religion today are both sophisticated and varied. In the modern academic world, the general study of religion, which is implicitly comparative because of the great variety of religious traditions found around the world, goes by a number of different names—religious studies, the history of religions, or comparative religion. Comparative work is further broken down into very specific fields such as anthropology of religion, sociology of religion, psychology of religion, and on and on. Sometimes the traditional topic of theology is another means of comparative study, although not all religious traditions are theistic in the sense of familiar traditions like Islam, Christianity, Judaism, and Hinduism.

Studying a single religious tradition in isolation, if taken to a detailed level, will reveal that the tradition carries multiple points of view, for all religious traditions evolve over time and place. Comparative work of a certain kind is needed to engage the length and breadth, as it were, of the tradition being studied. To study a single religious tradition, one needs to enlist the aid of many other disciplines such as history, sociology, ethics, politics, and ecology to grasp how religion works in the lives of individuals and communities.

Another approach requires even greater interdisciplinary work—comparing multiple religious traditions. Such work calls upon an astonishing number of disciplines such as psychology, literature, ritual studies, archaeology, anthropology, and geography to ferret out how religious traditions are at times alike even as they are also fundamentally different.

A third approach attempts to describe how religious traditions have impacted views of the more-than-human world. Some have analyzed claims about the world's living beings found in religious texts or other traditional literature and stories prized within a tradition. For example, the series of documents known as the Hebrew Bible or Old Testament features, by one count, 113 different kinds of animals.[20] Naming the number and identity of species mentioned in a particular religious text is a relatively simple task, of course, compared to describing the implications of what was claimed about nonhuman animals in these texts. Even more complicated is studying what human individuals and communities knew or claimed about the actual realities of other animals. Some individual religious communities exhibit detailed awareness of the lives of real animals even when the larger religious tradition is thought to be generally negative about animals. This fact illustrates an important task—recognizing that asking whether a tradition as a whole is open or closed, friendly or unfriendly to nonhuman animals is different than asking whether religious believers in daily life hold accurate, detailed information about other animals' actual lives.

Such important distinctions reveal that inquiries about religion and animals need to be more than superficial generalizations. One of the classic negative evaluations of an entire tradition is Lynn White's 1967 thesis that "especially in its Western form, Christianity is the most anthropocentric religion the world has ever seen."[21] White argued this because, in his view, "Christianity, in absolute contrast to ancient paganism and Asia's religions...not only established a dualism of man and nature but also insisted that it is God's will that man exploit nature for his proper ends." Many have argued that White's claims are wrong in important specifics. For example, many scholars have debated the meaning of the often-cited verse 28 of the first chapter of Genesis. In the Revised Standard Version, this passage reads, "Be fruitful and increase, fill the earth and subdue it; and have dominion over the fish in the sea, over the birds of the air, and over every other living thing that moves on the earth." One scholar has observed that this passage was not taken by ancient and medieval readers as any sort of license "selfishly to exploit the environment or to undermine its pristine integrity."[22]

Animal Studies provides a conceptual framework for interrelating both environmental and animal protection issues even though many scholars today continue to distinguish claims about nonhuman animals from claims about "the environment," "the world," or "nature." A religious tradition's mainline figures and institutions could conceivably be deeply committed to the holistic approaches of protecting ecosystems or the earth as a whole even though they are dismissive or otherwise apathetic regarding the moral issues regarding individual nonhuman animals, or vice versa. Relatedly, even a religious tradition that promotes an overlapping dismissal of the environment and subordination of all other-than-human animals may include individuals or entire subtraditions that put into practice altogether more positive responses to other animals, the environment, or both.

Such important qualifications reveal that generalizations about any religious tradition can be misleading. Each tradition characteristically features many options, some of which contradict each other. Good examples of the complexity of a large religious tradition are provided in recent books focusing on animal issues in Islam, a tradition that shares with Judaism and Christianity a historically important emphasis on humans as the centerpiece of creation.[23] Such work, which comes from both Muslim and non-Muslim scholars, suggests that, despite human-centered features, the overall tradition nonetheless has much room for claims that other animals have remarkable abilities, connections with Allah, and even their own revelation.

Critical thinking requires, then, that those discussing religious traditions and their impacts on other living beings assert more than mere generalizations underdetermined by facts about actual believers. Since exactly the same point applies to historical accounts, the lesson here is simple and humbling—any informed, nuanced view of a present or past religious tradition, as well as comparative work on several traditions, will tell a rich story full of diversity.[24]

It is possible today to study comparative religious issues most directly in the field known as "religion and animals," or in the allied approach known as "religion and ecology," which today is an astonishingly rich, international field.[25] It is also possible to do comparative study of religion and animal issues through the lens of religious ethics. In the indexes of leading textbooks on religious ethics, however, the issue of nonhuman animals is only rarely listed—the fact that a large majority of those who teach and publish about religious topics do not make humans' relationships with other-than-human animals a part of contemporary religious ethics is due to the human-centered traditions in scholarship and education, as well as to the relatively recent emergence (1990s) of religion and animals as a recognized topic in the modern academy.

Traditional education in universities also does not characteristically study how any single religious tradition creates opportunities and problems for living beings outside the human species—instead, most general courses about religion, like those about religious ethics, remain profoundly human-centered. Today, more and more general courses on religion begin to touch on the ecology issue, but even this important topic remains unexamined in many of the best-selling textbooks on religious traditions.

Comparing How Religious Traditions Impact Views of Animals

Both religion and animals and religion and ecology use comparative work to explore diverse religious voices that have played central roles in many historical eras and cultural settings where ecological and animal-related issues are discussed. These fields have already generated much scholarship about religious traditions' reflections on how and why humans might choose sustainable coexistence with other-than-human individuals and communities.

In addition, because both of these new fields are deeply invested in questions of ethics, they look creatively at the question of how religions have, on the one hand, historically supported attitudes that have created problems and, on the other hand, have been key players in proposing and putting into effect solutions to extremely complex problems in the modern world. Both fields study impacts of religious traditions on (1) adherents; (2) whole societies, cultures, and worldviews; and (3) individuals who view themselves as secular and uncharacterized by any religious or spiritual features of any kind. For example, some exclusively secular dismissals of animals seem to draw energy from various religious communities' dismissive points of view, human-centered language habits, ignorance of basic facts, and inherited failures to notice other animals.

The study of how religious traditions have impacted views of other animals has great potential for enriching any number of subfields in Animal Studies. Because both religious commitments and religion-originated views of the world are integral parts of so many humans' worldviews and lifeways today, the development of a critically sophisticated study of religion and animals can be crucial to the spread of healthy, historically informed and culturally sensitive forms of Animal Studies.

The Question of Realities

A key challenge in studying religious traditions will be how the issue of other animals' realities is handled. Concern for the actual realities and respect for the interests of other animals is deep and sophisticated in some religious traditions, and yet is lacking in many of the most influential and outspoken circles of the most populous religious traditions in the world today. Many mainline religious institutions play down the status of humans as animals, just as many dilute humans' obligation to include any nonhuman animals when individuals answer the root question at the heart of all ethics, Who are the others? Some religious traditions clearly are invested in the relevance of empirical (though not always science-based) investigations of other-than-human animals, while some are deeply apathetic about such matters.

By assessing how different human enterprises, such as science traditions and religious traditions, have engaged other-than-human animals, one can explore all the better how questions about humans' relationship to other animals have been an integral part of human existence. Further, every human endeavor, from religion to science to education, appears to include some spheres that ignore or otherwise dismiss some nonhumans' realities and some spheres that pay close attention to such matters. Discerning how and why this pattern emerges is a very special version of the comparative enterprises that Animal Studies can foster.

The answer to questions about whether religious traditions feature any commitment to awareness of other animals' realities, then, is only rarely, if ever, a simple yes or no. However, religious traditions and their adherents can promote such awareness if they choose. In effect, whether a religious tradition fosters awareness of other animals is a measure of that tradition's willingness to inquire about the actual world we live in, for religious traditions often feature a deep awareness that our world is clearly a shared world in which humans are not alone but, instead, accompanied by an astonishing array of other-than-human citizens.

Humans' religious impulses can actually help sustain a full engagement with other animals in their own communities. They can also provide sophisticated awareness of how and when human bias and self-interest play a distorting role. In this sense, religious traditions at times foster in adherents an engagement with the actual world. Comparative work in religion, then, can provide important perspectives regarding human societies' different answers to fundamental questions like, Who and what are living beings? Whether an answer fits any particular model—fellow animals, fellow created creatures, reincarnations of past lives, other communities, ancestors, souls, and on and on—religious traditions are the originators and, today, perpetuators of diverse viewpoints and claims regarding our fellow animals.

Comparative work in religion will also affirm that one human community after another has, as a practical matter, recognized the important task of naming our own limitations in interactions with other-than-human animals. There may in fact be no discipline that has recognized human finitude more often than religion has. Many religious communities have asserted in one way or another that humans' self-actualization is dependent upon humans' self-transcendence at the level of both individuals and the species as a whole.

An informed view about humans' many different religions, then, makes obvious how broad the range of human viewpoints regarding other animals can be. In particular, comparative religious studies can help answer the question of why Animal Studies is important. By providing carefully developed information about many human interactions with other animals, such comparative work confirms that detailed, specific, evidence-based information is all-important to understanding human interactions with other animals—this is an area where the devil is in the details (or are those the better angels of our nature?).

Animal Studies Deepening the Study of Religion

It may seem provocative to ask a question that reverses the claim that the study of religion helps enrich Animal Studies—how and in what ways can Animal Studies enrich the study of religion?

Animal Studies can assist anyone who wishes to assess whether a religious tradition pursues an ethic of inquiry. Similarly, Animal Studies can produce an informed account of a religious community's educational efforts by asking if such efforts enable members to see other animals well or blunt sensibilities because the community promotes humans in a way that dismisses other life forms. Human-centered accounts of a religious tradition that either explicitly or implicitly dismiss the more-than-human world fall short of providing a full perspective: "to engage in a history of the Christian tradition without considering the other-than-human animals that are part of this history is to deal with parts, not wholes. It results in a partial history, one that is incomplete and, in the long run, invalid."[26]

This frank comment reveals the risks taken by religious leaders and scholars who continue to suggest that religion and spirituality are naturally confined to human interests alone. Such claims are historically wrong, of course, but also existentially wanting for those believers who continue to be fascinated by other-than-human animals.

Questions about education also pertain to the ways different subtraditions prompt humans to notice or ignore the local world. Such education can be done in uniquely compassionate and ethical ways, such that these humans live and share community with other-than-human animals, or it can turn heads and hearts away from living beings across the species line. Answers to such questions illuminate the study of religion by providing perspectives that have long remained undeveloped.

Comparative Study of Cultures

The number of human cultures is staggering—the ten-volume *Encyclopedia of World Cultures* includes entries for more than 1,500 different culture groups, which is only a portion of those that now exist. From a historical perspective, these 1,500 cultures are only a small fraction of those that have ever existed. This is shown by the number of living languages today despite the loss of many languages in the recent past—in 2003 one estimate put the number of surviving languages at close to 7,000.[27]

Such numbers reveal that humans have developed startlingly diverse communities. While the discipline most commonly associated with the study of human cultures is anthropology (see chapter 9), many other disciplines also explore cultural issues. Here, general comparative issues are examined in a number of other fields that address different aspects and challenges of our species' multiculturalism.

Not surprisingly, the great differences among cultures have long fascinated people. Herodotus of Halicarnassus, the fifth-century-BCE Greek sometimes called the "Father of History," collected many accounts of different cultures and lands. It was an integral part of Herodotus's Greek heritage to claim that his own culture was superior to all others, but this early historian nonetheless assessed other cultures in relatively careful ways in his account of the Greco-Persian wars in his one surviving work, known as *The Histories*. Modern study of cultural differences has blossomed into distinct fields such as cultural studies, political science, sociology, and religious studies that go well beyond the historical forms of study heralded by Herodotus's work and the now-developed field of anthropology.

Cultural Chauvinism

In spite of our species' multifaceted sophistication in looking at multiple cultures, Animal Studies today proceeds in an environment full of discussions about the clash of cultures. Many people unreflectively assume the superiority of their own culture to others. Further, we are all heirs to historical and political problems spawned by attempts of one cultural group after another to dominate other cultures. There remain echoes of this tendency in twentieth- and early twenty-first-century accounts, such as Samuel Huntington's 1996 *The Clash of Civilizations and the Remaking of World Order*, in which the post–Cold War world is divided into antagonistic, armed camps of Western Christians, Muslims, Hindus, and a few other groups defined by cultural and religious heritage.

Although E. B. Tylor defined the concept of culture in print in 1871, there is no consensus today on how to define this key notion. Many contemporary approaches describe and interpret behavior in a culturally relativistic way. Such approaches refrain from judging cultural practices through application of a single, absolute standard and, instead, examine practices from vantage points within each culture and specific social contexts in which human behavior occurs.

Using this approach, questions such as "which cultures, if any, offer a paradigm for human behavior?" can be answered in ways gentler than clash-oriented approaches. It is possible, for example, to think of each culture as an example of human possibility. Inquiries can then focus on what we can learn from the many other cultures outside our own birth culture. Further a comparative view can say a great deal about human possibilities.

Such gentler, more inclusive questions are important in Animal Studies for any number of reasons, but perhaps most importantly because, in the manner of all critical thinking, they open up inquiry rather than narrow it down. Such an approach makes it clear how diversely members of our species have thought about other-than-human animals. By avoiding the assumption that any one culture's views are the norm, Animal Studies can enable students to explore the vast diversity in human-nonhuman interactions and possibilities and nurture wide consideration of different human cultures.

Work that memorializes languages and cultural differences is invested with a certain urgency. Each loss of a living language represents a diminishment of much more than language alone—such losses involve dislocation of people, the disappearance of a worldview and of human attainment, possibly even the permanent loss of an untold history that goes back centuries and even millennia. Such losses may not seem an Animal Studies issue per se, but merely an ethical issue anchored in the importance of each human and his or her group. But since each culture is a unique story of not only human-to-human interaction, but also human-to-nonhuman interaction and possibility, losses of living languages and dislocation of cultures diminish our ability to tell the whole story of human-nonhuman interactions.

Cultural Studies

The question of human-nonhuman interactions offers a chance to discuss the relevance to Animal Studies of a specialized field often known as cultural studies. Narrower and more politically engaged than anthropology, cultural studies is of recent origin.[28] It is also decidedly agenda laden because the field has roots in Critical Studies, which protests substantial harms to peoples and cultures. This heritage prompts cultural studies scholars characteristically to see their work as more than matter-of-fact description—"practitioners see cultural studies not simply as a chronicle of cultural change but as an intervention in it, and see themselves not simply as scholars providing an account but as politically engaged participants."[29]

Cultural studies, then, provides accounts of harms and political exclusions that are seen as unjust. These accounts, which are richly comparative, may be couched in a variety of theoretical frameworks—for example, some use Marxist notions to explain such harms, while others use one variety or another of critical theory, and still others might use race-based or gender-based theories to explain why some humans harm others. There are additional comparative issues beyond description and explanation of the severe problems of one culture dominating another, such as how cultures may cooperate or compete, clash or coexist. Such work is among the most complicated of human inquiries because no human stands completely outside of cultural influences (just as no human stands outside of animality).

Critical Theory

Like the forms of Critical Studies described in chapter 4, critical theory is highly specialized work characteristically driven by a concern to point out that liberating human beings from oppressive conditions is unfinished work. Paying attention to assumptions, value structures, and the role of ideologies (see chapter 1) helps immensely in seeing why this kind of oppression continues to be a feature of human society.

In his 1982 book *Critical Theory*, Max Horkheimer defined critical theory as concern "to liberate human beings from the circumstances that enslave them."[30] Because Horkheimer was part of a highly influential group of German philosophers and social theorists in the Western European Marxist tradition known as the Frankfurt School, some accounts capitalize Critical Theory to designate these European philosophers for whom a "critical" theory always had a very specific practical purpose, namely, promoting human emancipation.[31] A second common use of the term critical theory designates a broader group of theories that "have emerged in connection with the many social movements that identify varied dimensions of the domination of human beings in modern societies."[32]

The trio of Critical Studies, cultural studies, and critical theory immerse humans in descriptions of situations where domination is occurring, and they also propose prescriptions or norms by which that domination generally and its harms can be eliminated or at least decreased. The goal is an increase in freedom for the humans subjected to the domination. Given that there are many human-on-human dominations, there are many different versions of such critical theories. There are, thus, discussions guided by "critical race feminism," "critical race theory," "critical legal theory" (closely tied to critical legal studies), and cousin terms. These discussions can be wide ranging, as when critical studies delves deeply into communication studies and the arts.

The Relevance to Animal Studies

When humans notice and take seriously the domination of some humans by others, the inquiry has power of several kinds. Asking where and when domination exists and developing an eye for changing what we find in the way of injustice, oppression, or imbalance have more power than the human-side answers to such questions. Such approaches extend naturally and well to human harms to the more-than-human world. In sum, Critical Studies, cultural studies, and critical theory have traction in Animal Studies because each dauntlessly questions human privilege and uses a wide range of critical thinking tools to help open up questions about humans' intersection with other-than-human animals. Much power flows from these attempts to uncover assumptions, to describe practices frankly, and to call out privileges and various conditioned ethical blindnesses. In effect, the very openness of the inquiry allows the questioner to tap into the innate fascination humans have with our ethical potential. In turn, connections seen as one inquires into oppressions can nurture additional questions of how to prompt change and even an overthrow of privilege. Thereby, this style of critical inquiry throws light on harms that are buttressed by the way people speak and think. It also illuminates harms caused by particular practices of governments and private parties.

Because unmasking narrow and agenda-laden privileges in this way has inevitable social, cultural, political, legal, educational, intellectual, religious, and ecological implications, there characteristically is a confrontational tone in both the critiques and opposition to such challenges. Historically, such discussions often slip into highly polarized debates that seem unproductive. Nonetheless, such discussions can, if ably handled, offer fresh and fruitful insights into a wide range of contemporary oppressions and policy problems.

Beyond Noticing Exclusions and Oppressions: Reactions to Theoretical Pyrotechnics

Some writing in cultural studies, critical theory, and Critical Studies puts off a wide range of those interested in challenging harms beyond the species line. One often-cited reason is that much of this writing seems to be more complicated jargon than substantive analysis. Displays of erudition may tend to intellectual pyrotechnics. Another reason is that various people resist the novel terms in which specific problems are described. Still others are irritated by the oppositional and intellectualized tone of the challenges to privilege. Yet others struggle with both the style and level of generalization—as with all social movements, disputes arise as to how best to frame the underlying problems. Finally, some repudiate these fields because of the very breadth of their historical descriptions, preferring instead more fine-grained evaluations of specific problems.

Advocates of these approaches often make broad generalizations and sweeping historical judgments, expressing themselves with complex concepts drawn from some of humanity's most theoretical and complex intellectual traditions. The fact that they also challenge long-established claims from religious and cultural traditions about the place of humans on earth, the accuracy of our received history, and the morality of common practices in contemporary societies opens these views to close scrutiny and even disfavor.

Examples of Potential Problems

Passages from some of the most respected philosophers within what is generally called the European tradition of "continental philosophy" offer the chance to sample specific shortcomings in thinking about nonhuman animal issues. It must be added as well that the three examples given below, individually and collectively, reflect the altogether positive development of continental philosophers engaging issues that go well beyond a one-dimensional dismissal of lives outside our own species line. While the three passages considered by no means represent continental philosophy's full range and depth on Animal Studies issues, they do reflect well that continental philosophy has opened the door to the more-than-human world.[33]

Martin Heidegger, often considered among the three or four most influential philosophers of the twentieth century, addressed extensively the boundary that he deemed to exist between humans and other-than-human animals. His notorious statement that "existence" (*Existenz* in Heidegger's original German) is a capacity reserved exclusively for human beings must be read in context, for it includes negative-sounding claims (such as "horses are, but they do not exist") in order to make a very specific claim about human animals.

> The being that exists is man. Man alone exists. Rocks are, but they do not exist. Trees are, but they do not exist. Horses are, but they do not exist. Angels are, but they do not exist. God is, but he does not exist. The proposition "man alone exists" does not mean by any means that man alone is a real being while all other beings are unreal and mere appearances or human ideas. The proposition "man exists" means: man is that being whose Being is distinguished by the open-standing standing-in in the unconcealedness of Being, from Being, in Being.[34]

One scholar suggests, "Heidegger uses the term 'existence' to characterize the historical, linguistic, self-referential cultivation of meaning that is distinctive of human beings."[35] The shortcomings here are not caused by the fact that Heidegger takes a broad approach, for any number of thinkers have taken extremely broad, value-laden approaches when describing the human-nonhuman intersection—it is in fact a considerable strength that Heidegger and the continental approach in general have the virtue of mentioning, then analyzing, issues at the human-nonhuman intersection. Further, many of this tradition's most prominent philosophers describe features of this intersection in ways that are markedly less foolish than the radically dismissive views of Descartes and less narrow-minded and parochial than Kant's human exceptionalism.

One shortcoming in this passage stems from the fact that Heidegger's claim about humans' uniqueness appears to be empirical but is not grounded by extensive exploration of nonhuman animals generally—Heidegger neither sought time amid communities of the most socially and cognitively complex, nor engaged the humans who had done so. He was, in other words, content to accept generalizations about living beings that were as underdetermined by facts as they were overdetermined by the exceptionalist tradition.

Heidegger's generalities carry some superficial features. Horses no more represent all nonhuman animals than people in one town or a single institution of an unspecified nature in an unspecified country represent all humans. Heidegger's reasoning shares all the weakness of Descartes's choice of "flies and ants" to represent all nonhuman animals in a famous passage in his influential *Discourse on Method*:

> For next to the error of those who deny God...there is none which is more effectual in leading feeble spirits from the straight path of virtue, than to imagine that the soul of the brute is of the same nature as our own, and that in consequence, after this life we have nothing to fear or to hope for, any more than the flies and ants.[36]

Heidegger's claim plays to the human exceptionalism narrative that he inherited with his mother's milk, German language, and the philosophical tradition of the Greeks that he found so deeply compelling. Without question, Heidegger's work suggests many reasons one might use the adjective "exceptional" about humans, but his reasoning reveals how poorly even gifted intellectuals can reason about nonhuman animal issues. The result is disappointingly one-dimensional and ends up reinforcing the mentality of the exceptionalist tradition. While one might expect and excuse this level of imprecision in ordinary conversation, as part of an attempt to philosophize on fundamental issues it is as regrettable as it is vague and imprecise. Heidegger is, by consensus, a provocative thinker, but in this matter his reflections leave much to be desired.

Comparable problems appear again and again in other major thinkers in the continental philosophy tradition—the following analysis offers insights on the human side that are remarkably sensitive but, on matters beyond the species line, betray more shortcomings than strengths. Emmanuel Lévinas in 1974 published a story under the title "The Name of a Dog, or Natural Rights" in which he relates an encounter with Bobby, a "wandering dog" who approached Lévinas and other prisoners at a Nazi prison camp.[37] Lévinas reports that through interactions with the prisoners in friendly, very doglike ways, Bobby reaffirmed the prisoners' humanity because this dog's willing interaction contrasted so markedly with the intentional dehumanization of the concentration camp. While Lévinas's account makes it clear that he finds in the story the possibility that other-than-human animals are capable of certain forms of caring, his ideas about nonhuman animals are, at best, minimalistic and, at worst, an affirmation of the exceptionalist tradition's willingness to make nonhuman animals disappear by homogenizing them into a single group (for example, merely assuming that Bobby represents all nonhuman animals sets up problems not unlike Heidegger's use of horses to stand in for any and all nonhuman animals). The result is that many overtones in the account, such as his use of the adjective "subhuman" and his reference to dehumanization turning the prisoners into "a gang of apes," overshadow his evident gratitude for the dog's friendly reaffirmation.[38]

Ranked among the twentieth century's most influential thinkers about ethics, Lévinas's one-dimensionality on nonhuman animal issues rivals that of Heidegger. As a number of commentators have observed, Lévinas's view of nonhuman animals is not particularly informed—he is famous for his ethics of "the other," but in very few ways do these "others" include nonhumans.[39]

The net result is that Lévinas's story mentioning Bobby is mixed—a nonhuman animal is recognized in a profound way amid an implicit reaffirmation of traditional prioritizing of humans over any and all nonhumans. Thus, even as Lévinas honored Bobby, nonetheless this friendly dog and all other real biological nonhumans virtually disappear because nonhuman animals struggle to have any real life in Lévinas's extensive body of work.[40]

Similarly mixed effects occur in a more recent and heavily cited work published at the end of the career of the influential but notoriously challenging French philosopher Jacques Derrida. In the frankly titled "The Animal That Therefore I Am," Derrida addresses Bentham's classic question "not, Can they reason?...but, Can they suffer?": "The response to 'can they suffer?' leaves no doubt. In fact it has never left any room for doubt; that is why the experience that we have of it is not even indubitable; it precedes the indubitable, it is older than it. No doubt either, then, about our giving vent to a surge of compassion, even if it is then misunderstood, repressed, or denied."[41]

This essay in one passage after another reflects how encounters with nonhuman others put questions to humans' special ethical abilities. Derrida reaches across the species line and appears to focus on a single individual of a long-time domesticated species with which he has a meaningful personal relationship. But Derrida's essay reveals the risks embedded in conceptual thinking that depends on one-dimensional images of nonhumans. One can ask yet again, for example, why a single nonhuman animal (here, Derrida's domesticated cat) might be thought to represent all nonhumans. It is not at all clear how a relationship with a cat living in our carpentered world throws light on relationship possibilities with wild animals living in their own communities. In fact, unreflectively asserting that one human's dominating relationship with a single cat illuminates essential features of the human-nonhuman intersection can, if handled poorly, serve the exceptionalist tradition. That tradition is invested in making different nonhuman animals disappear into the "subhuman" category, thereby discounting actual realities and real differences. Thus, even though Derrida creatively identifies a range of serious problems, he fails to extricate himself from some of the most debilitating features of the exceptionalist tradition—the upshot is that this essay has limited power despite being full of keen insights that can act as signposts pointing the way to a nonanthropocentric world.

Two themes reveal this long essay's limitations and show that Animal Studies' journey has just begun. The first is how Derrida thinks about his cat, and the second is how Derrida talks about history.

Derrida on His Cat

At one point, Derrida standing naked and seeing his owned cat look toward him, muses about "the gaze of an animal, for example the eyes of a cat" (p. 372). Cats' eyes are, importantly, different than human eyes—it is possible, then, that cats may piece together whatever comes into their field of vision differently than we do. Derrida quite rightly assumes his cat has a point of view, but his comments seem to assume that his cat's vision works much like a human's gaze. Since this may or may not be the case, critical thinking pushes one to keep a basic question from disappearing—do cats see as we do or in fundamentally different ways? Did the cat in this case even notice Derrida's nakedness?

One can fairly ask another question—does talking about "the gaze of an animal, for example the eyes of a cat" assume that cats stand in for some other nonhuman animals, perhaps even for all of them? Phrases like "humans and animals" sometimes trick people into thinking that one can easily talk about "animals" (meaning here "all nonhumans, but no humans"). Does Derrida's "gaze of an animal" trick him into distilling lots of different animalities into but one image of "an animal"?

This problem arises whenever one talks about a particular nonhuman, as Derrida seems to be doing in this essay. But in what ways does referring to one kind of nonhuman animal illuminate issues regarding all nonhuman animals? How does one who lives with an owned cat in carpentered space learn anything at all about nondominated animals in other-than-human-dominated spaces? While the gaze of Derrida's own cat may help him begin such an inquiry, he appears to assume that his domination-filled relationship with a single cat opens up vistas for understanding a wide range of nonhuman animals. This is an inherited, culture-based assumption, as is the assumption that the breadth and depth of human thinking are such that humans can, without complications, easily describe well what this cat is doing in this particular instance or what other animals are doing in general.

It is an equally questionable assumption for Derrida to believe that, on the basis of perceiving his cat's gaze as he describes it, he understood anything at all about what his cat was thinking on that occasion. Does a cat's casual look in the direction of a naked human suggest in any way that the cat sees what humans might see? What if the cat is doing something along the lines of what a human does when, as we say, "she stared blankly ahead"? In other words, cats may look in our direction for many reasons, and it is quite possible they are not focusing on anything at all, let alone some highly specific feature of the person standing before them which we assume them to be noticing. So while one can hazard a guess that Derrida's self-awareness in this case is part of human sexual dynamics, we guess much more feebly about what a cat might notice when glancing at a nearby human.

Since it is a factual question how a cat's eyes, perception, and attention work generally, Derrida's musing about the cat noticing his nakedness was obviously at best a wild guess. This kind of speculation is backed by the standard assumption of Western culture that human minds are so remarkable that we can confidently estimate what an individual cat is doing, even thinking, when it looks our way. But Derrida in this piece purports to be talking about nonhuman animal issues even as he makes facile assumptions that serve yet another form of human-centeredness.

Talking in such a facile way about a specific encounter serves human-exceptionalist themes. Derrida in this essay is merely using the fact that he noticed and became reflective about his own cat to talk about an eminently human concern. To be sure, it would pose no problem whatsoever to use a cat's look as a literary device if the device is called out explicitly. But Derrida never reaches the deeper question of what his cat was actually doing on the occasion he describes.

It is this failure to question that reveals Derrida continues to work within an anthropo-centric horizon. A similar conclusion can be drawn from the fact that he does not problema-tize the implication that we are such remarkable animals that we can speak confidently of the inaccessible features of another kind of animal's gaze. Animal Studies constantly reminds us, though, that our human community needs, precisely because of the impoverished views we have inherited, a major effort to think out together the important, independent issue of what the particular cats we know might be thinking, feeling, and doing. We need commu-nal, interdisciplinary work in order to recognize the diversity of other-than-human animals. We especially need to foreground humility about the plain fact that we are guessing at what other animals are doing when they are in our presence and glancing our way—whether we are clothed or naked.

Further, our speculations about their awarenesses need the deepest commitment to communal discussion and humility. In such settings, we can work together to assess how our own thinking has been impacted by habits of speech and thinking that we were born into or constantly immersed within as we matured into reflective adults. Above all, we have a chance to problematize ingrained habits like the long, debilitating tradition of essentializing other animals as "subhuman" and inferior. In community, we stand a chance of seeing the anthro-pocentric horizon that needs to be called out far better than Derrida does in this essay.

Since many human circles are only now renewing such communal efforts, perhaps the obviously talented and sensitive Derrida is to be forgiven for not taking us all the way on this journey. His work suggests that philosophers cannot make this journey alone—many other people are needed as we explore this mysterious "more-than-human world" and then learn how to communicate far better about it.

Derrida on History

Embedded within this essay is a related problem. Derrida's answer to Bentham's question puts before readers of the essay the absolute dismissal of nonhuman animals most characteristically associated with Descartes. Because the modern world anchors the exceptionalist tradition through its mainline discussions in law, science, economics, religion, and education, Derrida takes a chance when he claims that Bentham's question has "never left any room for doubt."

The bluntness of Derrida's unequivocal affirmation that at least some nonhuman ani-mals suffer, however, helps everyone recognize that dismissal of nonhuman animals in so many circles of the contemporary world might be questioned as obviously self-serving. It can also, as this volume suggests, be questioned because it is ignorance driven and repeatedly begs questions about what many take to be intolerable immoralities.

But Derrida's forthright, even courageous framing of this issue keeps his readers squarely within the Western philosophical tradition. At a number of different points, as Derrida dis-cusses general issues, he problematizes thinking "from Aristotle to Lacan" and "from Aristotle to Heidegger, from Descartes to Kant, from Lévinas to Lacan."[42] While Derrida's queries are squarely within the best tradition of philosophy, thereby prompting each of us to exam-ine how we think on the question of nonhuman animals, it remains important to examine Derrida's framing of the historical issues. Is the tradition we must operate within when think-ing about other animals really the tradition "from Aristotle to Lacan"?

An interdisciplinary approach makes it obvious that "Aristotle to Lacan" is only a small part of what Animal Studies must explore. Other traditions, both philosophical and nonphilosophical, are very useful in addressing Derrida's concerns. Even if one chooses to work within this single philosophical trajectory (the Western world's mainline philosophical tradition), one must recognize plenty of subtraditions where ordinary people and talented, educated minds challenged the mainline tradition to open up minds and hearts about issues involving other-than-human animals.

Derrida's essay "The Animal That Therefore I Am" has much to commend it. In particular, it reveals how far Derrida and others, including Lévinas and Heidegger, have moved beyond Descartes's views on nonhuman animals. Derrida in particular talks about nonhuman animals with passion even though his essay remains within the narrow confines of modern thinking.

Minor Problems and the Prospects of Communal Change

The provocative work of Heidegger, Lévinas, and Derrida can be seen to have shortcomings, but these are problems that can be fixed by an attentive community in constant dialogue with other disciplines. The shortcomings merely reflect that humans are traditional animals, a fact which brings us both riches and challenges. The traditional nature of human animals causes even humans' most remarkable thinkers to advance views that are, upon careful consideration, subject to shortcomings, including conditioned ethical blindnesses anchored in traditional practices or ways of speaking reliant on specific phrases; they also involve human-centered ideas and categories that continue to buttress peculiar and debilitating claims like human exceptionalism.

As history has so often shown, humans as traditional animals struggle with change that threatens their privileges. The history of science offers any number of examples whereby the most educated people refused even commonsense changes. Perhaps the most notorious of these examples is the difficulties Galileo faced when academics and church authorities refused to even look in telescopes that revealed features of the planets, moon, and sun that prevailing theories said should not exist.[43] These people of "vested learning, unimpressed by the new discoveries" judged Galileo and other followers of Copernicus to be "men of no intellect." Those who refused even to peer through Galileo's telescopes did not have a background of thinking telescopes might help to see the obvious. Unfamiliar with a technology that was not part of the common sense anchored in "vested learning," they refused what was simply unfamiliar. As Santillana puts it, "It would be possibly more accurate to say that they were the first bewildered victims of the scientific age. They had come into collision with a force of which they had not the faintest notion."[44]

Very much like sixteenth-century academics and religious leaders, some people refuse to take seriously the possibility not only that other animals might have significant qualities, but also that humans have been plagued by strikingly inadequate ways of thinking and speaking about them. The key is finding ways to increase awareness, and this is most richly done in community. In other words, the very trait (communal opinion) that gives people confidence they can turn a blind eye to those realities is the means by which change becomes possible. Building communal acknowledgment takes time, to be sure. Galileo knew that Copernicus's ideas had been circulating for a half century when he asked people to look into his telescopes to confirm that Copernicus's theory demanded attention. Galileo even wrote, "It is a great sweetness to go wandering and discoursing together amid truths."[45] This could well be the mantra of Animal Studies.

Surely, Heidegger, Lévinas, and Derrida in their work reflect in a variety of ways the importance of humans admitting that we live in a more-than-human world. It is not surprising, then, that many today experience these philosophers as opening doors. One can often see key issues when one looks at their writings not as isolated philosophical ideas but instead as personal expressions of insights lived in reaction to real oppressions. Thus, even when the language these thinkers use is difficult to follow and seems remote from the immediacy of the issue of other animals' realities, readers can often experience something positive within or underneath, as it were, their writing about nonhuman animal issues. Such a positive impression can also be taken from work by scholars within Critical Studies, cultural studies, and critical theory. If one spends, for example, a few hours with a morals-oriented, sensitive, and people-skilled advocate of Critical Studies, cultural studies, or critical theory, one will likely sense that advocates of these approaches can be quite humble about human abilities even as they are at the same time both interested and powerful as they interrogate past and present practices in the finest tradition of philosophical and historical inquiry. This positive impression is connected to such advocates' ability to notice harms to both humans and nonhumans alike, as well as their commitment to getting beyond the justifications for human privilege found within the exceptionalist tradition.

Beyond Jargon to Theory

The perception that technicalities and jargon obscure the issues offers Animal Studies the opportunity to ask whether highly theoretical approaches risk certain forms of human-centeredness. Clearly, the use of a theory that has been structured so its generalized claims track verifiable facts closely can be helpful. What troubles many people is something quite different—some theories about nonhuman animal issues not only seem in no way determined by actual facts, but instead are overdetermined by something unrelated to nonhuman animals. In a very real way, the exceptionalist tradition is such a theory.

One of the twentieth century's most insightful philosophers of biology observed how "our understanding of the world is achieved more effectively by conceptual improvements than by the discovery of new facts, even though the two are not mutually exclusive" (chapter 4). Critical Studies, cultural studies, and critical theory regularly have created perspective at the level of theory in order to attain such "conceptual improvements" regarding the source of the oppressions they challenge.

The sheer number of explanatory theories, of course, makes it clear that some individual theories can—and do—go awry in a variety of ways. Some mislead, others caricature, and others are dominated by evaluations that seem driven by something like fashion, political correctness, or preference for an arid intellectuality. Human-centerednesses are, like the dismissals of the exceptionalist tradition, premised either explicitly or implicitly on theoretical constructs about the humans they favor; what can be problematic, of course, is that they characteristically are completely underdetermined by the verifiable realities of nonhuman animals. Instead, they work with either caricatures or a complete absence of information about most nonhuman animals. There are other ways in which a theory can obscure living beings—it can be based on intentional falsehoods or on ancient misconceptions that have become widely accepted simply because they have persisted over centuries. A theory can fall short simply because it is too one-dimensional, relying for example on one obvious trait but missing far more relevant features. Recall for a moment Bentham's famous comment, "the question is not, Can they reason? nor Can they talk? but, Can they suffer?"—Bentham was addressing those who accepted as valid the historically important generalization (a theory) that other animals do not (really, should not) come within our moral circle because they do not reason or talk. Such claims ignored many features other than various macroanimals' obvious capacity for suffering—they ignored, for example, some macroanimals' apparent intelligence, ability to understand human language, communication other than language, emotional displays, and much more.

In other contexts, some widely accepted and seemingly specific theories used in human situations (for example, legal rights) can in subtle ways impoverish a culture. When outsiders trying to solve a problem in an indigenous society propose a solution that focuses heavily on public remedies of the kind used in industrialized societies ("explicit rights formalized and implemented by the state"), such an approach can in fact impoverish the local people's way of thinking and talking, because the formal, state-based solution eclipses less formal but potentially more effective ways of resolving problems. In other words, when approaches that are familiar and more effective than formalized rights-based approaches are available, outsiders' preferred theory of legal rights sometimes does not resonate with local people.[46]

In light of such problems, one can question whether Critical Studies, cultural studies, and critical theory frame specific issues through lenses (theories) that are not helpful for one reason or another. This is an important question for many reasons, not the least of which is that these agenda-laden approaches themselves foreground the kind of thinking that prompts such critical self-examination.

Wandering Away from Other Animals?

Are some forms of theorizing at risk of refusing to explore or otherwise ignoring other animals' realities? Some theories dismiss other animals because they are controlled by a one-dimensionality in favor of humans. These can be found in certain circles of the law, public policy discussions, the veterinary profession, natural and social sciences, education, and religion.

Other forms of theorizing that are decidedly more animal friendly can, ironically, also be at risk of ignoring other animals' realities. This is because the abstractions of theory can also be pursued in ways that distract people from, rather than immerse them in, other animals' realities. Recall that Critical Studies, cultural studies, and critical theory offer vistas on various oppressions. By no means does this automatically produce concern for other animals, for any of these three approaches can be pursued with no mention at all of the more-than-human world. For example, one might recommend greening human practices not to preserve nonhuman communities but, instead, as resources for the poor or future human generations.[47]

Advocates of animal causes using the ideas and vocabulary of Critical Studies, cultural studies, or critical theory can produce an approach that falls short of concern for actual biological animals. This can happen when an analysis of general problems is expressed in theory, jargonized vocabulary, and intellectual fashion and pyrotechnics that are vague and unconnected to actual nonhuman animals. When theories purport to deal with other animals but in effect do not, they indirectly maintain an element of human-centeredness. Under such theories, humans talk to humans with concepts that intrigue the participants but have no traction either in the real world or for biological animals.

A 2010 academic conference convened to discuss the place of "real animals" in the humanities and critical animal studies was an attempt to address an important tension that arises from the exceptionalist tradition that dominates the educational megafield we name after ourselves—the humanities. The conference was convened with the purpose of exploring what these fields have to say about the real nonhuman animals in our shared world (as opposed to, say, the merely imagined animals that can easily be found throughout the arts).

During the conference, one discussion focused on the classic short story "Red Peter" by Franz Kafka, about an imaginary gorilla captured in Africa and transported to Europe. During the voyage from Africa to Europe, it is discovered that the gorilla has the capacity to learn human speech. Eventually the gorilla becomes a famous speaker, dressing as a human and providing his audiences with extremely interesting observations.

The discussion group tasked to engage this literary work's relevance to real animals was attended by more than a dozen scholars. One participant disclosed that Kafka's background prior to writing the story included exposure to native Africans (humans) on exhibit in a local zoo—interestingly, although these Africans wore clothes in their home habitat, they were instructed to be naked while on exhibit. Similarly, they were provided huts within the zoo exhibit very unlike their actual homes in Africa. Using the oppression-sensitive lens of critical animal studies, a scholar observed poignantly, "This zoo exhibit animalized these Africans." The group proceeded to focus exclusively on this oppression, which was not too surprising since Kafka's story is only nominally about a gorilla. Kafka's interesting fiction is from beginning to end about human issues (such as what it means to be a member of a human community). Actual gorillas are no significant part of the story—what is ironic, though, is that during the hour-long discussion, not one of the critical animal studies scholars suggested anything about the realities of gorillas or any other nonhuman animals.

The story represents how easily those attending Animal Studies gatherings—including those who pursue Critical Studies, cultural studies, and critical theory—can slip into comments exclusively about human issues even when the unequivocally expressed purpose of the gathering is to talk about the real biological individuals outside our own species. The realities of modern scholarship about nonhuman animals are such that a great variety of human-centerednesses continue to dominate, leading humans away from rather than toward nonhuman animals. Many scholars and students in both the humanities and the sciences have been so habituated to the justifications advanced by the exceptionalist tradition that they barely notice the ironies of ignoring nonhuman animals in this way.

To be sure, talking about only humans is, in fact, talking about real animals. So someone might argue that the discussion described above came within the conference theme of real animals. But clearly, the discussion ignored the task of saying something meaningful about the implications of Kafka's story about the nonhuman animals who share our world. Animal Studies does go forward in human-centered ways and needs to always remain cognizant of the human side of the human-nonhuman intersection. But the ironic imbalance in the Kafka discussion shows that those pursuing Animal Studies can wander away from other animals and, instead, pursue one form or another of the exceptionalist tradition.

Animal Studies offers great prospects for discussing the humanities' impacts on non-human animals, and not merely those literary animals meant to teach us lessons about our own species' achievements and foibles. Humanities-based discussions can promote pro-found, even unique explorations of other animals' individual and social realities. Further, Critical Studies, cultural studies, and critical theory have much to offer about the nature of oppressions within and beyond the species line. These fields can fully and insightfully explore oppressions of both human animals and other-than-human animals if they choose.

More General Challenges

While other theories have distorting or agenda-laden approaches, some analysts have stated general challenges to certain kinds of theorizing. In 2003, Terry Eagleton published his widely discussed *After Theory*, which argues that critical theory and cultural studies have failed to deliver on their promise to grapple with fundamental problems.[48] But most chal-lengers recognize that these efforts offer some important insights about how language works and how difficult it is to find an interpretation that is truly neutral or free of a controlling, narrow ideology.

Animal Studies, Realities, and Theory

Animal Studies pushes everyone to examine both (1) each specific theory for its adequacy, and (2) the very notion of theory as a valuable exercise. Theory that prompts detailed exami-nation of grassroots realities clearly provides the benefits of increased awareness of specific facts. It can help many people understand the overall significance and fine-grained details of actual interactions with other living beings, just as it can elucidate specific social movements that address local problems with real animals.

Animal Studies has many features that make it ideal for a range of tasks regarding the-ory-level thinking—it can help people construct the general features of a highly theoretical account about animal-related issues, just as it can prompt people to pursue deep questions about the value of theory making more generally.[49] It can explain why a theory that com-mends activism will seem to have virtues if one agrees that the problem addressed commands the attention of responsible citizens. Equally, Animal Studies can explain why the same theory will seem to others a problem (perhaps there is disagreement with the general prescriptions of the theory, or with the way it characterizes the facts, or the expression of the theory is so jargonized as to create a bad impression).

But since any theory is, in essence, a generalization, Animal Studies must stand ready to assess whether it provides a reasonably complete account of past and present facts. Additionally, Animal Studies needs to make plain the guiding assumptions of any general-ized account that is offered to help people understand the human-nonhuman intersection better. Above all, given how imaginative and limited human knowledge claims are, Animal Studies needs to keep everyone aware that even when theoretical abstractions are reasonably accurate and perhaps even indispensable to a good understanding of a problem, theory never exhausts our understanding of our general situation.

Human abilities and limitations being what they are, it would be tough for any single theory to account well for all past and present realities, let alone all future possibilities. Further, given that human creativity and imagination are so remarkably capacious, generalizing in terms of one single theory very likely puts one at risk of oversimplifying the astonishingly diverse and complex problems that have long characterized this intersection. Individual humans and different cultures over the millennia of human history have constantly suggested theoretical perspectives on this intersection, and will no doubt continue to do so. Animal Studies needs to handle these simple facts well—the open-ended feature of our creativity and imagination in no way invalidates individual attempts to create theory-level understandings of ethical or scientific or sociological issues. But the very volume of theoretical proposals strongly suggests two important conclusions. First, any single theory that does not provide room for other theories may be at risk of becoming just another example of theoretical over-reaching or totalization that curtails in some way a full exploration of human practices and other animals' realities. Second, the sheer volume of competing theories also suggests strongly that multifaceted, interdisciplinary accounts will always be necessary if Animal Studies is to help people see the human-nonhuman intersection well.

A Role for Critical Animal Studies

As Animal Studies attempts to work with many disciplines in the humanities, it will undoubtedly run into additional problems such as the following regarding the role that actual biological individuals of other species might play in different kinds of research:

> Reynard the Fox is therefore a much more frequent subject of scholarly inquiry than are medieval foxes, and research by literary historians is necessarily one of the key pathways that the cultural historian in pursuit of medieval animals must follow. But alongside the abundantly evidenced popularity of animals in elite cultural contexts, the investigation of the role of animals in everyday and popular culture in a broader sense has proved harder to sustain, not least because of the difficulties in finding source material.[50]

Making meaningful statements about "medieval animals" or, more specifically, foxes that medieval people would have encountered, is an obvious challenge. Animal Studies must constantly work with the fact that we are heirs to long-standing fascination with our own literature and its stories (which, of course, contain many extreme and inaccurate caricatures of other animals). Focusing on our own texts, constructions, and interpretations, or even the role of specific nonhuman animals "in everyday and popular culture" is important. Still, such an entry point is surely not free of risk that some who focus on "popular culture" will wander away from the nonhumans they purport to study and, instead, perpetuate some form of human self-centeredness.

Our genius for word-based constructions of reality may cause us to miss some of the simplest facts about human-to-nonhuman encounters. For example, our fascination with written and oral words about such encounters may cause some to ignore features of such encounters that are not word dominated. Arguably, such encounters have several heart-beats, so to speak, beyond the biological ones of the humans and nonhumans involved. Such encounters are impacted greatly by a third and fourth heartbeat—the nonhuman animal's realities, and the human's preconceptions (such as social constructions).

Because sorting out such complexities is an issue for human knowledge quests, just as it is an issue for ethics, encounters with nonhuman animals need to be both touchstone and gauge of Animal Studies. We can easily affirm that our minds do not construct reality itself—in other words, we are justified in being confident that the world exists independently of human awareness (see chapters 8 and 10). Our minds nonetheless actively structure in truly profound ways our perceptions of whatever we discern of the world's multifaceted chaos.

Given careful, humility-inspired input from sciences and other empirical traditions, and given that our ethical and critical thinking abilities can be marshaled as we seek to maximize what we can know of the world, however dimly perceived, Animal Studies has much hard, highly interdisciplinary work ahead. This is the thrust of a caveat offered by the scholar of medieval studies quoted above: "The excessively simplistic comparison of sources that are completely different in nature and [which] emanate from separate discourses has lain at the root of many stereotypical perceptions of animals in the Middle Ages."[51]

Animal Studies has the task of piercing through such "stereotypical perceptions" no matter what their source. It needs to look for insights even in jargonized, vague, and fashion-able expressions, and in discussions that are overwhelmingly human-centered. Its goal must be the presentation of lucid, substantive discourse that can help people from multiple disciplines maximize their own skills and findings. It must also work with theorizing that fails with regard to, as Eagleton suggested, a "rather a large slice of human existence."

An Aside on Cultural Studies and Indigenous Insights

A number of remarkable benefits can be derived from working in the subdivision of cultural studies sometimes called "indigenous studies." This approach focuses on the several hundred million people who live in the thousands of remaining small-scale societies around the world. The term "indigenous" works in a number of different ways depending on the area of study—for example, in the study of cultural and religious traditions it refers to human groups that live in small-scale societies and whose lives feature a heavy emphasis on ancestral memories, local and specialized mythologies, special commitments to their own kinship systems, distinct languages, and deep commitments to place or homelands. When used in this sense, "indigenous" has implications to be distinguished from a literal sense of the term, such as "originating where it is found." Many of the peoples we call indigenous originated where they now live, but many have been displaced from their homelands and yet still keep their cultural traditions alive.

In some small-scale societies, the most basic idea of what it means to be a person "flows from being creatively attuned to the surrounding world of animals."[52] Personhood, then, is not primarily a separateness from others but "a relationality," that is, it comes about by paying attention to and being in relation to the living beings, human and nonhuman alike, in one's local world. Most interesting from the vantage point of Animal Studies is that it is "work toward relatedness, especially in knowing animals" that makes such relation-built personhood an "achievement."[53] What humans can learn is significant if we "join the dance of life in the knowledge that all of us—humans and nonhumans—are bound together by networks that feel more like kinship than we can ever have imagined before we encountered the sacred ecology of our Native American predecessors."[54]

Comparative work of this kind supplies an obvious piece of the answer to What is Animal Studies?—it is the field that engages multiple styles of thinking about other-than-human beings. At one end of the spectrum identified by comparative work are an astonishing number of cultures that view other animals as, in their actual lives, "reactive social others, alternately collaborating in and obstructing the designs" of our own species.[55] Such insights have much in common with certain definitions of ecology, such as "the study of earth's household of life."[56]

At the other end of the spectrum are the mainline institutions of modern industrialized cultures that treat nonhuman animals like machines intended for human use. In the busy middle are some who are deeply troubled by those who conclude that there are no moral problems when humans dominate other animals. Some claim that whatever use they choose to make of nonhuman animals is a natural arrangement intended by a human-favoring deity, while others insist that duties, not privileges and special human rights, follow from humans' greater intelligence, power, and moral capacities.

The variety of cultures, religious traditions, and legal systems on the question of human-nonhuman relationships helps immensely in seeing why Animal Studies is important. Humans have a variety of self-interested reasons for learning about the harms done to nonhuman animals since such learning helps us understand our potential for oppression of any kind of living being, including humans. Additionally the possibilities of self-actualization are, for humans as social and moral animals, integrally related to human individuals' capacity for self-transcendence. Studies of different cultures around the world as well as religious traditions suggest strongly that Frankl's candid claim that "self-actualization is possible only as a side-effect of self-transcendence" applies to humans as a species as fully as it applies to each of us as individuals. In other words, the human species is at its best not when we claim superiority, but when we humbly work at community. The species self-actualizes when it self-transcends, not when it excludes all other living being from moral considerations.

Further, comparative cultural studies foreground more than diversity, integrity, and domination—by opening up the mind and facilitating recognition of the centrality of ethics in human life, they open up the search for truth and raise awareness of our species' history of having broadly interacted with and delighted in other-than-human animals. Such awareness fosters better historical studies, more accurate sociological analyses, more creative artistic activity, and even richer ethical and spiritual breadth and depth.

Above all, such comparative studies confirm the day-to-day realities in our present world. Some people live in societies that openly delight in a mixed community of life; others may live in societies impoverished by industrialization and the exceptionalist tradition, but they can, through study of other cultures, recognize that life is ubiquitous and can be nurtured in a great variety of ways. Comparative studies help us recognize the existentially significant dimensions of personal connection in interactions beyond the species line.

Such work thus helps us see the importance of telling a more-than-human story. It also helps us discover preexisting stories that support the insight that the larger story is our story. Such enlargement of the human spirit helps the entire human community explore the realities of other animals.

Finally, since comparative work on cultures helps everyone see human possibilities, it also makes clear that every generation has a hand, as it were, in choosing the future. That this future is glimpsed not only through science and literature and religion, but also through humans' extraordinary abilities to care and even love can be seen in a 2011 volume titled *Loving Animals: Toward a New Animal Advocacy* by the feminist Kathy Rudy. She suggests the importance of emotions as the basis for focusing on other animals.[57] This claim echoes not only arguments about the place of "love for animals" in veterinary medicine but also the passionate commitments of animal protectionists and the relationality arguments of indigenous peoples described above.[58]

Because comparative work is both implicitly enabling and humbling, it underscores the importance of calling out our limits adequately. The collection of human cultures is impressive in a way that no one culture is, and this collection reveals that humans have regularly recognized that other-than-human individuals and communities are an integral part of the more-than-human world we occupy. Attempts to turn such a rich, diverse world into a human-centered realm are perilous not only for nonhuman animals, but for humans as well, since as noted regularly in this book, human possibility is best achieved through humility rather than arrogance.

Animals and Modern Social Realities

In this chapter we look at ways in which humans' social nature plays a key role in Animal Studies. While the field does not aim to make humans' social nature its primary focus, Animal Studies deepens perspectives on ways in which humans are very animal in our social dimensions, just like orcas, elephants, bonobos, and so on.

Realism about Humans as Social Animals

Some people unreflectively assume that humans not only are, but must be, the most social of animals. Thus, it is not uncommon to read claims like "man is the social animal."[1] If this way of speaking is taken to imply that humans are more social than all other animals, it is misleading. A reflective individual might question such claims, for example, because she senses that our human abilities to verify them are surprisingly limited. Scientists who inquire about nonhuman animals' individual and social realities recognize that confirming the nature and extent of many other animals' mental, cognitive, and emotion-based abilities is a formidable challenge. Since these and other difficult-to-confirm capacities play key roles in other animals' social lives, it is an unavoidable conclusion that humans simply do not know which animals are, in fact, the most social. For this reason alone, then, it is hard to be confident in the judgment that humans are, in fact, the most social of animals.

Further, even casual observations of one's local world—backyards, patches of overgrown or isolated ground, street medians, local woods, and fallow fields—provide abundant bases for wondering whether some other animals are so thoroughly social that they are our equals and perhaps even our superiors in this facet of animal existence. Children often notice that insects are fundamentally social creatures. In both formal and informal education, reflective humans eventually encounter the science-based view that many social insects are genetically inclined to give their individual lives readily for the sake of the community. Some scientists have even suggested that each local society of such insects can be understood as a single organism rather than a family or society of related but distinct individuals as is the case with mammal and bird social groups.[2] Social insects, then, are an example of living beings that might properly be called "the most social animals."

Some other mammals with big brains are extremely social, such as orcas, elephants, gorillas, bonobos, and chimpanzees, although they may not be as social as the schooling dolphins who are described as "utterly alone outside their schools" (chapter 3). Captive gorillas or dolphins, separated from the social group into which they were born, have died from sheer loneliness.[3] Many people wonder if perhaps dogs have, in a minor sense, a more social nature than humans—dogs can become members of human families far more easily than humans become members of dog groups. If one values such social skills highly, one might then assert that dogs, not humans, are better described as "the most social animals."

Nonhuman Societies as Illuminating Our Own

Observations regarding the social dynamics of other living beings for millennia have shed light on humans' fascinating social realities. Today, many researchers and authors encourage humans to explore other animals' lives and then reconsider our complex human social realities in light of what we learn about other animals. The primatologist Frans de Waal provides abundant suggestions about how insights from biology help one recognize features of all primate societies, humans included.[4] Humans feature any number of social skills that appear in many nonprimates as well—for example, our deep-seated biological inclinations to nurture our young are part of a heritage we and other primates share with all other mammals.

Familiarity with contemporary research about the widespread occurrence of social needs and skills evident in different animal communities prompts one to see how work in the natural sciences can illuminate not only our social sciences, but also literature. Like present authors pursuing interdisciplinary approaches, many humans in the past have also attempted to bring multiple perspectives to bear when addressing our connections and similarities to some other animals. Those breakthrough thinkers we call the *philosophes*, who helped nurture the eighteenth-century European movement widely known as the Enlightenment, were enchanted by reports from explorers in Africa and Southeast Asia regarding various apes now known to be humans' close evolutionary cousins. Many centuries before, various pioneer thinkers, including the remarkable early fourth-century-BCE synthesizer Aristotle, wondered about other-than-human animals' actual realities. Throughout many cultures, the overlap of human and nonhuman traits has been easily and often noticed.

Humans as Eminently Social Animals

Even if one forgoes comparing humans to our animal cousins, it is only too clear that humans are astonishingly social creatures. Such realities led Aristotle to comment famously that humans should be defined as "political animals."[5] An upshot of our status as unambiguously social animals is that we must, if we are to understand ourselves well, recognize how our social dimensions impact us. Like some other complex mammals, each human is an extremely distinctive, potentially independent individual who is nonetheless overwhelmingly drawn to group life. How such individuals are raised deeply impacts personality and social skills. Humans are, in political matters, just one of the many primate species who contend with each other in their local groups in altogether Machiavellian ways.

Because these social realities impact greatly our claims to know the more-than-human world, Animal Studies must constantly explore how our sociological dimensions, our social psychologies, and our social constructions of realities (defined below) shape our relationship possibilities with other animals. Cultural factors, for example, can open up or shut down individuals' ability to pay attention to "the larger community" that "constitutes our greater self" (chapter 1).

Ecological Reality and Nested Social Lives

The ecological reality for humans is that each of us lives in a series of nested communities—we are members of our own family community, even as we are members of the local community (village or neighborhood or town or city) in which we live. Many humans also think of themselves as members of a national community, an ethnic community, and a cultural community that crosses political borders. Of great historical and ethical importance has been the emergence of humans' willingness to recognize the whole "human race" as a special community as well. We are also members of the local ecosystem community, although one prominent figure in the history of ecology had to urge humans to again recognize themselves as "plain member and citizen" of "the land-community" rather than "conqueror" (chapter 10). As science, common sense, and Animal Studies show, we are also members of the entire animal community and the Earth community. We are, then, members of a series of nested communities that create opportunities for humans as perceptive, intelligent, moral animals.

The Root Question

This series of social realities presents each of us a range of possible answers to the root question of ethics—who, in fact, are the others about whom I will choose to care? Different cultural and individual answers reveal startling variety in how humans perceive membership in these communities. It is deeply mistaken to claim that our sole community is and can be humans alone. Shared, multispecies communities of life are, plain and simple, ecological givens. Many humans choose lifestyles that foster sharing and thereby create special, more-than-human communities—for example, many households and families include nonhuman companion animals. Other lifestyles, traditions, practices, and even worldviews are premised on radical dismissals of all living beings beyond the species line.

How does any individual human choose among these options? Why in some cultures does one option become pervasive socially and culturally? Public policy is often remarkably human-centered—how does that sort of option, in the face of human membership in so many human and more-than-human communities, come to dominate entire human communities and eras in history? What do the humans making such decisions know about the living beings beyond the species line? The question of what we know about other-than-human animals is intimately related to "the most ancient of questions not only of philosophical inquiry proper, but of human thought as such."[6] Humans have for millennia recognized the tensions between, on the one hand, claims to see reality "as it is" and, on the other hand, social processes impacting our awareness. These tensions are closely related to the most basic questions that philosophers have long been asking, namely, What is real? and How is one to know?

These knowledge-seeking inquiries have often been individual-focused—for example, many first-year philosophy students are confronted with the question of how an individual might answer the question, How do I know what is real? Philosophers have, however, long been aware that an individual's membership in a particular society impacts greatly what that individual claims to know. This important insight has been discussed in countless ways, and the terms "sociology of knowledge" and "social construction of reality" are two contemporary ways of raising a whole battery of fundamental issues about how individual humans claim knowledge about the realities we live amid. Berger and Luckman in 1966 described "social construction" as seeking to understand "the observable differences between societies in terms of what is taken for granted as 'knowledge' in them."[7] The French philosopher Blaise Pascal illustrated the problem wonderfully as he mused about human knowledge and the well-known problem of great variation in claims about truth and justice: "We see neither justice nor injustice which does not change its nature with change in climate. Three degrees of latitude reverse all jurisprudence; a meridian decides the truth....A strange justice that is bounded by a river! Truth on this side of the Pyrenees, error on the other side."[8]

Inquiries into both the sociology of knowledge and the social construction of reality, then, attempt to explain the processes "by which any body of 'knowledge' comes to be socially established and 'reality.'"[9]

Humans' opinions about many subjects, of course, exhibit astonishing diversity—so it is by no means only with regard to the realities and significance of other living beings that tensions exist between claims about reality "as it is" and claims impacted by social processes. But in the area of nonhumans animals, these tensions play out in particularly powerful ways. Sociology of knowledge inquiries, then, provide important insights that can illuminate why researchers find such great variations in the thoughts, attitudes, and dispositions in different groups regarding nonhuman animals.

Also of particular relevance to views about nonhuman animals are attempts within the field of sociology of knowledge to explain the astonishing variety of what societies hold to be everyday knowledge. Some cultures prominently feature ethical views that contrast markedly with modern dismissals of other animals, holding it a matter of common sense that humans are but one animal community among others.

Because so much education in our industrialized societies is dominated by the exceptionalist tradition, some educated people find it controversial to suggest that social factors decisively shape, even control, the content and structure of such claims—some people are so certain that what they know is "the absolute truth" about human identity and human relations to other-than-human animals that they resent any implication that their claims may not be completely true. A respected commentator on biblical religion raises an important issue that goes "to the root" (thus being, in one sense, radical) regarding what at first might seem a simple issue: "What I am thinking of is what the anthropologist Foucault calls 'the politics of truth'—that is, that what each of us perceives and acts upon as true has much to do with our situation, social, political, cultural, religious, or philosophical."[10]

That any human's search for truth might have a "politics" is less than obvious to some—many people unreflectively assume that any human can know the world easily and fully simply by looking at "the facts." All the humans in a room may perceive some simple situations quite accurately—for example, whether there are four, five, or six books on a table. But more complicated fact situations are regularly disputed. Even more controversial are difficult subjects like the meaning of history, the place of ethics in our human lives, the complexities of humans compared to those of other living beings—such issues are decidedly difficult to determine and thus "know" in any simple sense of that important word.

An inquiry into how knowledge might in practice be subject to a "sociology," or into what it means for reality to be socially constructed, will seek out factors that lead to great differences in the way humans assess the world's many complicated features. The spirit of such inquiries seeks to identify the social and psychological sources and consequences of particular knowledge claims. Sociologists, psychologists, anthropologists, philosophers, political scientists, comparative religion scholars, and many others today are deeply interested in how social relationships operate as a foundation for different ways of thinking. What makes their work plausible is that many people for centuries have, like Pascal, observed not only differences of opinion, but that the very structure of a society can impact the categories of thought that prevail in that society. The claim is that social organization can shape not only the content of thought, but also the very structure of knowledge. One implication of this feature of human social life is that different social conditions, which include cultural and political realities, can also hide the truth from members of a society.

"Sociology of knowledge" and "social construction of reality," then, are terms used by reflective human thinkers pursuing questions about the way different societies and cultures foster specific knowledge systems. Such thinkers seek to identify how humans' rich and complicated psychological realities impact belief and even consciousness of the local world. They also explore how our everyday language can impact what people feel they confidently "know."

This introduction makes numerous references to the place of "self-inflicted ignorance" about other-than-human animals. In many cases, even though people talk easily and often about nonhuman animals, what is in fact being passed along may be inaccurate, caricature-like information. Further, authorities such as government officials, teachers, and perhaps religious leaders may speak with just as much certainty as they pass along the same errors. Finally, the education system may even teach students to hold such popular views to be scientific, rational, religious, ethical, and a matter of common sense.

Since each of us receives much information second-hand, its quality is not always easy to assess. Through inquiries about the sociology of knowledge, one can begin to assess the quality of what people claim to know. One can work at looking closely at the quality, depth, and breadth of humans' attachment to many different kinds of claims not based on firsthand knowledge, such as beliefs about other humans or the more-than-human world. One can also inquire about the ways in which group viewpoints impact perception, recollection, claims to certainty, and thus knowledge. One can examine pressures to conform of the kinds discussed below, as well as other factors that shape or blunt diverse beliefs, lifeways, and imagination about future possibilities found among humans in different social and cultural settings.

Sociology of knowledge analyses direct attention, then, to the fact that knowledge claims come through social mechanisms that impart definite features to what is claimed as "knowledge." As different examples of the phenomenon are explored, the idea that knowledge claims have a sociology begins to seem a kind of common sense, especially to people who live in pluralistic societies. One learns from meeting people of different cultures and religions that, in altogether important ways, both subtle and overt, different social conditions cause people to see "the world" and its "facts" differently and thereby impact societies in startlingly powerful ways.

Putting the Tension in Context

There are other reasons some people react skeptically when they first encounter the notions of sociology of knowledge and the social construction of reality. Many people confidently intuit that the physical world and its myriad realities exist completely independent of humans' views of the world—it is common sense to them that whenever we speak of a mountain, a tree in the deepest remove of a forest, or a pod of whales in the mid-ocean away from any landmass, these realities and countless others exist apart from whether humans know about them.

The prevalence and commonsense features of these intuitions, known in philosophy as "realist" or "objectivist" views, cause some to balk at the claim that we socially construct either knowledge or reality. Inquiries about the sociological aspects of knowledge claims and the social dimensions impacting how reality is viewed push everyone to consider recurring, familiar experiences as a reason to be open to the possibility that knowledge has a "politics" or "sociology." For example, a very high percentage of children raised in any single religion end up following, as adults, that same religion and, importantly, holding it to be the single most important orientation to the world. Similarly, most people hold their birth culture's stories to have an obviousness, as it were, and thus to be a valuable guide. This commonsense-like "obviousness" is often anchored by daily language and practices that constantly reaffirm how basic the underlying point of view is for anyone who wishes to understand the "truth."

This underlying point of view, which characteristically features some wisdom about the importance of one or more groups of humans, also carries moral overtones: "one's ethical, as well as one's ontological framework is determined by what entities one is prepared to notice or take seriously."[11] In other words, any young human's familial, educational, and communal environments prepare this impressionable human animal to expect and value certain features in the universe. These features might be brought home most poignantly by the stories of a community of faith, or by examples of personal compassion, or, alternatively, by bigotry, racism, and other exclusions. The child learns through this early training that some features of the world, but not others, have moral importance.

Such views often assume that realist or objectivist intuitions are completely correct, and some even maintain that humans' perception of the world's reality creates a mirror image of reality.[12] Both critical thinking and experience reveal that even if such views have an intuitive appeal, they present obvious problems. Ordinary language, for example, contains many concepts that warn us that what we see and what seems to be the case may not upon further investigation track reality. Some who have described this aspect of the world have gone considerably beyond William James's "buzzing, blooming confusion" to describe "interminable Appearance, mistress of endless disguises,... the kingdom of lies."[13] Every human, thus, is familiar with the risk of our senses deceiving us, and words like "illusion" and "mirage" alert us to the fact that we need flexibility in both negotiating reality and asserting certainty about some states of affairs.

In this way, the very common sense that tempts us to assume without argument that humans readily and easily track reality "as it is" also offers us reasons to pay attention to the findings of social psychologists and sociologists of knowledge. Simply said, our claims about the world are complicated—it is especially with claims about complex, ambiguous phenomena like "nature" or "history" that social construction of both knowledge and reality become major factors.

More relevantly, as this book in many ways reveals, social construction plays a key role in each human group's claims about both nonhuman animals and the more-than-human world. As animals trying to learn about other animals, humans have long noticed that some other animals, such as the other great apes, are very much like us, even as others are radically different. Animal Studies requires us at one and the same time to study other beings that may be like us in a great many respects, in only some respects, or completely different than we are.

Further, the idea of "construction" may confuse some because it is commonly invested with an intentional sense—one usually intends to build whatever is being constructed. Yet the complex processes of sociology of knowledge and social construction of reality include the emergence of views that are not chosen in the normal sense. For example, human individuals find themselves using inherited images, ideas, and languages they clearly did not construct themselves. Each human inherits distinctive and historically conditioned ways of understanding the world by way of the particular family, clan, local authorities, culture, religion, ethnic heritage, language, and national identity into which they happened to be born.

In addition, some humans as creative, meaning-making animals actively work at embellishing or evolving their cultural inheritance. Some even contribute entirely new visions, such as theoretical explanations of the world. In a number of ways, then, individuals can further or even alter sociologically impacted knowledge and even the very processes of social construction. Typically, though, social construction operates at the group level where passive, even unconscious acceptance of the inherited constructions dominates. Below we ponder implications of the widespread failure by so many modern citizens to question culturally transmitted ways of seeing other animals. The upshot is that inherited forms of bias, caricature, and self-inflicted ignorance remain the prevailing ways of "constructing" the world in a human-centered manner.

Raising Awareness, Creating Ferment

It is possible, of course, to raise awareness of the processes by which social constructions operate—historically, raising awareness has become particularly important with regard to social constructions that discriminate, exclude, and harm, as with gender or race bias, religious intolerance, or exclusions based on ethnic identity or social class. Identifying such harmful social constructions is facilitated greatly by comparative work across cultures—this is the thrust of Pascal's musing about the peculiar differences created by merely "three degrees of latitude" or the fact that two neighboring cultures happen to be divided from one another by a river or a mountain range. Above all, comparative work has been particularly helpful in identifying how certain practices thought to be the unalterable order of nature are, in reality, mere custom that has been invested with psychological power by virtue of long-standing tradition.

Those whose learning takes place solely within a single cultural tradition may struggle to see how the images they have inherited implicitly order their universe. Such ordering is often so basic that it is unseen. The result is that inherited views of other animals continue to be taken by many as entirely natural and accurate. Before turning more directly to the impact of sociologies of knowledge and social construction of nonhumans' realities on nonhuman animal issues, it will help to examine pertinent features of two major fields dealing with the extraordinary complexities of humans' social dimension—sociology and social psychology. Both of these fields make it clear why and how Animal Studies has power to illuminate not only inherited social constructions and their power to distort or honor other animals' realities, but also the crucial role of human heritage. It is possible to see each person's heritage as a particularized, often idiosyncratic frame that they themselves can adjust at the level of fine-grained particulars to be consistent with the discernible realities of the nonhuman animals with whom they share a local world. When sociologies of knowledge and social constructions of realities are understood in this way, each individual human is enabled and can choose to honor the fact that each human group is but one among others in the more-than-human world.

Sociology

While the separate discipline known today as sociology took form in the nineteenth century, the approach has roots in the ancient recognition of the central role of social realities in our individual lives. Ancient philosophers long before Aristotle wrote of humans' social natures (for example, the Chinese sage known in Western culture as Confucius). Techniques anticipating the empirical approaches of modern sociology were pioneered in different spheres by creative thinkers, such as the fourteenth-century Islamic scholar Ibn Khaldun, an Arab from North Africa, who was fascinated by both cohesion and conflict in different societies. Thus while the French word *sociologie* was not created until 1780 by a French essayist, the substance of this field was long developing prior to its formalization in the nineteenth century by the French philosopher of science August Comte.

The field, which has developed dramatically, now can illuminate fundamental issues and key concerns of Animal Studies. In particular, sociologists of many different kinds now provide detailed descriptions and explanations of realities and trends that comprise the rich diversity apparent today in the worldwide animal protection and environmental movements.

As both of these social movements address the astonishingly varied set of facts of the human-nonhuman intersection, they continue to evolve and will certainly change further. With some trends already obvious—for example, the increasing number of people interested in protecting nonhuman animals (whether this takes a classic animal protection form, an ecologically framed form, or some other approach altogether)—sociology will surely provide important insights for Animal Studies. In addition, sociology-based efforts stand to benefit as different subfields of Animal Studies illuminate previously ignored dimensions of the human-nonhuman intersection, such as the idea of nested communities. Further, understanding animal protection and environmental protection efforts as they impact and are impacted by socioeconomic situations—such as poverty, which impacts people's ability to protect unnamed "others"—is crucial to assessing what is happening and what might happen in the future with regard to mixed-species communities.

Sociologists can also offer perspectives on and information about ways that humans have been reasserting deeper, background values that traditionally fostered animal protection across societies and cultures. Sociologists also provide detailed studies about varieties of opposition to such protections, resistance by privileged elites who do not wish to give up special advantages, or apathy by general publics. Such research has explored consumerism, interlocking oppressions called "the Link" (see chapter 11), and the treatment of workers in dangerous sites such as slaughterhouses.

The Animal Issue in Contemporary Sociology

Today university-based researchers, scholars, and educators in sociology often discuss the human-nonhuman intersection as part of the modern field's developed body of literature. In 1997, sociologists began efforts to have the American Sociological Association create a formalized group to focus on issues arising at the human-nonhuman intersection. After some resistance from the professional organization, in 2002 these efforts led to official recognition of the Section on Animals and Society.[14]

Ferment on animal issues is particularly well reflected in the self-consciousness that many scholars and theoreticians of sociology display today in their work. An observation made in 2008 by a pioneer of the subfield of sociology and other animals goes to the very heart of what sociologists do:

> While many sociologists bring the power of sociological analysis to a range of social issues and to different forms of oppression, challenging tradition, convention, and existing political-economic arrangements, as a rule most sociologists in the U.S. accept human treatment of other animals as normal and natural. The reluctance of most sociologists to recognize the elite-driven arrangements that oppress other animals and to bring them into scholarly and public focus highlights the question ... "Sociology for Whom?"[15]

The self-consciousness about human-centeredness so evident in this passage today extends to teaching methods, research questions, and findings—this increased openness to sociology-related issues beyond the species line has now produced degree-granting programs at colleges and even some professional schools.[16] Sociology today has moved well beyond its nineteenth-century origins to offer a developed battery of concepts and explanations that can be used creatively in combination with the insights of other fields to reveal unnoticed features of any number of issues that impact humans' interactions with other animals.

Specialized Sociologies

Because sociology contains many approaches and subjects, there exist today any number of impressively developed subfields, such as sociology of religion, sociology of law, sociology of education, sociology of science, and on and on. The relevance of sociology-developed information is immense and will no doubt inform much of the Animal Studies work on each of these topics.

The sociology of religion is itself a sophisticated field that offers detailed information on social dynamics within contemporary religious communities and perspectives on factors that have long shaped religious traditions. To date, this work has been quintessentially human-centered, but it need not be exclusively so as it goes forward. Similarly, the sociology of law has been focused on law in human societies, which has resulted historically in this subfield being completely dominated by the exceptionalist tradition. Given the rapid emergence of animal law, especially the companion animal paradigm, sociology of law today clearly can involve nonhuman animals. Today's sociology-focused circles offer many quantitative and qualitative methods for studying enforcement realities, the emergence of different aspects of animal law, voting patterns, and consumption practices.

Sociological studies of formalized education reveal many battles over what subjects will be included at various levels of formal education. Where students will primarily meet other animals—at home, in the woods, at the zoo, on their plate—is an important question that sociologists help everyone see better. Similarly, sociologically focused studies of science directed at understanding the practical features of modern science face the question of which priorities and other values drive modern research practices regarding nonhuman animals, and whether the prevailing practices call into question core values of the scientific tradition itself.

In many ways, then, sociology and its many subfields have been raising awareness of different dimensions of humans' inevitable interactions with other animals. This work has in turn opened up general study in other disciplines of the communal implications of nested communities that go well beyond the species line. Sociologists of many kinds can, then, offer invaluable insights into our own species' great diversity of attitudes toward nonhuman animals of all kinds even as they develop information regarding both positive and negative interactions with other-than-human animals—in a word, the work of sociologists is essential to developing Animal Studies.

Social Psychology

The respected psychologist Gordon Allport in 1985 defined social psychology as "an attempt to understand and explain how the thought, feeling, and behavior of individuals are influenced by the actual, imagined, or implied presence of others."[17] Given the overwhelmingly human-centered features of industrialized societies in the middle of the twentieth century, the "others" referenced by Allport may well have been solely human individuals. Allport's definition on its face, however, clearly encompasses many practices and attitudes impacting nonhuman others. This field, then, which emerged only in the early twentieth century as an interdisciplinary intersection between psychology and sociology, provides some of the most valuable insights regarding the manner in which humans perceive other animals. Social psychologists have regularly shown that a great many human individuals' views of the world and its specifics are decisively impacted by the views of other people in their community. As noted below, one eminent psychologist suggests that what is at stake, then, is "the nature of truth itself." Even when we, as individuals, are demonstrably wrong, we may nonetheless have, as cultural and traditional animals, incredibly strong inclinations to resist change.

Humans' social dimensions play astonishingly decisive roles in our accounts regarding the human-nonhuman connections and interactions we experience. As importantly, our social dimensions also greatly impact the connection and interactions we might in the future experience. Of great relevance to Animal Studies is the fact that careful examination of these social dimensions within our own species provides valuable perspectives on the quality and nature of the knowledge, facts, and experiences each of us personally claims.[18]

The decisive role of social pressures in the life of the average human as he or she explores, and then makes claims about, the world has been studied in many different ways. Cumulatively, such work calls into question any view that facilely assumes each human is capable of dispassionately evaluating the world and its parts simply by looking earnestly and then speaking plainly about one's conclusions. For example, Philip Zimbardo in his 2007 *The Lucifer Effect: How Good People Turn Evil* reveals that the way each individual thinks and draws conclusions can be greatly influenced, at times even governed, by social pressures. Zimbardo's perspective is grounded in experiments that strongly suggest that "other people's views, when crystallized into a group consensus, can actually affect how we perceive important aspects of the external world, thus calling into question the nature of truth itself."[19]

Zimbardo refers to a range of research, but concentrates most fully on two classic experiments—Solomon Asch's 1950s research revealing how a wide range of normal human beings become subservient to peer pressure and thereby claim something that is obviously incorrect, and Stanley Milgram's subsequent and better-known demonstrations in the 1960s of blind obedience to authority.[20] These experimental results—which suggest that a startlingly high percentage of humans will defer to social influence such as peer pressure and authority and thereby change their own individual judgment—have often been confirmed. In summary, experimental data consistently confirm that many humans, when confronted with a social consensus that conflicts with their basic perceptions of the world, will go along with peers or authorities.

The results of such experiments often astound people. In the experimental data itself, however, is what amounts to a saving grace. Sometimes when peer pressure is brought to bear on human individuals, the simple fact that at least one person dissents plays an important liberating role for many others. Asch's research showed, for example, that the presence of merely one dissenter increased greatly the chance that other individuals would not give into peer pressure but would, instead, voice their own independent judgment.

Such phenomena are of obvious interest in realms like Animal Studies where knowledge, good communication, and critical thinking are prized. Further, these results underscore the importance of creating environments in which questions and, especially, dissent can be expressed. When dissent (expressed explicitly or more implicitly through the power of questions) is part of a social situation, judgment independent of peer pressure is facilitated and even nurtured, thereby prompting educational opportunities of many kinds. A culture of tolerance and creativity can emerge, which in turn nurtures more exploration in more areas.

As a noncanonical academic discipline, of course, Animal Studies benefits greatly when freedom of inquiry is nurtured, for much of the academic world remains squarely within the exceptionalist tradition. The great relevance to Animal Studies of social psychology (as this term is used here), then, is that such studies help each of us recognize that what humans claim to know, be, see, experience, and value can be decisively impacted by group views and even pathologies (these include, of course, cultural and familial heritages). Experiments like those of Asch and Milgram can help explain why whole groups seem to tolerate fictions and deceptions. Such experiments also suggest the importance in debate and education situations of creating realistic opportunities for exploration and dissent through raising new, alternative ways of viewing or explaining a situation. If such creativity is allowed and colleagues or students are encouraged to identify possible problems with any group's prevailing explanations or reasoning patterns, then responses of individuals in the group can avoid automatic, superficial conformity. These are real problems today in the matter of marginalized groups of both human and nonhuman kinds.

Such benefits help explain why Zimbardo suggested that what is at issue is "the nature of truth itself." In Animal Studies pursued in contexts where the exceptionalist tradition is politically dominant, inquiries about other animals can easily become extremely complicated (even politically risky) because of vested interests of a human group not inclined to ask open-minded questions. Science and ethics play critical roles in protecting those who create optimal conditions for fair, integrity-driven research. Education can also play this crucial role because it pushes young children to engage the important human endeavor of searching for the truth with an "interested mind" buttressed by "the disinterested motive."

The Knowledge Issue in Animal Studies

The ancient question of what humans can really know is a multifaceted, recurring issue in Animal Studies for two basic reasons—first there are basic limits in knowing other living beings. Second, the problem of knowledge in general has from time immemorial prompted questions about who humans are, how our minds and hearts work, and how the ecologically interwoven world we find ourselves amid is both part of us and more than human.

As to any single human's possession of actual, abiding knowledge about all other animals, the number and daunting diversity of competing claims to possess such "truth" conspire to suggest that knowledge of this caliber is likely a rare thing for any one human. When one factors in individual humans' psychological complexity in social settings and the fact that today's prevailing ignorance about nonhuman animals has features that are self-inflicted, what then counts as knowledge in Animal Studies will be a particularly complicated issue to address.

Constructing Others

It is particularly our (human) construction of them (nonhuman animals) that reveals how social construction of alien beings can be, for all practical purposes, immune to correction whenever the humans involved do not care to assess the quality of their own claims. Some social constructions of other animals persist over time even though they are wildly inaccurate and even harmful—recall from chapter 3, for example, Linnaeus's anxieties about revealing that he could identify no "generic character...by which to distinguish between Man and Ape." Linnaeus was courageous enough to classify humans together with monkeys and non-human apes in the seminal (for taxonomic classification) tenth edition of his *Systema Naturae* published in 1758. But like so many classification schemes purporting to describe the natural world's continuum of living beings, Linnaeus's framework featured social construction. He elevated some humans (notoriously, Europeans) even as he demeaned others (the mysterious African human groups he knew as "Hottentotti") by placing them below chimpanzees on the "Chain of Being."[21]

Different Kinds of Social Construction

Engaging how different fundamental processes and limits shape humans' convictions and exclusions about nonhuman others fosters recognition that social construction comes in a variety of forms. For example, the "cognitive revolution" mentioned in chapter 3 moved away from a sociology of knowledge that had fostered a decidedly one-dimensional scientific view of nonhuman animals' intelligence and other cognitive abilities. The upshot was that the cognitive revolution fostered much more responsible exploration of other animals' realities. Exploring both sociologies of knowledge and social constructions of reality illuminates equally well the driving spirit of the dismissals of living beings outside the human species line that one finds in the work of intelligent but nonetheless myopic philosophers like Descartes and Kant. Such exploration also helps one recognize that education and theorizing based on self-inflicted ignorance have social construction features as well. In many areas, then, one can see the relevance of the complicated but important idea "that reality is socially constructed and that the sociology of knowledge must analyze the processes in which this occurs."[22]

The prevalence of some humans favoring only a few nonhumans but dismissing most others also provides opportunities to identify contrasting types of social construction. Recall the distortions of nonhuman animals cited by the scholar quoted in chapter 1 who suggested that pet keeping in the Victorian tradition "appears as a phantasmagoria, a fantasy relationship of human and animal most visible in the trope of the animal as child, the pet as a member of the family" in which "the pet...is a de-animalized animal" in ways that foster humans' aggression and domination.[23] This construction of dogs and cats does far more than lead to protection of these animals—it shapes and distorts the realities of the nonhuman animals involved. Such a preoccupation with companion animals has a place in what chapter 4 described as first-wave animal law. The significance of this is easier to appreciate if one takes the dynamics of sociology of knowledge and social construction seriously—some people who think of themselves as the leading edge of animal protection in actual practice favor only some owned dogs and cats but neither feral or unwanted dogs and certainly not wild animals more generally.

Another sort of social construction exists when nonhuman animals hold significance primarily as metaphorical symbols rather than fellow biological beings. Sociology of knowledge helps one recognize that entire eras have been dominated by inquiries into other animals' symbolic importance far more than their actual abilities—such as the emblematic tradition of the Renaissance. Such information in turn prompts one to ask how a group's metaphors, paradigms, classification schemes, and other highly generalized ways of thinking about the world shape, mediate, or create what the authorities within that group present to the society's children as knowledge. It can also remind students of Animal Studies to stay cognizant of the ways that institutions channel our perception, shape our thought, and thereby play a decisive role in what has appealed to each of us as "certainty" of the kind we are confident calling "true knowledge."

The tendency to impose constructed meaning upon certain species of nonhuman animals need not harm the biological beings. In some cases the biological individuals of a nonhuman animal species held to be a culturally significant symbol are not impacted in any way. Indeed, such status may even result in additional protections for members of that species. It is impossible to deny, however, that some animals held to be symbolically important are harmed greatly even though humans believe they are protecting and honoring the biological animals. Recall the story of the elephant Pra Barom Nakkot, a white elephant "honored" with captivity by certain Buddhists (chapter 5).

There are yet other, even more basic forms of social construction that, like the distortions of captivity, impact other animals' actual realities. One of the most basic and distorting of social constructions involving the animal world is the dualism "humans and animals" that is repeatedly described, analyzed, and challenged in this book. The fact that this dualism still prevails in so many scientific circles reveals how a social construction from another era can be persistent—it also reveals that persistence requires the participation of theoreticians and rank-and-file scientists, for if either group opted to use accurate scientific terminology, this antiscientific habit would quickly fade away. There are, of course, other forms of social construction that project distortions and thereby cause individual humans to lack accurate factual information about a group of nonhuman animals—such as distorting social constructions regarding wolves (chapter 2).

Reprise: The Question of Discernible Realities

In many of these cases, we can recognize that our social construction of other animals contends with observable realities. Humans' different forms of intelligence offer our species a rich even if admittedly limited set of abilities to perceive some of other animals' realities. Even when we refuse to follow one form or another of discredited, naive realism that assumes humans know all essential features and realities of other animals, we can still attempt to work out the truth to the extent it is available to us. We can honor the intuitively obvious fact that other animals have their own realities that are in no way dependent on human perception of such realities even as we attend to the complex dynamics of the different sociologies of knowledge and the social constructions of other animals' realities.

Animal Studies must contend with realist points of view in animal-focused discussions and yet also soar to questions about what other animals know and teach us. Realism is the implicit point of view that drives the following passage from C. S. Lewis, but so is a hope that we might attend to other animals well enough to learn something important in our own lives.

> The man who is contented to be only himself, and therefore less a self, is in prison. My own eyes are not enough for me, I will see through those of others. Reality, even seen through the eyes of many, is not enough.... Even the eyes of all humanity are not enough. I regret that the brutes cannot write books. Very gladly would I learn what face things present to a mouse or a bee; more gladly still would I perceive the olfactory world charged with all the information and emotion it carries for a dog.[24]

Such a combination of realism and hope is an unspoken foundation of much attention paid to other animals—for example, book titles reveal that humans regularly hope we can know key dimensions of birds' actual lives:

> *The Way Birds Live*
> *Birds as Individuals*
> *Mind of the Raven: Investigations and Adventures with Wolf-Birds*
> *The Bird Detective: Investigating the Secret Life of Birds*[25]

Frankness about Inevitable, Powerful Social Construction by Humans

It is not only normal but inevitable that we construct meaning as we try to make sense from our important but limited vantage points. Living as an animal in a world full of unknowable realities, random events, and occasional life-destroying chaos elicits sense making and meaning making from us. These require interpretative abilities, for humans are fully capable of recognizing their own and their culture's social constructions. We are also fully capable of recognizing as well that, important though they be, social constructions are not uniquely real—in other words, unlike the world we inhabit, they disappear altogether whenever we abandon them.

Just as we have a responsibility to describe as fully as possible the history we have inherited (chapters 2 and 10), we have a responsibility for our constructions in a world we share with others. We can, if we choose, distinguish distorting constructions from more benign ones. Indeed, both our science and our ethical abilities tell us that our own interpretations of reality, and those we have inherited from family or culture, are not the only ones that count.

Power

We must also recognize that our social constructions carry awesome power of several kinds. They shape reality, especially given that humans are now using their role as the earth's most powerful animals to shape the earth to exclusively human ends. Thus, social construction is much more than thinking that is "independent of the proposed determinative factors."[26]

It has long had power to construct values, ethics, choices, and human imagination about possible futures. Thus, just as social constructions often have reduced women to merely men's property and thereby created myriad injustices and much suffering, so, too, social constructions that demean nonhuman animals foster one problem after another. Social constructions, then, have enormous potential to carry psychological and cultural power for individuals. Such observations also imply that social constructions can carry prospective ethical power whenever we come home to the fact that we are social animals capable of membership in a number of nested communities that go well beyond our families, clans, nations, and species.

Politically, we recognize today that sociologies of knowledge produce forms of power and control over others. Most discussions of sociology of knowledge, however, focus solely on human-on-human, intraspecies implications of knowledge forms, and thereby ignore entirely interspecies implications. The tendency to focus on human-on-human problems has been very productive—it has made clear that knowledge claims have often been used as mechanisms of social control (for example, in some religious traditions, women or outsiders have not been given access to texts claimed to be revealed and essential for understanding reality—the consequence of this is exclusion from leadership). The sociology of knowledge thus often features detailed, conscience-driven discussions about complex intraspecies problems, such as the use of knowledge claims to implement gender or class discrimination or to confer legitimacy or identity on some humans but not others.

But as a source of insights about human knowing, the sociology of knowledge critique has special power to help us see, and then take full responsibility for, views about other-than-human animals. This has the potential to change our character, for "caring both within and across the species line is...the form of self-transcendence that prompts the richest, fullest, most human forms of making community" (chapter 1). In other words, recognizing our memberships in a number of nested communities advances our own moral development and, in turn, "making morals means making community."[27] The good news, then, is that social construction carries as much power to develop an inclusive community as it does to dominate, kill, and extinguish.

Turning to Real Animals

If one watches carefully how a humpback whale mother supports her newborn calf near the surface of the ocean in order to facilitate the calf's breathing, or how the mother is alert to protect her calf from dangers of many other kinds, then one easily notices that these realities have their own integrity (that is, they exist whether we think about them or not). Nonetheless, our willingness to attend to such a relationship at all, rather than our own lives, is impacted by the sociologies of knowledge and social constructions of reality amid which we have been raised. Further, the willingness to take seriously this relationship, even to look at it as possibly within the realm of morality, is similarly impacted by the socially constructed knowledge imbibed along with mother's milk, familial interactions, educational experiences, and earliest developing loyalties to family. As noted above, many different environments prepare each of us when we are young, impressionable primates to expect and value certain features of our surroundings.

Some cultures enable their maturing children to notice how mammalian mothers normally nurture, socialize, and protect their offspring, but some do not. Some encourage sensitive observers to add their own observations and thus deepen their own and others' appreciation of individual differences that each mother has by virtue of her own distinctive personal skills when nurturing her offspring. Some societies encourage their youth to put away any fascination with nonhumans and from that point forward to ignore actual realities (in such cases, individuals may fail to notice what is right before them, or see something but not understand it because the inherited notions simply do not alert anyone to such events). Because social constructions and other expectations can enable or disable, help one perceive or ignore, each person needs to pay attention to the key roles of not only social construction but also unlearning if one is to understand one's heritage fully.

Discovering One's Inheritance

Each individual's personal connection with other living beings is often a powerful motivation for involvement in Animal Studies. Each of us can, much as archaeologists dig into the past layer by layer, probe his or her own history of encounter with living beings beyond our species. Such work uncovers, by definition, personal aspects of the multiple and diverse ways in which each of us has learned about other living beings. Deeper in our personal history, we encounter experiences with animals other than the humans who populated our early life. Some of these experiences will be actual encounters, but most of them involve claims about other-than-human animals by authority figures (such as parents).

This kind of exploration is enabled by question after question—Which individual nonhumans did you meet early in your life? Who taught you about them? Were you taught to observe them carefully? To dismiss them? To eat them? To protect them? What did those teachers really know when they taught you about the living beings outside our species? Were your teachers passing along good information or bad? Were you taught to treat what you were told as absolute truth, or as your teachers' careful guess at the realities of other animals? How often did you learn on your own about other animals? And when you did learn, how was it that you knew how to learn about them?

Such a search reveals layer after layer of experiences upon which each of us has built our present understanding of other-than-human animals. A sensitive exploration will reveal not only one's unique personal history with other living beings, but also a wide range of complexities as well. One will encounter both good and bad modes of learning. Above all, though, one will learn that each of us encounters a truly unique set of experiences when learning about nonhuman animals. This is so because each of us has met different nonhuman animals than anyone else. Take as an example the dogs and cats you have met in your life. Each of us has met, and been affected by, a unique set of real dogs and, similarly, a unique set of real cats. Some of us will have had positive experiences with family member dogs or cats that were very bright and affectionate. Others will have been marked by experiences with vicious or less intelligent animals of these kinds. Each of us has also met a unique subset of wildlife, research animals, and food animals.

There is much to be learned from doing such a personal archaeology—one eventually recognizes that the process of learning about animals is a very unique, richly constructed event for each person. How we perceived living beings early in life, and then later, depends a great deal on what each of us was told to expect, who taught us, and certain predispositions in our personality such as the presence or absence of strong "biophilia," a term used by Erich Fromm in 1973 but made famous by E. O. Wilson.[28] Finally, because conducting a personal archaeology of this kind gives one a sense of one's own past, it also implicitly opens up the future—one recognizes that it is possible as an adult to take full responsibility for one's own views and thereby get beyond reliance on what "authorities" have claimed.

Personal archaeology provides a tailor-made example of how knowledge claims about other-than-human animals are constructed over time, shaped by each person's unique reaction to personal, familial, and cultural experiences. It is often easier to grasp the power of such influences in one's own life by studying a social system with features that are noticeably distinct from those within which one was nurtured. For individuals raised within, say, a Western cultural tradition that gives primary importance to humans, exposure to the Jain tradition provides an opportunity to learn how children in a Jain family and community are from the very beginning nurtured by language, practices, and teachings centered on the importance of the ahimsa tradition (literally, "not harming"—see chapters 11 and 13). The same point can be made by reference to countless other human societies. Studying different social realities, one gets a glimmer of how factors as diverse as diets, prayers, stories, and ethical concerns impact a group's idea of other living beings. Yet it is, of course, a much deeper kind of immersion—namely, living for decades in a specific society—that offers one the best chance of grasping how deeply and personally a particular social situation can produce a very powerful sociology of knowledge.

Animal Studies must constantly engage the fact (not merely the hypothesis) that any culture includes great differences in individuals' preparation to notice other living beings (these will be macroanimals, of course). Some individuals are fully invited by their educators and community leaders to take other animals seriously, while some live amid adults who simply do not care to notice other animals. These differences can cash out as competing claims about whether certain nonhuman animals have certain abilities (say, intelligence or emotions) or for some other reasons are important or not.

Education—which is, of course, a primary means by which any human group prompts its children to use preferred social constructions—can take forms that not only cause people to shy away from seeking the best information but can—and often does—reinforce ignorance about nonhuman animals by influencing what one is willing to consider ethics and morality. Education can thereby move those "educated" away from questions about, encounters with, and compassion for nonhuman animals. Education can foster social conditions that control and produce ethical and scientific ghettos. But education can do the opposite, of course—it can prompt robust attentiveness to lives beyond the species line.

Human societies have often featured ignorance-driven and broad claims about the world that, even generously construed, have nothing at all to do with reality. When one studies our species' abundant racisms and ethnic slurs, sex-based subordinations, multiple claims about miracles or pantheons of locally active divinities and saints, or persistent beliefs like astrology, one might well agree with the philosopher Nietzsche, who once observed, "In individuals, insanity is rare; but in groups, parties, nations and epoch, it is the rule."[29] Such

212 | Animal Studies

society-level problems are legendary, of course. In 1852, Charles Mackay's book *Extraordinary Popular Delusions and the Madness of Crowds* provided one chapter after another of astonishing events that suggest strongly how carried away entire societies can be by fads and collective insanity.

In the face of diverse, recurring phenomena of this kind, the explanatory potential of sociologies of knowledge and social construction of reality is obvious. Given the human-centered history and contemporary emphases of a number of the most affluent societies today, Animal Studies faces major challenges as it tries to call attention to the distortions found in popular caricatures, misleading folk tales, and claims such as "all other animals were made for humans' benefit" in some of humans' most influential cultural and religious traditions. If education, however, permits students to explore Muir's observation (quoted in chapter 6) that he "never yet happened upon a trace of evidence that seemed to show that any one animal was ever made for another as much as it was made for itself," they may recognize that the exceptionalist tradition has shortcomings comparable to those of the obviously false claims that the tides were created to float our boats.[30]

If well-intentioned humans observe a mother humpback whale and her calf, they can record sounds, gather as much data as possible about observed actions, and correlate sounds with those actions. They must work with imaginative constructs of some kind when considering the purpose of communications, the question of intentions, and the nature of these two individuals' connection with each other. Through doing such work, the humans will inevitably encounter these animals as individual presences, as a pair, and as social mammals in an environment very unlike our own. Whether the ideas these investigators generate about the realities of these whales are in any way a full measure of these two nonhumans' lives is a rich and complicated question we only now begin to address in any detail.

Humans as land-based animals need more than big brains and clever technology to maximize our own exploratory abilities—we also need a willingness to explore as fairly and fully as possible both the nonhuman animals we encounter and the role that social constructions play in science and nonscience realms. Our answers will be partial, for we are imperfectly situated to know every relevant fact. Yet even when limited access and our finite abilities force us to construct only partial answers to profound questions about their intelligence, emotions, suffering, and awarenesses, we can still aspire to the goal of dispassionately describing their realities without regard for our own wishful thinking, prejudice, and bias.

Critical Thinking Tasks

The very attempt to identify sociologies of knowledge and social construction of realities offers abundant opportunities for employment of critical thinking skills. The personal archaeology exercise is a kind of critical thinking that recognizes that we are social animals who have been nurtured by meaning-making authorities within our own culture. Critical thinking in the Animal Studies classroom helps evaluate different ways of finding, then integrating, the best of humans' perspectives on other animals. It can help scholars and students combine the power of science's willingness to question everything with the power of ethics-sensitive questions about humans' place in the more-than-human world—each requires critical thinking skills for multiple reasons. First, biological phenomena are exceedingly complex. Second, human limits contend with human arrogance, but we can journey especially far even with these limited abilities because we are capable of communal, humble, and imaginative reflection that nurtures truly open debate. These are the benefits of robust, relentless employment of the widest range of critical thinking skills about all of the earth's animals.

Critical thinking skills also are needed for us to notice how our categories impact our thinking—how we sort living beings into categories is a telling feature about our social species. Categories are, by their very nature, constructed—whether they are honest reflections of what our sciences tell us (such as that we are animals as fully as any other animals), or continue to be human-biased will say much about us. How we deal with our current domination of farmed animals, a stunningly large category obviously socially constructed for our purposes alone, will be as revealing as whether we are able to shape a world in which many different animal communities are healthy and respected as free-living individuals. We can wonder how our human societies will handle the ethical sleights of hand and character by which some still contend that food animals were "made for us" by the universe's creator, and others newly contend that that since "we made them," we therefore "can do what we want with them." If the same reasoning was applied to purpose-bred human animals, it would be condemned immediately. Which nonhuman others will these same voices protect as vigorously?

Questions to Make Science-Literate Moderns Think

While broad, ignorance-driven dismissals of nonhuman animals may seem to create freedom to imagine a future that advantages humans, dismissal of our larger community harms humans and nonhumans alike. What are the practical consequences of such dismissals? What social consequences prevail when whole societies fail to help their children notice other animals? Are the consequences worse when an education system regularly fails to mention them, or when it promotes deficient caricatures? What happens when children are not exposed to people willing to speak about animal-related issues as ethically charged? Sometimes children break through to animal protection issues on their own, but what happens to those who do not? In societies where apathy, ignorance, and caricature prevail, what are their chances of becoming educators, scientists, media reporters, and even secular or religious leaders who trade on the society's indifference and myopia? What happens in law and public policy circles when human-centeredness is seen as the leading form of morality? Who will then develop and pass along what counts as knowledge? Who will raise the kinds of challenges that keep societies thinking humbly and creatively?

If we are willing to question everything, we can wonder about the risk that our children may languish in a sterile, one-species world. We may bequeath them science done poorly and dishonestly and habitats destroyed so mindlessly that they live in a less beautiful, more dangerous and unpredictable world. It is altogether possible that we and our children are taught to see our scientific practices honestly, to wonder if current practices are dominated by human-centered social ethics rather than the basic integrities of the scientific enterprise. We can question, too, whether our education system is more self-affirmation than unbiased pursuit of the truth.

To answer such questions, Animal Studies must nurture historical inquiries, grapple with philosophy of science issues, and delve into disciplines committed to describing the subtleties of social construction in human lives.

The Vantage Point of Time

Just as social construction is particularly easy to see when one looks at different cultures, it is easier to identify when looking across large swaths of time. For example, it is easy to see across the Western intellectual tradition where the dominant views of nonhuman animals changed again and again—ancient views were supplanted in Western Europe during the medieval period by heavily symbolic approaches, which in turn gave way to modern views that were much more representational. Hargrove explains,

> When someone in the Middle Ages looked at an image of...[an] animal (for example,...a lamb), that person normally began mentally spinning through a host of passages in the Bible in search of the most appropriate line or parable. A lily would trigger a thought about the "lilies of the field," the lamb a passage from John, "Behold the lamb of God...." Once the modern period began, however, any person shown similar images automatically thought instead of a real lily or a real lamb.[31]

Hargrove adds that the modern period has been supplanted by what he tentatively calls "postmodern," although he suggests, "we still do not have a very clear picture of the new suppositions that are supposed to" supplant those of the modern period.[32]

The Vantage Point of Religion

Social construction is also easily recognized in different religious traditions—whether non-human animals have a soul, for example, is answered affirmatively by virtually all of the sub-traditions in the Islamic and Hindu traditions, but negatively by many Christian traditions. Similarly, if we use sociological tools to survey adherents of different religious traditions, we find varying views of different animals—dogs, cats, pigs, snakes, and wolves, for example, have each elicited an extraordinary range of positive and negative views.

A consequence of these differences is a bewildering multiplicity of stories—sometimes human origins are linked to this animal or that, while sometimes humans have refused categorically to connect humans in any way to other-than-human animals. Such great variety has produced multiple ways of talking ("discourse traditions"), immense variety in iconographic traditions, a vast array of ethical approaches, and innumerable stories. One role for Animal Studies is sorting out the history and genealogies of these many different views of other-than-human animals.

Returning to a Realistic Underpinning: We Are Animals

Some will view the claim that humans are animals as a form of social construction meant to demean humans. However, it has abundant anchors in hard evidence, such as the almost countless physiological overlaps that sciences have mapped. The best known may be the startlingly great genetic similarities between humans and our closest relatives (the other great apes—different methods for measuring genetic overlap of humans with chimpanzees and bonobos produce figures ranging from 93 percent to as high as 99 percent).

One claim that lacks such hard evidence anchors, and thus is a prime candidate for the category of misleading and dysfunctional social construction, is the dualistic framing that isolates human animals into their own category and then relegates nonhuman animals into another, catchall category. This framing is a particularly exaggerated and dysfunctional social construction, much the same way that calling the earth flat is exaggerated—though the earth is, in fact, flat in places, the claim "the earth is flat" is completely wrong as a generalization because it is underdetermined by all the relevant facts.

Similarly, "humans and animals" is a dysfunctional social construction because it misleads, implicitly supporting the exceptionalist tradition's dismissals of other animals as different in kind. Naming humans as members of the animal world supports the claims that humans are, along with tigers, birds, microorganisms, and so many more living beings, members of a single community. The generalization of all animals as members of some shared category is also a social construction, albeit one that tracks facts widely accepted by not only science and many cultures, but also by many religious traditions, many forms of ethics, and that quotidian common sense that recognizes how fully humans are members of their local ecosystem even when our ideologies try to deny this plain truth. Said another way, pulling humans alone out of the animal category is the sort of social construction that distorts and produces a wide range of social pathologies.

We Are, and Will Remain, Meaning-Making and Meaning-Driven Animals

We are animals who are born into, and live within, envelopes of culturally specific storytelling and meaning making that dramatically impact any search for the truth. As symbolically fluent creatures, we work with inherited visions not only in our arts, but in our sciences, ethics, religions, and even purely secular political visions. Individuals may attempt to vary and embellish any or all of these (for they change even if at very slow rates), or even attempt to create entirely new social constructions based on some combination of fact and our human imaginations.

As social creatures, we live amid an astonishing range of meaning, for we are not only meaning-making animals, but meaning-driven animals as well. A great challenge is to have our chosen meanings be more than mere construction, that is, to have them in some reasonable way engage realities outside ourselves. If we fail to do our best to discern other animals' actual lives, our meanings are mere construction—and "mere" can be a misleading word: "When we blame a man for being 'a mere animal,' we mean not that he displays animal characteristics (we all do) but that he displays these, and only these, on occasions where the specifically human was demanded. (When we call him 'brutal' we usually mean that he commits cruelties impossible to most real brutes; they're not clever enough.)"[33]

Our different skills as imaginative social animals open up individual awareness of how inherited meanings work. We foster creative meaning making whenever we ask questions about any received understanding, and when we test old and new claims against observable facts. Social constructions change because they belong to the human imagination, which means they belong to each new generation.

Reprise: Our Social Abilities to Care

Social ethics is a major issue in Animal Studies. The worldwide animal protection movement is a telling social manifestation of humans' ability to care in collective ways. For many people today, this movement is an important community-making effort. Such communal concern reaching beyond the species line has been at the center of countless human cultures for millennia. The modern animal protection movement, then, is just a recent manifestation of Berry's insight that "the larger community constitutes our greater self." This core ability to care about both human and nonhuman others has been called ethics or morality, or love for one's neighbor, biophilia, justice, right conduct, community building, spirituality, or any of dozens of other names.

Animal Studies asks how humans might answer the question at the heart of our ethical capabilities, Who are the others? Animal protection advocates have had to do a complicated dance of possibility and limit, challenging those who harm nonhuman others even as they recognize and solicit the support of the many modern citizens who have long been doing a great deal with this gift. Any answer to the question, Who are the others? must always be given by an individual in daily life. So even when governments, education systems, religions, legal systems, and other human institutions ask it, the answers come through individuals making day-to-day choices.

Much can be learned from exploring the range of our abilities to care—for example, where on the following continuum of possible others does one's caring stop? How far can one go past oneself as an individual actor? What issues arise when one goes further along the continuum than one's fellow humans?

- Immediate family
- Extended family
- Local community group
- Clan
- Tribe
- Ethnic group
- Nation
- Regional group
- Race

- Species
- Taxonomic genus (*Homo*)
- Group you perceive to have general physical or cognitive similarities to you and your social circle
- Your clade (group of species or taxa sharing features inherited from a common ancestor—humans are, for example, members of the great apes clade)
- Taxonomic family (Hominidae, humans and all other apes)
- Taxonomic order (primates)
- Taxonomic class (Mammalia)
- Taxonomic phylum (Chordata)
- Taxonomic kingdom (Animalia)
- Sentient beings
- All recognizable living beings in one's ecosystem
- All forms and groups of life in one's ecosystem or bioregion
- All forms and groups of life on this planet
- The entire earth
- The solar system
- Our galaxy
- Our group of galaxies
- The entire universe

This continuum contains, of course, some nonbiological candidates. Is it really meaningful to ask if nonbiological entities can be the subject of ethics? For some people, the answer is clearly yes. An anthropologist named Hallowell studied the Ojibwe language. After learning that, grammatically speaking, rocks were animate nouns in this language, Hallowell asked a native speaker if each and every stone was alive. The native speaker, after reflecting on the issue, responded, "No! But some are."[34]

Many people, of course, are raised in cultures that do not view the world in this way. No matter where we are raised, each of us faces a standard dilemma—as we move along the continuum above, it becomes progressively harder to know what can be made of our capacities to care. How do we care, for example, about the last candidates, "our group of galaxies" or "the entire universe"? How is caring about an entire ecosystem in any way similar to caring about a recognizable individual who lives nearby?

Importantly, the ability to care about others is by no means solely a human trait—it is, in fact, a mammalian trait (it seems to appear in some other animals as well, but clearly all mammals have this ability in some form). We as human mammals have a very special set of abilities to care. Perhaps some other animals have similar abilities, but that is not important when we answer the basic question we wake up to each day of our lives, namely, Who are the others whom I, as a human, will choose to care about today? How anyone answers this inevitable question (that is, strikes a balance among self-interest, our inherited cultural views, and additional interests) is truly an individual matter.

Our answer always has a practical side, that is, practical judgments about what we are capable of doing for the others we choose to put inside the protected circle. Some people think it is practical to care about insects, while others don't (in the sense that this level of care is impossible). Some include microorganisms, plants, and other entities even as most of us do not recognize these living beings as others it is practical to protect. Others choose to protect only macroanimals. Many people, of course, assert that it is common sense or even religiously mandated to care about only human beings.

No matter what position one takes, we can care about many familiar, easily recognized others like individual dogs, cats, elephants, dolphins, and various wildlife in and near our homes. In our daily lives, it takes practice, self-awareness, and courage to follow convictions and compassion to live out our choices regarding which beings receive our attention and protection.

Societies often censor individuals who go beyond socially approved forms of ethics. Those who first protested racism in the virulently racist societies of the early nineteenth century were ridiculed, like those who championed the equality of women in the eighteenth and nineteenth centuries, continuing today in societies that subordinate women. Some humans are unable to shed the biases they were taught. Some justify refusals to change based on their belief in divine favor, while others refuse to change as a way of honoring tradition or simply because they believe their superiority is the order of nature.

Expanding the circle of protected beings creates not only opportunities but also risks. History is full of stories about people who challenged traditional privileges and domination and were punished severely, whether the excluded beings were humans or not.

Sociobiology

Sociobiology attempts to provide a comprehensive explanation of all animal behavior from an evolutionary, science-based perspective. Proposed by E. O. Wilson, this theory has drawn much attention.[35] Some have found this theory adequate for describing other-than-human animals, but inadequate to explain certain human behavior. Wilson's suggestion that ethics be "removed temporarily from the hands of the philosophers" because it needs, in effect, to be grounded in biology, has caused an outcry. While critics have focused on the inability of this approach to account for humans' life of the mind and related ethics-based abilities, they have been less apt to see such an approach to nonhuman animals and social systems in a negative light.[36] Hailed by some respected pioneers of scientific research as a "giant stride forward" because it offers an imaginative, science-based general explanation of all social instincts,[37] sociobiology's attempt at comprehensiveness has also been criticized. De Waal, for example, observed that "the sociobiological idiom is almost derisive in its characterization of [nonhuman] animals."[38] Some argue that the theory is "essentially reductive," that is, it errs by reducing complex phenomena to component parts and thereby obscuring much of great importance.[39]

Dissatisfaction with some features of this broad theory does not require one to deny that sociobiological approaches can illuminate some features of the world. The theory goes forward on the assumption that natural selection and competition between animals have worked in powerful ways that help us understand certain features of many animals' lives and social realities. But whether the theory is comprehensive, calling out "an all-powerful, incessant pressure, a quasi-physical force explaining every development" is subject to debate.[40] In effect, critics challenge the theory's explanatory monism—a one-dimensional approach that "behavior is nothing but a product of evolutionary pressures and reproductive strategies."

Relatedly, some view sociobiology as tending to the scientistic, that is, to the view that only conclusions anchored solely in the natural sciences offer a valid, authoritative worldview and form of human knowledge.

These criticisms are significant in light of Animal Studies' organizing commitment to explore as carefully as possible the realities of any and all animals. This commitment requires regular, confident use of different scientific methods, but it also repudiates exclusivist approaches that dismiss other, nonscientific approaches. Many inside and outside science believe it shallow to hold scientistic views, to advance solely reductive explanations, or to insist upon explanatory monism. Arguably, such narrow ways of thinking lack a truly scientific perspective, which uses many different methods.

Thus Animal Studies includes a range of inquiry that goes beyond sociobiology. Further, Animal Studies can offer resources for seeing ways in which sociobiology's reductionism is like, but also different from, other science-based reductions. Ethology tends to focus on individual examples of instinctive behavior but uses language and science-based measurements that pull the life out of, rather than illuminate, other animals' realities. Ethology's reductionism can thus be contrasted with sociobiology's focus on living beings as shaped by their environment or their fellow beings in a rich social milieu. A more aggressive reductionism is Skinnerian behaviorism, which reduces living beings to entities measurable by their parts.[41] In its effort to surmount single-minded reductions or one-dimensional explanations of other animals' realities or of humans' complex, often dynamic and evolving interactions with other living beings, Animal Studies thus offers a wide range of resources to see the complexities beyond (as well as within) the species line.

Reflection on Our Social and Moral Natures

Today we know that what counted in the past as knowledge about human groups as well as nonhuman animals was often biased and misleading. Changes in prevailing views can continue—scientific revolutions have taken place, knowledge of new peoples and nonhumans has increased, new religious messages have come along, some political processes have opened up, and around the world there is today a pronounced ferment, even reexamination, of humans' possibilities with other-than-human animals.

Sociology-related disciplines offer abundant concepts and survey tools by which such issues can be identified, just as they illustrate well how important it is for humans to stay aware of the values driving their claims and inquiry. With such tools, Animal Studies can highlight the need to unlearn preconceptions at critical junctures even though they carry powerful psychological and social force. This is important because the average modern citizen seems less knowledgeable about nonhuman animals than were hunter-gatherer peoples. Modern citizens are, on average, far less attentive to the daily habits of many nonhuman neighbors. What it means to know another living being is a very complex issue and can vary greatly from context to context:

> A knowing of knowing… would mean knowing how an artist thinks, putting a
> thing together; knowing how a scientist thinks, taking a thing apart; knowing how

a practical [human] thinks, sizing up a situation; knowing how a [human] of understanding thinks, grasping the principle of a thing; knowing how a [human] of wisdom thinks, reflecting upon human experience. It could mean being able to think in all these ways...all in one.[42]

Our social natures constantly prompt us to notice what others do and suffer. They prompt us to care about others, thereby creating myriad social interaction possibilities, which may go beyond the species line. For this reason alone, Animal Studies can elucidate contemporary interactions with other-than-human animals and thereby plumb the very meaning of community and society. In this way, Animal Studies creates depth and breadth in human lives.

The Special Roles of Anthropology, Archaeology, and Geography

Citizens of any city may hold their urban surroundings to be the most baffling and complicated of all realities, perhaps more so than the larger, more pristine natural world amid which the city and its suburbs are placed. While urban complexities can surely be challenging, in this chapter we explore whether they are, in fact, less baffling and complex than the natural world's complexity.

Urban noise and passing crowds invade one's space, providing shifting forms of intensity that disturb and disconcert; layer upon layer of personal, social, and political challenge and expectation overwhelm those who want to live simply and freely; many enterprises and institutions supply ever-morphing demands that appear to have no known analogues in the natural world. Perhaps most strikingly, cities are carpentered. Cities are layered with abundant human meanings, of each urbanite's own making or long ago constructed by others. In this last respect, a cityscape is not personal, but rather a world that past humans imagined for today's and tomorrow's inhabitants. Inhabitants and visitors in modern cities can find the urban environment not only alien and cold but teeming in challenges. It is little wonder, then, that our layered, altogether human urban worlds could be judged far more complex than a simpler natural world.

Yet citizens daily see others who appear to share common features—dress, hairstyles, consumer products—and there are long-standing buildings, familiar vehicles, designated paths, and much more that provide a certain uniformity and predictability. Urban complexities, then, exist among powerful stability and comfortable fluencies such as shared language, subcultures, and even political hopes or woes that provide connections with other urban citizens.

The natural, more-than-human world, on the other hand, is in essential ways far more diverse and baffling than any city environment. Citizens may no longer be aware of the natural world's particular complexities, but a little reflection allows them to emerge. Spending a night in the natural world can be a frightening experience for humans used to well-lit, warm places and the easily negotiated right angles and consistent lines and curves of the carpentered world. Further, the cityscape is defined by the exclusion of most forms of life. Thus cities may well be far simpler than the natural, more-than-human world where so much more life abounds and competes.

The notion that the city is supremely complex may merely be a human assumption that, upon reflection, is yet another human-centered construction, an overgeneralization rather than a fine-grained distinction. The simplicity of our social constructions has its benefits, for it allows us to map the world as a human-centered realm. That many social constructions are also often dramatically misleading counts little in the city, for its business is human-centered and thus unfocused on the more-than-human world.

So urban citizens proceed with misplaced confidence if they assume that their single-species world is more complex than the multispecies world beyond the city limits. Urban citizens may assume that the animals they see—humans and their companion dogs, carriage horses, pigeons, and the occasional scurrying squirrel or rat—are all that exist in the urban environment. Humans' cities host, in fact, many types of nonhuman biological lives. In cities' interstices and margins, underground, within the walls of buildings, in the trees and medians of landscaped subworlds are, in fact, myriad creatures of diverse kinds. What the vast majority of human urban dwellers see, however, is much more desertlike—many nonhumans keep a low profile or otherwise avoid passing humans to survive. The result is predictable—an artificial, unrealistic view of who and what lives in the city supports an ironic and continuing failure to notice any lives but humans, their companion animals, and occasionally glimpsed nonhuman inhabitants in a local park.

Many humans prefer unduly simple, even antiscientific social constructions such as "humans and animals." Such simplistic mapping of the living world fosters unreflective acceptance of the constructions of the natural world that foreground humans. This book's distinction between micro and macroanimals plays to our familiarity with dualistic notions. Humans are capable of noticing differences among several thousand species. These macroanimals are but the tiniest fraction of the earth's countless trillions of other living beings. Millions of other species comprise individuals we cannot notice as unique because they are simply indistinguishable from one another or too small to be seen (hence micro as their description). The micro/macro division is a useful construction because it can be used to underscore a practical point about humans' limited abilities to notice the astonishing complexity of the natural world.

Another common but overly simple construction is our sorting of the natural world's complex living beings into the dualism "plants and animals." In reality, the border between these two groups is less determinate than commonly supposed, for there are living beings that have characteristics of each.[1] The division, however, is for all practical purposes correct at the level humans experience the world, which makes this social construction a valuable tool that helps us sort the complex biological world into two basic categories.

The social construction "humans and animals" is, however, qualitatively less benign. It divides the world's macroanimals unrealistically, separating humans for the purpose of elevating them even as it homogenizes all other macro and microanimals. Its prevalence misleads many about humans' connection to other life, and it obscures the important historical and cultural fact that thousands of human cultures have recognized many nonhuman macroanimals as fellow citizens of the world.

City and Civilization as Simpler

Because a cityscape banishes so much nonhuman life, it is a simplified world shorn of many complex features of life on earth. Admittedly, the removal of complexity can be, in the case of the city, replaced by a new set of complexities, namely, those arising from humans' wonderfully rich civilizing abilities. But since these special abilities can surely thrive in communities that are far more mixed than human cityscapes, the city is for this reason alone not the most complex world one can imagine.

Further, given humans' need for engagement with the complexities of the biological world in order to reach their fullest selves, urban environments end up fundamentally simpler than environments with a richer biological mix. The absence of biological complexity could be potentially detrimental to those humans whose cognitive, ethical, and imaginative abilities are developed within such an isolated and impoverishing cocoon.

The so-called urban animals issue, then, prompts many questions: which living beings actually are in the city, what benefits would noticing them potentially provide for humans, and are there ways to live in a multispecies cityscape? These questions provide a bridge from the carpentered and aesthetic precincts of human society to the more-than-human world. As noted below, geographers travel this bridge most often as they address both cities and nonurban issues—they contribute important observations and ideas about the problems of all urban denizens, that is, both humans and their many other-than-human neighbors. This field has a long and distinguished history of engaging past, present, and even future forms of life at human-nonhuman intersections.

The problem of urban animals is particularly important to the work of Animal Studies because it displays so well a recurring pattern—studying animal issues begins in human precincts simply because it is humans who are pursuing the research. But studying other-than-human animals forces us out of comfortable, not-too-complex human circles to the more complex world beyond. Finally, humans who happen to be urban animals have a special range of opportunities, for a discerning eye will note that the city includes many other-than-human lives.

We look first at the classic study of cultures (anthropology), then at a form of history (archaeology) focused on times and places where ancient humans lived in circumstances less removed from the more-than-human world. We then return to geography, looking first at its global understanding of our shared more-than-human world and, second, how geography brings us back across the bridge to humans' urban setting.

This discussion of anthropology, archaeology, and geography will confirm the ubiquity of other living beings in human lives and thought. Additionally, discussion of these three fields models how Animal Studies journeys through the humanities, social sciences, and natural sciences in order to weave together information and insights.

Anthropology

When anthropology was first developing, the more mature field of sociology dominated the study of larger, more urbanized human societies. Early anthropologists thus concentrated on small-scale, indigenous societies rather than their own larger, industrialized societies. Today, however, anthropology has broadened its scope to cover social and cultural behavior in all human societies. For Animal Studies, early anthropologists' work on small-scale societies was serendipitous—through the study of these cultures, many of which were threatened by change, a large number and an extraordinarily wide range of lifeways, worldviews, and notions about humans' relationships with other-than-human animals were memorialized. The upshot is that the field of anthropology offers a deep and wide corpus of materials revealing how richly and diversely human societies have been connected to other-than-human animals.

This body of research has been seminal because of the now-well-recognized problem of biases and preconceptions distorting early anthropologists' framing of indigenous peoples' thinking about other-than-human animals—recall, for example, the dismissive comments quoted in chapter 2 regarding views of animals found in "the lower stages of culture." Ironically, today's Animal Studies turns the table on such dismissals, pointing out that it was European thinkers who failed when they belittled the views of "savages" who expressed "reverence...towards the animal creation" and openness to other animals having "a vastly more complex set of thoughts and feelings, and a much greater range of knowledge and power" than European philosophers, explorers, scientists, writers, and educators had assumed.

What now is easily understood as arrogance was a cultural problem as much as an individual one—educational and cultural institutions failed to supply early anthropologists with forms of critical reflection by which they might examine their own assumptions. These shortcomings not only caused the European founders of modern anthropology to miss central features of other cultures but also indirectly supported European colonial powers intent on dominating many non-European cultures as well.

Humility and Modesty: The Question of Culture among Nonhuman Groups

Anthropologists, historians of this field, and other scholars now help everyone recognize the importance of approaching other cultures with more openness to variety. Our growing ability to identify more and more about our own species' "earliest recorded history," for example, commends a certain humility about assuming that today's prevailing views will never change. Collectively humans continue to learn about and refine the story we know as human history.

Defining culture in exclusively human terms has long prevailed in the Euro-American tradition. For example, "culture" is generally thought of as the way of life of a community of people (humans alone), and includes all of a particular human group's learned behavior patterns, attitudes, and material goods, such as tools. But refusals to even consider cultural possibilities beyond the species line are unscientific and myopic because critical thinking readily prompts wide-ranging inquiries about whether the notion of culture might apply to some nonhuman groups. This is a key issue in Animal Studies for multiple reasons. First, the question follows from the definition of culture used above because notions such as "a group's learned behavior patterns" and "tools" on their face apply to some nonhuman groups. Second, findings about the complexities of social life among a number of other animals also beg the question of whether any animals other than humans exhibit culture or other developed traditions. Third, Animal Studies needs to assess this question on its own merits, not ignore it simply because the very notion of culture among other animals has often been ridiculed and even deemed politically incorrect in the academic world.

One highly respected field scientist suggested in 1992 after reviewing the evidence regarding chimpanzee tool use, "If the same findings cited…came from a range of human societies across Africa, we would not hesitate to call the differences cultural." Addressing the issue of culture among some other animals, the researcher adds, "if the contents of this chapter were reported unchanged except for a single independent variable—species—then the answer to the question…would be taken for granted as positive."[2] Such findings help explain why, for several decades now, detailed discussion about culture as a feature of a number of nonhuman societies has been common. While the best-known examples come from primatology, this term has been used for some cetaceans, elephants, and other animals as well.

Whether people will even ask about culture in some nonhuman group is a test of critical thinking skills that seek to make sense of, or problematize, definitions of culture in the Western cultural and scientific tradition. One set of scholars noted in 1994, however, "Virtually every definition of culture in the social sciences premises human uniqueness. Even a book entitled *The Evolution of Culture* claims that 'man and culture originated simultaneously; this by definition,' thus barring any thought of continuity with other species."[3]

The contributors to *Chimpanzee Cultures* have more than expertise in primatology; they also feature a deep commitment to describing other animals' realities honestly and dispassionately in the tradition of good, careful, rigorous science. That book and many since reveal that many scientists now regularly use the prized word "culture" for all chimpanzee groups.

Refusal to use the notion of culture regarding *any* nonhuman communities arguably violates critical thinking canons that require an examination of all available evidence. It also supports the exceptionalist tradition's perpetuation of ignorance about nonhuman animals. Beyond their compromising of scientific values, such refusals distort humans' ethical abilities and support traditional legal concepts that refuse to consider any nonhuman lives as candidates for fundamental legal protections.

A similar reluctance prevails with regard to admitting that other traits long claimed as unique human features might appear in some nonhuman animal individuals. There has long been, for example, great reluctance to even consider the claim that various nonhuman animals have communication abilities that feature some of the complexities found in human language. Refusals to examine this possibility or others (such as the existence of emotions) must be subjected to critical thinking for the reasons suggested by an English philosopher:

> We remain doubtful that animals could be said to have a language. In part, this doubt is a mere device of philosophy: it is not that we have discovered them to lack a language but rather that we define, and redefine, what Language is by discovering what beasts do not have. If they should turn out to have the very thing we have hitherto supposed language to be, we will simply conclude that language is something else again.[4]

We must also examine how simple assumptions may be grounded in language—consider, for example, the common habit of generalizing about a particular nonhuman species with terms like "the horse" or "the wolf." One researcher described how claims about chimpanzees, like claims about human uniqueness, must be scrutinized carefully: "There is no such creature as 'The Chimpanzee.'... [There is] enough variance in the data to make any attempt to generalise about the whole species a nonsense."[5]

Animal Studies is congenial to such observations not only about culture but also about continuing strategies to maintain human uniqueness or to deny the uniqueness of individuals in other species. Above all, it is a deep commitment to know the realities of other animals that prompts many today to inquire about cultures beyond the species line.

Anthropology offers immense bodies of information about hundreds of other cultures that have paid close attention to a great variety of other-than-human animals and have viewed some nonhuman macroanimals as members of their larger community. Anthropology also prompts thinking about the insights and techniques needed when attempting to fathom unfamiliar cultures of other living beings, be they human or other-than-human.

An Anthropology-Inspired Comparison

Ruth Benedict's 1946 *The Chrysanthemum and the Sword: Patterns of Japanese Culture* was not only a best-selling book, but also an anthropological work that provided insights about how an outsider might describe a community (Benedict wrote the book during World War II and from the United States because it was not feasible for an American to visit wartime Japan).[6] Benedict's task was to help Americans and others understand as much as possible about a society widely thought to be radically different from the Euro-American tradition. Benedict, who displayed sympathy for some of the marginalized people in Japanese society, concluded that the cultural phenomena she witnessed were both complex and hard to reduce in simple ways. Although her observations, methods, and conclusions were criticized, her book skillfully describes some general features of the anthropologist's task—after observing, "The job requires a certain tough-mindedness and a certain generosity," she famously added, "The tough-minded are content that differences should exist. They respect differences. Their goal is a world made safe for differences."[7]

Because Benedict's phrase "a world made safe for differences" makes it clear that a human born in one culture can, even from afar, develop general interest in a completely alien social community, it is possible to draw a limited analogy to an important task in Animal Studies—humans also have the capacity to nurture interest in, and appreciation for, unfamiliar contexts like entire ecosystems, other animals as a local population, or even a group of nonhumans as an integrated community with its own special features. The analogy is limited because the problems of trying to understand, say, a troupe of bonobos or an elephant matriarch's family-based group are not the same as those of trying to understand another human culture. But there are overlaps—all of the groups are communicative, intelligent, highly social mammals that form communities that work together as they face the challenges of surviving in a world full of powerful competitors.

The overlap is, in fact, interesting enough to prompt one to ask preliminarily, Do these social situations have similarities? If we notice similarities, we can then ask, Can we compare these societies in ways that illuminate each? If we then conclude that there are nonhuman animals whose lives deserve protection, who carries the obligation to create "a world made safe for [these] differences"? Benedict's insight calls for us to extend caring enough to notice and respect differences. The target of caring for Benedict was another human group, of course, but further along that same trajectory (which in Benedict's case involved describing respectfully a foreign culture at war with her own) is caring across the species line—such reasoning could apply, of course, to any living beings and communities outside our own.

Such expansions of care clearly are happening widely now. But an especially important anthropology-based lesson is this—if you take other peoples seriously, you will also notice that some care across the species line in a variety of ways. Such caring will be culturally constructed, mediated, and expressed, and one of the most important future tasks of anthropology is to develop even greater sophistication at seeing such differences and working with scholars and students in Animal Studies and other disciplines to imagine what it means to have "a world made safe for differences" across the species line.

Interdisciplinary Anthropology

Anthropology is often said to be more than merely a battery of inquiries about cultures—which is the primary aim of the subdivision called cultural anthropology, ethnography, or social anthropology. Physical anthropology pursues an understanding of the genesis of and variation among human beings, and the subdivisions of archaeology, psychological anthropology, and linguistics provide additional breadth to this multifaceted discipline.

From its very beginnings, the field has been deeply immersed in comparative work—for example, early figures in anthropology worked on comparative law and eventually focused on languages and linguistics. The central role of religion in different cultures prompted early anthropologists to think about this complex phenomenon, thereby spurring comparative religion and the anthropology of religion. As one observer suggests, "Anthropology as a field has long been open to interdisciplinary work and, indeed, may even be conceptualized as a kind of transdiscipline."[8]

Theoretical Turns

Multifaceted challenges await anthropological researchers, who characteristically have been born in literate cultures and who may not surmount the biased images and other social constructions they learn through education. Although a key purpose of the field is to be descriptive, another is to provide enough information for comparative work.

Through decades of dialogue, the field's practitioners have identified the biases that dominated much of early anthropology. Such reflective work has prompted contemporary anthropologists to increase their sensitivity to the problem of overinterpreting or importing alien notions when describing a specific culture. As the following list of generalized, theoretical frameworks reveals, there is no single approach that everyone agrees upon: "Anthropologists have also, over time, variously emphasized evolutionist, rationalist, functionalist, social structural, structuralist, symbolic, interpretive, political, Marxist, social constructionist, phenomenological, psychoanalytic, poststructuralist, cognitive, aesthetic, and ethical approaches or modes of understanding."[9]

Some skepticism about the value of theory has arisen, as a reaction to the sheer volume of competing theories, or to various theorists claiming that only their own theory is needed to explain the field's data. Also, from an early point in the field's history, skepticism about religious claims was a distinct feature of anthropological scholarship. "Anthropology became what it is with, and in a definite sense because of, the great historical dissent from religion. Its debt to skepticism is profound."[10]

This skepticism was limited, for the founders of anthropology were less willing to challenge their own claims and generalizations, which often reflected personal and cultural biases. Despite these early problems, today's cultural anthropologists recognize that thinking about human cultures is inherently complex and challenging: "Reflection on the partiality of past interpretations demands reflection on the partiality of the present."[11]

Critical Turns

Today anthropology is recognized as a lively tradition of intellectual inquiry with a great capacity for discussion of possible connections within and across cultures. The vibrant quality of current discussion is a function not of any theoretical consensus, but of its willingness to embrace diversity as we think about our vast cultural abilities. Cultural anthropology today includes approaches that can be not only comparative, but universalistic, holistic, dialogical and sensitive to context, critical, historically sophisticated, and respectful of the views of the people in the subject cultures.

"Universalistic" approaches arise because contemporary cultural anthropology takes "the whole range of human societies, past and present, as its subject matter and attempts not to privilege the western tradition or literate societies."[12] This has had important practical effects, for it "has meant it was the only discipline to take seriously the existence of small-scale societies without traditions of literacy."[13]

Cultural anthropology's commitment to context-sensitive approaches brings perhaps the greatest challenges, for it is difficult for cultural outsiders to be precise about the many nuances apparent to those born into and living wholly within a particular culture or subculture. It is, in a way, the very diversity of human cultures that prompts anthropologists' commitment to understanding phenomena and "facts" in terms of the culture in which they are found. An additional, obvious challenge for anthropological interpretations is to call out linguistic nuances accurately, for without attention to this particularly fertile dimension of a social group's life outsiders have almost no chance to appreciate cultural, social, religious, and political significance. One of the methods used to achieve such a lofty goal is being "dialogical," that is, using methods that are "rooted in conversation with, and especially listening to, those whose practices, knowledge, and experience we attempt to understand."[14] Yet again, this is no simple task, for how does an outsider determine which facts are to be emphasized? Further, how is one to understand the significance of diversity? Of course, it is helpful to be in dialogue about such issues with the people of the culture being studied.

Modern anthropology conversations also fit several of the definitions of "critical" (chapter 2):

> the anthropological conversation can be characterized as critical, meaning by this both "critical" in the sense of literary critical and "critical" in the political sense of concern with power and its subterfuges and abuses. The best anthropology is also self-critical;

here our concern with overcoming the various and multiple forms of ethnocentrism and intellectual narrowness remains a characteristic feature of any contribution that wishes seriously to be taken as anthropological.[15]

A Model for Animal Studies

Because of its universalistic, comparative, holistic, context-sensitive, dialogical, and critical thinking commitments, cultural anthropology models important virtues that must play out if Animal Studies is to meet its promise. Just as anthropology seeks to be universalistic by taking the entire range of "human societies, past and present" as its subject matter, Animal Studies aspires to focus on the widest possible range of living beings. Thereby, the umbrella of Animal Studies can include the great variety of human efforts to understand individual lives, notice nonhuman communities, and factor the astounding diversity of life into attempts to make meaningful, scientifically accurate observations about the human-nonhuman intersection.

Animal Studies is, like anthropology, necessarily comparative. Further, comparative work in Animal Studies is among the most difficult of human endeavors—it must approach a range of differences that is qualitatively more complex than that which anthropologists study. For this reason, Animal Studies will undoubtedly continue to feature an astonishing range of generalizations and theories.

Animal Studies obviously also has an investment in developing context-sensitive approaches. Just as individual sciences, the field of ethics, and so many other searches for truth require a deep commitment to understanding "facts" in full context, Animal Studies necessarily foregrounds as a core commitment the investigation and fair reporting of other animals' realities and capabilities to the maximum of human ability.

The dialogical features of cultural anthropology have a parallel in Animal Studies—namely, listening to and observing as carefully as possible the lives of other-than-human animals. Human ingenuity is stretched to its fullest capacity as humans attempt to listen and otherwise attend as carefully as possible to other animals' "practices, knowledge, and experience." Animal Studies often aspires to operate at the very edge of human capabilities.

Lessons in Relativism

Animal Studies benefits from anthropology's well of experience in dealing with the key problem of relativism, which is the question of how to handle the common assertion that, because it is so difficult to say definitively which one of many competing points of view is the truth, all points of view therefore have equal validity. While important insights can be gained from contrasting "objectivism" with "relativism," many advocates of relativism oversimplify statements of the underlying problem and thereby produce debates that mislead and lack realism.

Anthropology may seem at first to be on the side of relativism, for anthropologists shed light on how individual cultural viewpoints deserve respect. The claims and beliefs that prevail in a particular culture are characteristically taken by its members to be a definitive, realistic account of the world. Because of the investment that members of a culture have in their own worldview, anthropologists acknowledge the value and power of any individual culture's worldview. Anthropologists at the same time celebrate the number and variety of such worldviews. As one scholar suggests about religious views, "anthropology must understand them as so many means for acting, asking, shaping, and thinking, rather than as a set of fixed answers whose validity either can be independently assessed (objectivism) or must be accepted as such (relativism)."[16]

With such perspectives, anthropologists have helped comparativists in other fields work with the claim that each worldview must be respected on its own terms. In the study of religion, for example, comparativists work with multiple religious points of view as they ask how each different claim reflects an underlying vision or worldview that is generated, structured, and legitimated in unique ways. With such tools, comparativists can also help explain how and why members of a particular culture hold certain features of the world to be fully "natural" rather than humanly (socially) constructed and thus hold their own culture's claims to be true rather than merely mythological or convenient and self-affirming.

These lessons in how to deal with differences supply important insights about crucial distinctions regarding types of relativism. One version of relativism is sometimes called skeptical, cynical, or complete relativism. This is the belief that what a person thinks is right for herself constitutes the sole criterion of such views. Some even hold that such personal views should be the only measure when judging human morality.

Most people sense the risks of taking seriously the claim that each and every opinion on good and evil is just as valid, or invalid, as any other. If it truly does not matter which views people hold, then outsiders have no way of condemning murders, injustice, and other harms that have become traditional practices within a particular culture. Such claims trouble many people because they believe that "there are rules of moral reasoning that are inviolable, however free spirited one may be."[17]

Cynical, skeptical, complete relativism is plagued by more than these doubts—there are important logical problems with such a full, uncompromising form of relativism. Those who advance cynicism of this kind, for example, have no obvious answer to the following line of argument—if all views are equally valid, then nonrelativist views are just as valid as cynical relativism. Relativists of the cynical stripe advance their position as if it was the only obvious choice, but their own claims imply that their position is only as good as its opposite, namely, denial of their relativist view. This logical conundrum—advancing a position that implies that it is not any better than positions which cancel it out—bothers many people.

Nonhuman animals do not fare well under cynical relativism—as one philosopher observes, "doing as you please" with other living beings, which cynical relativism permits because it provides no way to challenge any practice, is in fact a "moral position."[18] However, cynical relativism effectively implies that "doing as you please" is immune to challenge of any kind. This, too, bothers many people (for the harms such an approach would permit to humans as much as nonhumans).

Another form of relativism, sometimes called principled relativism, is anchored in the practicalities, humilities, and open-minded commitments that abound in critical thinking circles. This kind of relativism, which has important applications in Animal Studies, has both historical and practical roots related to the problems faced by pluralistic societies. Principled relativism suggests that in some very basic human situations, it is reasonable and practical, and therefore realistic and common, to judge definitively that some views are better than others. Only some actions fit into this category—such as murder of innocent people, taking sexual advantage of innocent children, advancing Nazi-like privileges for some while advocating ethnic or other irrational exclusions, and perhaps a few other assorted other harms. In most cases, though, individuals cannot be so confident that the vast majority of fellow humans would agree with the judgment that a problem exists—these are the cases in which facts are complicated, too vague, or suffer from some other feature that makes judgment difficult. In these more common and complicated cases, most humans recognize the value of choosing what amounts to a principled relativism.

Many examples could be offered—such as a group that agrees to tolerate religious diversity or political pluralism. In such cases, while people might be confident that they are entitled to hold their own views and even opine that others should follow their example (and therefore judge the matter in the same way), most of us are not confident that we can definitively say that others must hold our interpretations of complicated issues, like which religion to join or which political candidate to favor. As noted below, views of other living beings fit into this larger category—in other words, judging whether macroanimals or microanimals are worthy of protection is a complicated matter subject to important but difficult factual inquiries.

The point of principled relativism is simple—human life, and certainly the big issues like love and loyalty to a group, are extremely complicated. Such complications are best worked out by each individual in light of experience and belief. Thus, as a matter of principle and as a way to allow people to live together peacefully, people in groups often agree that there are competing options among which each person should be able to choose. The appeal of this principle is simple enough—in some situations, the underlying judgment is complicated, and judgments are best made by isolated individuals precisely because no one is competent to claim that everyone else must judge in the same way. The major issues like murder and harms to innocent beings do provide exceptions, of course, but these are rare enough. For the bulk of life's decisions, each person, for practical, personal, and political reasons, needs to judge each issue's complexities and then choose a personal response.

The need for a principled relativism became especially obvious during the religious wars that dominated sixteenth-century European history—as one analyst suggests, "Cultural relativism is in part a fresh expression of the peace formula which followed the end of religious conflict in Europe."[19] An important operative assumption of principled relativism is that when people disagree, "it is both possible and necessary to keep talking."[20] This practice helps avoids problem-fostering fundamentalism, that is, insistence that one individual "has all the answers and therefore [has the right] to drown everything anyone else has to say," which is the "position of every form of fundamentalism, religious or political."[21]

Protracted disagreements also have made clear the problem of cynical relativism because that position "expresses not a generous tolerance but a weary or cynical giving up on truth."[22] One thinker who concluded that a "new ethic, embracing…animals as well as people, is needed," helps us understand why he drew this conclusion: "We have to see that moral relativism represents not a value position but an abdication from holding a value position."[23]

In the twentieth century, many people began "holding a value position" as they challenged existing power structures and cultural customs supporting exclusions based on race, ethnic origin, gender, sexual preference, age, and even consumer and environmental practices. Interfaith dialogue again began to flourish—the upshot was that awareness grew regarding the peace-generating prospects of principled, generous relativism that made political and cultural space for different points of view. It is in relation to this background that one needs to see the ferment and calls for a "new ethic, embracing…animals as well as people."

Clearly, the emergence of principled relativism has not eclipsed cynical relativism—in the twentieth century, for example, such skepticism continued to thrive along with nationalism, racism, and greed in high-profile circles. Consumerism also prevailed as relatively affluent segments of human society became preoccupied with money and privilege. Consumerism that leaves behind segments of the human population (and, of course, impacts the more-than-human world so adversely) is connected to the larger phenomenon of "economism," which some analysts suggested had replaced nationalism as the principal means of structuring public life. A by-product of this development was that not only public policies but also human interactions more generally were measured in economic terms. Economic growth for its own sake emerged as the organizing principle of many industrialized societies.[24]

Through such changes, nonhuman animals were reduced to commodities as industrialized or factory farming emerged and was advanced technologically by "animal science" (chapter 3). Awareness of the harms that such developments annually cause to tens of billions of nonhuman animals has, as noted previously, been one of the driving factors in the emergence of animal protection and its natural ally, environmental concern.

These increasingly popular movements can be seen as holding a value position, that is, as repudiating cynical approaches. Both animal and environmental protection stood on the shoulders of other social movements and benefited from their predecessors' critique of greed, exclusion, and relativism.

Animal Studies is, in this regard, clearly the beneficiary of other liberation movements—feminist and race-related critiques, for example, provided key insights into the ways privileged groups that benefit from an oppression defend the status quo as "tradition," "the natural order," "God's design," or "business as usual." Further, because critiques developed in other social movements model a willingness to look at issues from new vantage points, they foster habits of mind that open up even further issues, all of which is conducive to Animal Studies.

Robust, Cross-Species Multiculturalisms

Given the great failures within the human community in the narrowest of multicultural challenges (that is, between one human group and another), it is not surprising that understanding the more complex multicultural challenges between humans and the nonhuman animal communities whose social realities include culture-like features has lagged in human-centered circles. However, some human cultures have featured a fundamental openness to this possibility. Recall that some human cultures have long featured "integral traditions of human intimacy with the earth, with the entire range of natural phenomena, and with the many living beings which constitute the life community" (chapter 5). In fact, many human cultures have found both community and "souls" in other animals. Consider, for example, what an outsider could learn by visiting certain American Indian ceremonies where the Dakota word *omatakwiase* is chanted. Literally, this word means something like the English phrase "to all my relations," although this translation lacks the resonance for English speakers that *omatakwiase* has for native Dakota speakers—perhaps the closest analog in contemporary English is religious use of the word "amen." One scholar describes how *omatakwiase* functions in context: "Traditionally, chanters offer prayers to the heavenly grandfather. At the end of each invocation, the speaker chants: *Omatakwiase*...The Sioux say that this word is not to be spoken lightly. Intuitively, the sound symbol seems to focus its speaker back into the earth and to all the creatures."[25]

Through focusing "its speaker back into the earth and to all the creatures," *omatakwiase* reveals that an inherited language tradition can open up people within a culture to connection and relation. For those living in industrialized, large-scale societies dominated by the dualistic and exceptionalist thinking and speaking so characteristic of law, education, and "caring," the challenges of understanding lifeways, virtues, symbols, and narratives found in societies who see other-than-human animals as "relations" are daunting. These societies are, to be sure, no less representative of human possibility than today's most populous societies. Such small-scale, eminently human societies are, however, still being destroyed by insensitivities, greed, and other harms that originate in the industrialized world. Notice how the fate of both nonhuman and human animals is connected, as both are impacted by the exceptionalist tradition.

Animal Studies needs a general awareness of such trends within modern industrialized societies, just as it needs frankness about how these trends help or harm small-scale human societies and other-than-human communities. Through its commitments to address such matters, Animal Studies is thrust into the midst of the complex problems and debates that anthropologists and others address.

Animal Studies can, in turn, model for anthropology a style of responsible, full inquiry into perspectives on cross-species cultural diversity, as well as frank recognition of existing biases and inevitable limitations that may influence any researcher's work. In a very real way, comparing human culture with nonhuman cultures might be seen as a principled multispecies multiculturalism. Most debates about multiculturalism concern, not surprisingly, only human cultures. These debates sometimes touch on harms to nonhuman animals—such debates are not in principle too different from debates about harms to humans. Some people prefer a form of multiculturalism that disfavors any criticism of a culture's practice by outsiders (that is, members of another culture). The implications of deferring in this manner, of course, include tolerating practices now widely held to be offensive, such as murder, virulent racism leading to genocide, or sexism that leads people to kill young girls because they would rather have male offspring.

Other forms of multiculturalism allow outsiders to criticize some cultural practices—this position has its risks, too, such as critics imposing their own values on the culture in question. Despite such risks, in the international community today there is a large majority willing to walk in this direction on some issues, such as senseless murder of innocent people and child rape. A smaller majority is also willing to condemn certain traditional cultural practices that obviously produce irrevocable harms to innocent human beings, such as Indian suttee (a practice in which a widow immolates herself on her husband's funeral pyre), Chinese foot binding, or genital mutilation.[26]

However, the human species has reached no consensus as to the principles, values, or justifications that allow outsiders to criticize cultural practices that harm nonhuman animals. There are almost countless real-world situations where traditional cultural practices of one kind or another produce serious harms to nonhuman animals. Whenever outsiders value the nonhuman animals that a long-standing cultural tradition is harming (as many people throughout the world value dogs harmed in certain places, or as many people in India value cows killed in so many other cultures), challenges to such harms raise thorny issues. In the early twenty-first century, the realities are that human-on-nonhuman oppressions of almost any kind are tolerated under the notion of multiculturalism.

If one takes the idea of nonhuman cultures seriously, a qualitatively different set of issues can also be called multiculturalism—this is the principled multispecies multiculturalism referred to above. These ideas are anchored in the notion that humans might choose to refrain from harming nonhumans because the humans recognize nonhuman "cultures" as valuable. This more encompassing form of multiculturalism will, through the lens of the exceptionalist tradition, seem far more radical than the important but human-focused notions generally thought of as multiculturalism. Some might even be tempted to suggest that merely speaking of the possibility of a multispecies multiculturalism trivializes human differences, but this ignores the views of those many human cultures that would have found such a robust, multispecies multiculturalism to be reasonable.

Questions of such breadth have a power that prompts one to identify both the complex questions of human-centered multiculturalism and the conceptually humbling notion of a form of multiculturalism that reaches beyond the species line. Animal Studies has significant tools by which one can take into account a wide range of views on the busy continuum stretching from individuals to small-scale cultures to industrialized societies and their transnational corporations. This continuum offers a wide variety of claims about what counts as a virtue when engaging other-than-human individuals and communities, and how one determines ethics in such cases. This continuum also features a bewildering array of symbols, narratives, and beliefs, all of which reveal both humans' fecund creativity and extensive abilities. Animal Studies in the future will be further challenged to develop tools that allow scholars and students to get beyond the mild form of human-centeredness that is implicit in recognizing only human cultures. But this is precisely the sort of challenge that prompts a field to self-reflection and growth.

As both anthropology and Animal Studies proceed with their cultural investigations, they can make clear why careful, open-minded work prompts again and again the observation, "what a thing is the interested mind with the disinterested motive."

Physical Anthropology's Contributions to "the Animal Question"

Physical anthropology has played an important role regarding one relationship that humans have with other animals, namely ancestry. The search for human origins has led many researchers to examine tools used by our nonhuman relatives, and while the motivation for such research has often been a preoccupation with discovering why and how humans are unique and superior, the results have confirmed that our closest relatives feature significant intelligence, culture-like transmission of learning and tool making, social skills, and other fascinating complexities.

It can be argued that cultural anthropology has played an even more important role for Animal Studies because of its ethnographic accounts. Such accounts of our own species' stunning cultural diversity are a window on the abilities of one kind of animal to create astonishing variety. They also underscore the great variety of human groups' impressions of the inevitable human-nonhuman intersection. Ethnographic accounts make only too clear how many human cultures have framed this intersection through themes of kinship, common origin, or, as the Islamic tradition suggests, neighboring communities.

Through such valuable information and broad perspectives, anthropology has shown how metanarratives can enable or mislead. "I aim to show that the story we tell in the West about the human exploitation and eventual domestication of animals is part of a more encompassing story about how humans have risen above, and have sought to bring under control, a world of nature that includes their own animality."[27] This proposed generalization about why Western culture is invested in control themes goes beyond nonhuman animals, of course, and helps one see features of human lives as well. In the following is an explanation of why some humans demean other humans whose relationship with nonhuman animals is disdained: "In this story a special place is created for that category of human beings who have yet to achieve such emancipation from the natural world: known in the past as wild men or savages, they are now more politely designated as hunters and gatherers."[28]

Tim Ingold's account links stereotypical portrayals of certain humans and the subordination of nonhumans. This generalization illuminates some of the reasons why dualisms like "humans and animals" continue to prevail even in scientific circles that in so many other ways openly advance Darwin's insights regarding the fundamental continuity between humans and other species. As will be apparent to anyone who studies small-scale societies or the Indian or Chinese civilizations, this continuity was acknowledged in many other cultures that accepted implicitly and explicitly that humans are but one kind of animal related to other animals in a shared world.

Beyond illuminating why scientific circles might foreground (even require) antiscientific ways of talking, anthropological accounts of this kind model how a discipline may become more self-conscious and thoughtful about its traditional ways of thinking and talking. These are key developments in any form of Animal Studies because of the background human-centeredness that dominates educational circles. Further, anthropology can display very fully the roles that critical, reflexive thinking plays in any educational or research enterprise that deals with narratives and social constructions of the human-nonhuman intersection.

Domestication Seen Anew

Ingold speaks of an "alternative account of the transformation in human-animal relations that in western discourse comes under the rubric of 'domestication.'"[29] In nineteenth-century portrayals from European-based cultural traditions that contrast hunter-gatherers with those humans who created agriculture-based ways of life, the former are often depicted as "living like animals." Such portrayals, suggests Ingold, "carry force only in the context of a belief that the proper destiny of human beings is to overcome the condition of animality to which the life of all other creatures is confined."[30]

As this view suggests, domestication is a form of control premised on, and practiced through, domination and coercion. These realities undergird countless modern practices—including not only food production in both its intensive and less aggressive forms, but also private ownership of nonhuman animals and the zoos and circuses that so obviously pull these other-than-human living beings out of their natural contexts. Animal Studies also points out that domination and coercion are often unacknowledged elements of the companion animal paradigm.

The view that "the proper destiny of human beings is to overcome the condition of animality" also helps explain why habitat destruction is regularly framed as "progress." In fact, identification of these "overcome animality" and "humanity versus nature" metanarratives helps one see more clearly what anchors modern scientists' justifications (chapter 3) when they opt to use living beings as experimental tools rather than the various alternatives that are available today. All of these views repudiating our own animality, the benefits of being separate from the rest of nature, and our right to use other animals because they are "subhuman" are mainstays of the exceptionalist tradition.

Because some have argued that hunter-gatherers saw themselves as conservers or custodians of their environments, Ingold distinguishes the Western idea of conservation from the attitudes of hunter-gatherers: "[For us, conservation views are] rooted in the assumption that humans—as controllers of the natural world—bear full responsibility for the survival or extinction of wildlife species. For hunter-gatherers this responsibility is inverted. In the last resort, it is those powers that animate the environment that are responsible for the survival or extinction of humans."[31]

Many cultures have held that "the proper role of humankind is to serve a dominant nature."[32] Such views have been used to suggest that "rather than saying that hunter-gatherers exploit their environment, it might be better to say that they aim to keep up a dialogue with it."[33] Such a connection to one's local world requires not detachment, but involvement, because it implies that other animals "are not regarded as strange alien beings from another world, but as participants in the same world to which people also belong."[34] Ingold's description uses "trust" and other positive, value-laden words to convey hunter-gatherer societies' overall attitude, which "presupposes an active, prior engagement with the agencies of the environment on which we depend."[35] This anthropologist also suggests that "instead of attempting to control nature," hunter-gatherers take a responsibility-based approach as they instead "concentrate on controlling their relationship with it."[36]

Small-scale societies have often developed the view that their world was populated by "other-than-human persons."[37] But "it is quite otherwise with pastoralists" who exhibit care for the nonhumans they have domesticated, but "care of a quite different kind than that extended by hunters. For one thing, the animals are presumed to lack the capacity to reciprocate."[38] Hunting societies characteristically perceived nonhuman animals to have "full control over their own destiny": "Under pastoralism, that control has been relinquished to humans.

It is the herdsman who takes life-or-death decisions concerning what are now 'his' animals, and who controls every aspect of their welfare, acting as he does as protector, guardian and executioner. He sacrifices them; they do not sacrifice themselves to him."[39]

In summary, our human forebears, through the long process we now generalize under the term "domestication," changed their view of those living beings which they domesticated, from "trust" to "domination."[40] The result is a human-centered viewpoint that is not unlike the exceptionalist tradition—nonhuman animals under domestication become "like dependents in the household of a patriarch" and "their status is that of jural minors, subject to the authority of their human master."[41]

Culture as Home

Each human who identifies with a culture will grasp the implications of the claim that one's culture is like one's home, even one's family. Through a variety of means, birth culture shapes each person long before he or she becomes self-aware. As we learn to speak with our birth culture's established language choices and as we are taught by parents and other authorities in many other ways, we learn to live and honor our society's implicit and explicit ethics. We also hear, internalize, and repeat stories and narratives, and thereby fathom our world in terms of the social constructions that structure the perception of those who teach us.

This socially mediated learning has, to be sure, deep personal meaning. Those who grow up in a single intact culture (that is, without major disruptions imposed by outsiders) know that birth culture or subculture provides deeply formative experiences on virtually every major issue of human life. When humans who have grown up in such circumstances walk into nearby woods, local neighborhoods, or other human communities, they carry within them a cultural understanding of those surroundings' features and possibilities. Such culture-infused understandings do not necessarily bind us irrevocably, but they clearly impact what we initially expect and, often, what we ultimately believe we have experienced—and can experience—in our surroundings. Such socially and psychologically significant expectations are integral parts of what it means to be a cultural animal.

The concept of culture as home is relevant to other intelligent, social animals as well, because their awareness of the surrounding world's features is profoundly shaped by their mothers and others. Thus, in Animal Studies as it is part of humans' attempt to understand other living beings (and, of course, themselves as animals, too), paying attention to the micro and macro details of the ways culture and family shape living beings is crucial.

There are enabling and limiting factors other than culture—we know from our own personal experiences as embodied creatures, as well as from our inherited wisdom traditions, that our limited sense abilities shape what we can know. A good example of a technical term that calls out such constraints is *Umwelt*—this German word is commonly translated as "environment" or "surrounding world," but it was originally coined in the early twentieth century by Jakob von Uexküll, a seminal figure in the fields of theoretical biology and biosemiotics, to mean "appearance world."[42] Sometimes *Umwelt* is translated as "subjective universe" or "self-centered world."[43] Sometimes it is translated with longer phrases such as the "biological foundations that lie at the very epicenter of the study of both communication and signification in the human [and nonhuman] animal."[44]

Such natural constraints anchored in our own biological foundations, or imposed by some feature of the surrounding environment, are often obvious, unavoidable factors that influence any individual's understanding of the world. Thereby, they impact greatly what one can know about other living beings. Animal Studies takes as a key task identifying these and other limits encountered as we try to learn about other animals (chapter 1).

Failure to acknowledge that humans inherit social constructions can produce serious problems with our already limited abilities to understand other animals' realities. Even our most advanced research scientists glimpse but a few things about what can be considered the internal realities of other animals' awareness. They know even less, sometimes nothing at all, about the social and cultural realities of large-brained social mammals, such as cetaceans, in whom we might expect to find assorted traditions and complexities, some of which might even be understood as "cultural" in nature. Because we know that human animals are decisively shaped by their birth cultures, we can wonder if the cognitive abilities, communications, and personalities of some other animals are molded by their birth group or local community.

Anthropology and the Emergence of Animal Studies

Even though anthropology continues to be dominated by the view that human-to-human relations are far more important than human-to-nonhuman relations,[45] it has nonetheless made exceptional contributions to all humans' abilities to see the human-nonhuman intersection as it has played out in diverse cultures. In this alone, anthropology has been instrumental in the emergence of Animal Studies. Further, anthropology exemplifies how a discipline can become progressively more self-conscious and thoughtful about its own history of thinking and talking about complex subject matter. Thus anthropology has helped Animal Studies by making particularly clear the importance of critical, reflexive thinking in any educational or research enterprise.

Archaeology

The prevailing way of defining archaeological science is human-centered, as when the field has been called an "anthropology of extinct peoples" because it "concerns [humans] in the past."[46] But there are fuller definitions. A 2007 encyclopedia of nonhuman animal issues seamlessly adds nonhuman animals to this field: "Archaeology is the scientific study of physical or cultural remains left behind by peoples or animals who lived in the past (remote or recent)."[47] What makes this fuller definition more appropriate is that even though it continues the tradition of describing the science of anthropology by means of a nonscientific use of the word "animal," it confirms the prominent place of other animals in the lives of the peoples that one explores in archaeology. The human-nonhuman intersection is, not surprisingly, reflected often in ancient digs.

As with anthropology, our ability to identify earlier and earlier periods in human history continues to advance regularly, such that our ability to tell our own story (see chapter 10) is constantly enriched. Because it digs down deeper and deeper into the remote past, archaeological research, for obvious reasons, often must deal in approximations and surmise. Even with such limitations, though, two issues on the question of nonhuman animals are clear. First, many peoples were, in remote times, very riveted by certain nonhuman animals. Second, modern attitudes are far more dismissive of nonhuman animals' abilities and significance.

This movement away from considering nonhuman animals may seem the normal trajectory to many readers—indeed, a metanarrative of separation drives many mainline segments of Western culture today and the exceptionalist tradition more generally. There have been times in Western history, however, when "progress" away from humans' connections with their larger community was not the main storyline or narrative. Instead, a steady, long-term decline was assumed to be the human species' lot. For example, a classical Greek metanarrative was that in earliest times humans were a golden race that then declined—as Hesiod says in *Works and Days*, "First of all the deathless gods who dwell on Olympus made a golden race of mortal men who lived in the time of Cronos when he was reigning in heaven."[48] After the passing of this original race, "then they who dwell on Olympus made a second generation which was of silver and less noble by far. It was like the golden race neither in body nor in spirit." Finally, after the passing of the second, less spectacular silver generation, "Zeus the Father made a third generation of mortal men, a brazen race,... and it was in no way equal to the silver age, but was terrible and strong."

While today the preferred narrative in industrial societies reverses this story of decline (we commonly speak not only of progress, but also of evolution as producing higher stages), the progressive removal of humans from the natural world brings a number of different complex challenges and negative results. Scholars of comparative religion have long pointed out that ancient peoples often saw other animals as bringers of blessings or even divinities.[49] Many of the best-known religious traditions featured in their early periods views of other animals as deserving respect and compassion. It is interesting, for example, to compare the abundant references to other animals as nations in North American Indian materials with the Islamic claim in Qur'an 6:38 that all other animals have their "own communities." This special feature of ancient peoples' view of nonhuman animals by no means prevented them from hunting other animals, although it clearly entailed a fundamental respect for them. Strangely, though, "progress" has not made the average modern person more knowledgeable about the nonhuman animals that live in one's "neighborhood," however broadly or narrowly one construes that term.

Archaeology can offer revelations about the connections between ancient humans and the nonhuman animals in their local environment. The capacity to notice such issues, however, developed late in archaeology because, like so many of our sciences born in recent centuries, this field developed amid debilitating frailties in thought and action. For example, like many other sciences, archaeology struggled with sexism—it was "not until affirmative action policies were implemented in the late 1960s [that] colleges and universities [began] to hire women archaeologists in visible numbers."[50]

Like other sciences, then, archaeology passed through eras where dismissive and inaccurate preconceptions dominated. Similarly, many early archaeologists were not alert to the importance of even noticing, let alone taking seriously, nonhuman animal issues revealed in the sites they examined. Today, however, archaeology makes major contributions to understanding how peoples in ancient cultures understood other-than-human animals. Archaeology led the way in the mid-1980s with special meetings and publications on animal issues at world archaeology conferences, especially the One World Archaeology series beginning in 1986.[51] These publications show how critical thinking opens up tradition-bound fields.

Researchers today commonly cull information about humans' relationships with and possible views of other animals from evidence previously examined only for its relevance to exclusively human issues. Even human-centered archaeology can engage nonhuman animal issues, because one of the aims of archaeology is "construction of developmental sequences and the explication of the outlines of culture history"; another is the "discovery of the functioning of cultural systems at single points in time."[52] Both aims are enhanced when one ponders evidence in its full context—what is found at what levels, what is found together, what the sequences are, whether they have been disturbed, and much more—for such evidence is needed to develop "the value of complete contextual inference."[53]

Given the different kinds of ubiquity of nonhuman life in human affairs, it is obvious that past humans not only lived in the proximity of other animals, but also surely held views of at least some of them. The ability to speculate about such views and interactions requires, of course, that researchers have some concern for the relevance of the human-nonhuman intersection to the archaeological issues they pursue. Inquiries along such lines, however, occurred irregularly in many late nineteenth- and early twentieth-century archaeological digs, and research findings were not particularly attentive to the kinds of issues that are pursued today. However, it is possible today to use powerful techniques to examine the available evidence for previously ignored connections. Conclusions may be rudimentary and limited, but they can open up powerful questions that stimulate further research on human-nonhuman intersections. Evidence of the types and concentration of bones, teeth marks on such bones, plant pollen, burial patterns, and cooking remnants can suggest not only the presence of other animals, but also consumption and other relationships (even codomestication possibilities), all of which may reveal glimmers (or more) of how an ancient human group may have interacted with and understood other animals of various kinds.[54]

A particularly well-documented connection between humans and other animals comes from archaeological research on ancient Chinese communities.[55] Archaeological work on the capital of the Shang dynasty that ruled roughly 1550–1050 BCE, for example, reveals vast quantities of bones used in the ancient form of divination known as scapulimancy, in which the shoulder blade of different nonhuman animals, such as sheep, cattle, boars, horses, or deer, is burned in a fire until cracks appear. It was a cultural presupposition that the diviners could then use these cracks to foretell the future. Oracle bones, such as tortoise shells, are also commonly found in certain archaeological digs.

Findings from much older archaeological sites in the ancient Indus civilization settlements known as Mohenjo Daro and Harrapa include animal-based art in the form of small terra-cotta seals (only an inch or two in length and width, these possibly were used by merchants to stamp their goods). These seals are "mostly realistic pictures of animals" denoting that they were "apparently worshipped as sacred."[56] Very different, far more recent animal images in the shape of massive earth mounds in central North America fascinated late nineteenth-century archaeologists.[57]

Archaeologists have long recognized that people have "been burying or otherwise ritually disposing of dead dogs for a long time. They sometimes treat other animals in such a fashion, but not nearly as often as dogs. This presentation documents the consistent and worldwide distribution of this practice over about the past 12,000–14,000 years."[58]

Morey suggests that this phenomenon reflects "how people often have responded to the deaths of individual dogs much as they usually respond to the death of a family member." Similar evidence of the burial of three dogs in separate graves in the central United States "hints that an affectionate relationship between humans and dogs may have existed over 8,000 years ago in the North American Midwest."[59]

More recent evidence from Egypt reveals both an astonishing number of cat mummies buried together and certain ironies in the modern era.

> *National Geographic* reported in its November 2009 issue that after a mass grave of mummified cats was discovered in 1888, the volume of bodies was so great that they were sold for fertilizer, with one ship alone hauling more than 180,000 mummies to be spread on the fields of England. This story contains some ironies—it begins with cats so valuable at one period that they were mummified, and then ends with these cats' bodies, centuries later, being used as fertilizer in another country known for its animal sensibilities. In this way, the story helps one see that one era's valuation of certain animals can be quite different from that found in another era and another place.[60]

Archaeology, then, can bring contemporary humans face-to-face with ancient humans' connection to the two nonhuman species—dogs and cats—that are again very prominent in the concerns of citizens in modern industrialized societies. Such evidence, especially when put alongside the surprisingly realistic wildlife-focused cave art of Paleolithic peoples, invites humans to contemplate the wide span of time across which humans have been fascinated with some of the macroanimals outside our own species.

Geography: The Return Bridge to Urban Animals

An ancient and astonishingly wide-ranging pursuit, geography is sometimes thought of as merely the study of places and regions. But the modern academic field of geography, which was formalized as a discipline in the eighteenth and nineteenth centuries, is thoroughly committed to interdisciplinary approaches that have moved the field within only a few hundred years into a premier place in the academy. The field has been called "the world discipline" and "the bridge between the human and the physical sciences." Limits that characterized the field in its early years have been surmounted, in part because of new information about unfamiliar animals in the lands that Europeans were exploring and conquering from the sixteenth century onward. Recent publications reveal the significant number of humans living in these lands before they were "discovered."[61]

In the twentieth century, the field achieved a higher profile with a continued stream of books popular with the educated public such as the Pulitzer prize–winning *Guns, Germs and Steel*.[62] Yi-Fu Tuan developed the special subfield known as humanist geography. This approach, which focuses on the ways humans interact with space and place (meaning their social and physical environments), includes discussions of art, philosophy, religion, psychology, and more. This subfield also highlights the way human perception is impacted by creativity, personal beliefs, and experiences that, in turn, shape attitudes regarding local place and the larger environment.

Modern geographers have also focused heavily on altered landscapes. In his 1983 *Changes in the Land: Indians, Colonists, and the Ecology of New England*, geographer William Cronon describes Henry David Thoreau's reactions in the first half of the nineteenth century to the altered landscape in the vicinity of Concord, Massachusetts.[63] Cronon cites the observations of many nongeographers whose descriptions helped him piece together as full a picture as possible of the prechange landscape. Such creative, interdisciplinary work is aided greatly by advances in certain technical sciences that help geographers and others provide a more accurate and detailed story of humans' impact on their surroundings. The late twentieth-century emergence of pollen counting and other intensive techniques provide data that help reconstruct the composition of ancient landscapes.

Such blending of multiple disciplines and skills models yet again the special role of communal, interdisciplinary work as consortia of geographers, historians, ecologists, and so many others try to piece together past and present puzzles. A particularly complex example is the attempt to tell the history of landscapes that have changed again and again. As David Blackbourn's 2006 *The Conquest of Nature: Water, Landscape, and the Making of Modern Germany* reveals, the history of intentional landscape changes and unintended consequences along rivers running through long-developed parts of heavily industrialized countries is particularly complex.[64] Attitudes toward such profound changes are, understandably, diverse. At the time when many changes were proposed and accomplished, those with the power to control the local society's decision clearly favored the change. Cronon mentions a 1683 report by a historian who counted major changes in New England by which "remote, rocky, barren, bushy, wild-woody wilderness" had within a generation been made "a second England for fertileness."[65]

The centuries-long process of change has prompted many to ask poignant questions that often take human-centered forms, paying little or no attention to nonhuman animals. As readers may already surmise, nonhuman communities have been far more fully impacted by such changes than have local humans. Apart from the well-known, often-condemned extinction of many species in North America, "the entire body corporate of animate creation" suffered "horrendous diminishment" after Europeans came to the continent.[66]

Such work has made geographers principal contributors to humans' resurgent interest in other-than-human living beings and thereby greatly enabled Animal Studies. Geographers today often mention other animals because their discipline has long nurtured recognition that the earth is a more-than-human world. Biodiversity, animal communities, and the actual realities of other animals have long been implicit concerns in geography even when geography, like other social sciences, was captive to some form of human-centeredness.

It is not surprising, then, that geographers regularly include discussion of nonhuman animal-related issues. Cronon's 1983 book mentions an astonishing range of nonhuman animals—alewives, bass, bears, beavers, cardinals, cattle, clams, cod, cormorants, crows, deer, dogs, ducks, eagles, eels, elk, foxes, flounder, goats, geese, grouse, hawks, herring, horses, jays, martens, mice, minks, moose, muskrats, mourning doves, oysters, otters, owls, passenger pigeons, pigs, porcupines, porpoises, quail, rabbits, raccoons, rats, salmon, seals, shad, sheep, smelt, squirrels, sturgeon, swans, trout, turkeys, walruses, whales, wildcats, and wolves. Recall that the Bible mentions 113 different kinds of animal by one count.[67] Similarly impressive lists of different kinds of nonhuman animals mentioned by other authors who touch on geography-intensive issues could be compiled easily.[68]

Fascination with natural places and wild, free-living animals is a theme raised by geographer John Wright in a wide-ranging essay titled "Notes on Early American Geopiety."[69] Wright points to many religious sources that recognize that the human spirit is fascinated by the more-than-human natural world. This has much in common with the insights of the Axial Age sages (chapter 2).

Urban Animals

In an often unrecognized way, urban nonhumans also provide connections to other-than-human dimensions of life on earth. Recall how the development of civilization's roots within ancient Greek city walls left its "mark deep in the minds of men" by setting up a principle of "divide and rule" (chapter 3). Even though some of humans' most powerful twenty-first-century cultures bear the deep, dividing mark of city walls, the presence of nonhumans in urban environments provides lessons that can help those who passed through city-based education see how one-dimensional their understanding of life on earth is. They can then, *if they choose*, unlearn any sociology of knowledge and social constructions that honor only human-centered realities.

One way to prompt such unlearning is to recognize that even within city boundaries an astonishing array of other-than-human lives exist, well beyond invisible microanimals. As pointed out by geographer Jennifer Wolch, urban dwellers' neighbors include many macroanimals that are not humans. Urbanization in Western culture was historically tied to "a notion of progress rooted in the conquest and exploitation of nature by culture. The moral compass of city builders pointed toward the virtues of reason, progress and profit, leaving wild lands and wild things—as well as people deemed to be wild or 'savage'—beyond the scope of their reckoning."[70]

Morality in such environments would seem to be human-centered, but Wolch makes clear that it is business- and consumer-centered. "Today, the logic of capitalist urbanization still proceeds without regard to nonhuman animal life, except as cash-on-the-hoof headed for slaughter on the 'disassembly' line or commodities used to further the cycle of accumulation." A footnote adds, "Such commodified animals include those providing city dwellers with opportunities for 'nature consumption' and a vast array of captive and companion animals sold for profit." She underscores how law may slow the process in ways, but does not help with individual animals generally—"Development may be slowed by laws protecting endangered species, but you will rarely see the bulldozers stopping to gently place rabbits and reptiles out of harm's way."

Some forms of geography, such as "urban theory," feature an unabashed human-centeredness revealed through language that misleads. We read of land that is "improved" by "development" measured only by economics done squarely within the exceptionalist tradition. We read of laws that mandate "highest and best use" measured in one-dimensional ways that focus only on certain human groups.

It is significant that Wolch draws a conclusion about environmental issues, for these are clearly central themes in Animal Studies: "our theories and practices of urbanization have contributed to disastrous ecological effects."[71] Wolch prefaces her article with an observation of the ecologist Daniel Botkin suggesting how much is at stake when urbanized humans fail to see their larger community: "Without the recognition that the city is of and within the environment, the wilderness of the wolf and the moose, the nature that most of us think as natural cannot survive, and our own survival on the planet will come into question."[72]

Wolch points out how "theories" or broad, generalization-dominated approaches that seem friendly can be, in fact, debilitating and harsh—she mentions the familiar measure of "progressive environmentalism" under which "[other-than-human] animals have been objectified and/or backgrounded."[73] Recognizing that "progressive environmentalism" is not being driven by science, Wolch observes,

> Progressive environmental practice has conceptualized "the environment" as a scientifically defined system; as "natural resources" to be protected for human use; or as an active but unitary subject to be respected as an independent force with inherent value. The first two approaches are anthropocentric; the ecocentric third approach, common to several strands of green thought, is an improvement, but its ecological holism backgrounds interspecific difference among animals (human and nonhuman) as well as the difference between animate and inanimate nature.[74]

A truly science-based environmentalism would be less human-centered, for contemporary environmental and ecological sciences are grounded in "the Darwinian revolution [which] declared a fundamental continuity between the species."[75] Such a perspective ought to open up ethical questions about how "improved" land might in fact be measured, or how "highest and best use" might be assessed. Instead, progressive environmentalism prefers a social construction under which ethics has pre-Darwinian features drawn from the exceptionalist tradition.

Wolch adds observations about how progressive human-centered movements preoccupied with critiques of human-on-human harms often remain within the exceptionalist tradition:

> Animals have their own realities, their own worldviews; in short, they are subjects, not objects. This position is rarely reflected in ecosocialist, feminist, or anti-racist practice, however. Developed in direct opposition to a capitalist system riddled by divisions of class, race/ethnicity, and gender, and deeply destructive of nature, such practice ignores some sorts of animals altogether (for example, pets, livestock) or has embedded animals within holistic and/or anthropocentric conceptions of the environment and therefore avoided the question of animal subjectivity.[76]

Such problems are ignored as well by "various wings of the urban progressive environmental movement" which

> have avoided thinking about nonhumans and have left the ethical as well as pragmatic ecological, political, and economic questions regarding animals to be dealt with by those involved in the defense of endangered species or animal welfare. Such a division of labor privileges the rare and the tame, and ignores the lives and living spaces of the large number and variety of animals who dwell in cities…I argue that even common, everyday animals should matter.[77]

Wolch's essay reflects both the continuing human-centeredness of some widely used theory and the versatility other theory can have in the work of a geographer committed to interdisciplinary approaches that take seriously the realities of other-than-human animals. Her work also reflects that critical thinking has very important roles to play when values driving specific claims and generalizations that purport to be inclusive and progressive are far narrower than claimed. Such critical thinking fosters, in turn, a healthy dose of ethical reflection that is not human-centered, which only opens up more issues for exploration.[78]

Wolch's insights have been developed with other geographers who analyze various theories and practices of urbanization that seem to offer positive views of wildlife and nature generally. Because such theories and practice tend to romanticize—and thereby distort—other living beings, in the end they harm wildlife communities. Such caricatures serve a human-centered purpose of marketing wildlife themes in consumer cultures.[79] The point of countering romanticizations with "transspecies urban theory" is to challenge the way unreflective, inherited assumptions make human interests superior to the interests of other-than-human animals. Transspecies urban thinking, then, opens up discussion of how urban environments do, can, and should include other animals. In effect, urban nonhumans no longer are automatically deemed unwanted others. It may sound merely intellectual, but the change has ethical consequences that both share a certain spirit with the Axial Age and honor the realities of other animals: "Diligent efforts at mutual understanding and learning…are needed to create an environmental ethic that recognizes the fundamental linkages between human justice and justice for animals."[80]

Geography and the Urban Animals Bridge

As discussion of urban animal neighbors opens up minds, it bridges human precincts and the more-than-human world. The urban animal issue requires interdisciplinary work to introduce, identify, contextualize and make part of the community the nonhuman macroanimals that abound in city homes, businesses, undergrounds, parks, vacant lots, and myriad other interstices. Nonhumans are, of course, also overhead in migrating groups, and geography is well suited to identifying the far-flung places that such animals inhabit at different times of the year.

Like anthropology, geography as a mature discipline has an astonishingly large body of work that has memorialized a wide range of historical and cultural views about our species' place in the world. The field today also offers popular, journalism-oriented work that supplements more purely academic versions. Both the academic and popular versions reveal in great detail how interactions with other animals are a constant, inevitable feature of human life. Not surprisingly, scholars have formed a professional group that focuses on nonhuman animals.[81]

In all these ways and more, modern geography has developed insightful uses of critical thinking and theory, as well as diverse notions of place and meeting spaces, to help humans recognize different features of our inevitable encounter with other living beings. Thus, like archaeology and anthropology, geography has important roles to play as humans build a cooperative, ethically attuned and scientifically informed community that seeks a frank appraisal of our species' past and present. Such work is crucial to our species telling a full version of the larger story.

Telling the Larger Story

A primary task of Animal Studies is telling the full story of humans' history with other living beings. Of the partial versions of this story that have to date been told, some are more encompassing than others—in particular, accounts of this history written in the last half century reach further and wider than before. Contemporary versions, which still fall far short of the entire story, are being expanded as many scholars work to rectify this shortfall.

History as Exploration

"History from below," which also goes by the telling names of "social history" and "the people's history," foregrounds humans who have been oppressed, disenfranchised, forgotten, or otherwise marginalized because of their poverty, nonconforming beliefs or actions, or ethnic or cultural identity. Theory-level work has helped many people see fundamental exclusions that past histories have perpetrated—in fact, much of this human history remains to be told.

Another essential precinct of history is the larger story of human and nonhuman interaction. Thus, even if contemporary historians expand the human side of history from below, such human-centered accounts will still be impoverished. Whenever purported historical accounts fail to tell the story of humans in their multifaceted, inevitable interactions with other-than-human animals, what is passing for history is inadequate for understanding humans alone.

Historians have begun to utilize the open and encompassing spirit of history from below to venture across the species line. Such efforts have prospects of producing an even more radical history, and, in turn—because of the common association of the most marginalized humans with (other) "animals"—opening up an ever-wider range of human stories.

Opening Up the Human Side of the Intersection

In the 1980s, various historical studies showed that the modern academic discipline of history could illuminate explorations of the human-nonhuman story. When historians include in their narrative various features involving nonhuman animals that previously have been ignored, they challenge their own field's exclusivist tradition and that of the humanities more generally.

Narrow and blinkered versions of history that obscure human groups produce complex problems. While one subset of these problems concerns the human side of such histories, the focus here is the more extensive and virulent nonhuman side of these problems. Severe harms to nonhuman individuals remain obscure, accumulating to the point of depleting, then destroying, entire populations, communities, and, eventually, species. Even when such realities are known, they typically disappear from memory under the pressure of human-centric history. Even ecosystem-wide damage is often ignored, perpetuating ignorance and causing awareness of extraordinary harms to disappear from human consciousness.

When multiple and profound impacts on humans are ignored despite an ideology that "all humans matter," it is not surprising that human-centered institutions fail to describe the magnitude of harms to other-than-humans mentioned by a respected Canadian. "This is not a book about animal extinctions. It is about a massive diminution of the entire body corporate of animate creation.... the greater part of the book is about those species that still survive as distinct life forms but have suffered horrendous diminishment."[1]

The better-known tragedy of extinction has commanded much attention—"When the last individual of a race of beings breathes no more, another heaven and another earth must pass before such a one can be again."[2] Sadly, there are an astonishing number and range of harms to other-than-human animals that rival extinction tragedies, but those who learn the exceptionalist tradition version of history are rarely, if ever, alerted to the ways our shared, larger community has been emptied out and thereby deeply impoverished. Such issues are often framed in human-centered terms with questions like, What sort of world are we leaving to our children? Thereby, losses of individuals and communities are obscured even though our sciences and museums may catalog species extinctions. In modern societies, particularly at the day-to-day level, the depth and breadth of the problems are not well described.

The consequences of ignorance about harms to nonhuman communities ripple through education—jobs go to those who become expert in recognized (human-centered) problems, and programs center on recognized harms to our own species. Tenure is offered to those who work in our human-centered canon, making all too evident how education is not only rarely a place of daring but, worse, a place where ignorance and even apathy are perpetuated through silence or one-dimensional viewpoints about other-than-human animals.

In 1987 an influential historical study, Harriet Ritvo's *The Animal Estate: The English and Other Creatures in the Victorian Age,* examined ways in which ownership of certain non-human animals symbolized class relations in Victorian England. Ritvo offered a detailed social history of nineteenth-century English treatment of livestock, pets, and rabid dogs at home, and big game in the British Empire. Two earlier publications were the extraordinarily detailed *Man and the Natural World* (1983) by the British historian Keith Thomas, and *The Great Cat Massacre and Other Episodes in French Cultural History* (1984) by the eminent Harvard historian Robert Darnton. Although these and similar works often emphasize nonhumans as objects of human action (such as prize possessions, public health menaces, or exotic beasts), the net effect was to prompt other historians to look at sources with new eyes, in search of clues about human-nonhuman relationships in the past.

In the following decade, the publication of *Diamond's Guns, Germs and Steel* raised further awareness of why and how human-nonhuman intersections have influenced history—Diamond's subtitle is *The Fates of Human Societies*.[3] More and more details of how concern for other-than-human animals constantly has emerged and reemerged in daily life in different eras are evident in historical accounts such as Preece's detailed and nuanced 2002 study *Awe for the Tiger, Love for the Lamb: A Chronicle of Sensibility to Animals*.[4] Virginia Anderson in 2004 used nonhuman animal issues to provide impressive new insights about human history in a familar topic in her book *Creatures of Empire: How Domestic Animals Transformed Early America*.[5]

Exploring the Further Reaches of History from Below

Some may balk at any suggestion that we can tell a more-than-merely-human history. In 1984, a scholar respected for his considerable learning and very developed sense of ethics was decisive—and dismissive—about history in connection with nonhuman animals:

> If we look at a herd of cattle in a field, we can pick out individual cows from the mass. But no cow has a "history" in the sense that an individual human being does. Which is to say that although cattle, like human beings, live individuated lives which are extended through time, there is no particular significance which resides in the individual life-course of each. It does not constitute a "story." When Abraham entertained the three heavenly visitors by his tent at Mamre, he slaughtered a calf. Has anyone ever asked which calf? Yet you could not slaughter a human being without slaughtering some particular human being, someone with a name, of whom it would make sense to ask "Who was it that died?"... Individual humanity does not lose its significance when it is part of a multitude; rather the history of the multitude gains its significance from the fact that it is a multitude of persons, not of ants, each of whom has a significant history in him- or herself.[6]

Critical thinking can illuminate this passage's claims and conceptual structure. For example, it is not hard to notice that this theologian groups all nonhumans together into one amorphous category—the original topic is cows, but the ending comment (about living beings in a multitude) is about ants. Subtly, the analysis moves from a domesticated social mammal to an individual insect, and by implication to any and all nonhuman animals in the amorphous category "animals," which clearly excludes humans.

The passage focuses most heavily on a domesticated animal selected precisely for its docility and subservient response to human masters. It is a fallacy to set up a single animal as a representative of all nonhuman animals. Employment of this fallacy for the rhetorical effect of demeaning other-than-human animals calls to mind the fallacy of misplaced community. Both of these fallacies separate humans in order to distinguish us.

Such mental habits deceive this particular theologian into using the cow as a representative for other-than-human animals generally. The negative agenda behind this move is also evident in the odd implication that cows are like ants in some way—in fact, cows are normally very unlike ants in relevant respects since we easily perceive cows to have individuality, personality, and the ability to relate to others as individuals. It could also be argued that ants are a different form of life than cows because worker ants are genetically identical to each other, which is not normally true of cows—however, in factory farming today, cows, through genetic engineering, are characteristically mass produced as genetically identical individuals designed for optimum production.

This theologian has used what he thinks are trivializing examples of cows and ants to make the principal point of the story—that individual humans have a distinctive history. Surely this particular point is easily accepted even when one repudiates the passage's reasoning about nonhuman animals as flawed. A particularly important failure in the passage is the failure to signal that there are some nonhuman animals—elephants, orcas, orangutans, and so on—whose realities would make a better comparison to the admittedly wonderful complexities of human existence that give each of us a history. But cows, too, have rich complexities that have been honored by many humans even when our highly educated scholars fail to take note of them.

But critical thinking pushes us to confront this theologian's point that no human asked which calf was slaughtered by Abraham. If this claim is true (it is impossible to really know), the failure to ask would hardly lead to the conclusion that the calf had no story or history. The argument relies on a questionable rhetorical move—it confuses "no humans I know of have ever recognized this calf's story" with "there is no story." As a simple logical matter, humans' failure to acknowledge the story of any particular being in no way shows that that other being has no story. We would never argue that our failure to recognize the stories of other humans means those humans did not have a story. Why, then, does such reasoning apply to cows? The answer is that modern humans have been trained—by education, by language, by ethics—to ignore cows' individuality. But despite such failures, many humans in different times and places have noticed the history of cows.

To find such stories, in industrialized countries one can, for example, visit 4-H clubs of the kind that began in the United States over 100 hundred years ago and have spread to more than seventy countries in North, South, and Central America, Europe, Asia, and Africa. In these circles, one finds story after story involving cows and other nonhuman animals raised by young humans who named and still remember the specific animal they raised as an individual with a unique history.

Outside the industrialized world, the same phenomenon is ancient. In the Masai culture, for example, because of the intimacy and bonds between herders and their cattle, individual cows are named and remembered. This embeds in Masai culture a symbiosis that is part of the story of both humans and cows. When Masai greet each other, they say, "I hope your cattle are well." Such openness carries over culturally into general attitudes toward living beings outside our own species.[7] The Masai delight in telling stories with developed human ethical abilities that accept cows as the actual individuals they are who are both part of the Masai community and possessed of personal histories.

On the Importance of Recognizing Individuals

Humans have long recognized that it is both possible and good to tell stories about nonhuman animal individuals. In one sense, this entire book is an extended argument that everyone knows that such histories are not only possible but also important. Recognizing some other living beings outside one's own species as morally significant has been, at least since the Axial Age, in some people's opinion, a sine qua non of what it means to be a spiritually informed "moral animal."

Further, critical thinking, as well as common sense and intuition, make it clear that such stories (that is, histories) by no means must be, at their core, stories about human matters. Of course, many stories that appear to be about nonhuman animals are at heart about solely human issues. Such stories include Aesop's Fables, the *Jatakas* from the Indian subcontinent, and other popular literature around the world. Even when a story includes some information about nonhumans, most people recognize that the principal purpose, point, and substance are a human matter.

Stories truly about nonhuman individuals—that accurately relate their realities, personalities, and relationships—are more than merely conceptually possible. It is not difficult to imagine a story about a nonhuman animal that does not primarily serve human purposes. While such stories may be converted to human purposes, virtually everyone can relate a history of a biological individual that does not reflect human concerns. These may be tales of heroism or family loyalty or friendship, or some fantastic feat of travel or intelligence—the point is that histories of individual nonhuman animals abound in our society despite the ideology that history is a human affair.

Accounts of real animals are typically confined to the members of a limited number of species that impress themselves on humans. Dogs, for example, are obvious candidates today for such stories given the special relationship that so many people around the world have developed with them. One thus finds abundant examples of writing about real dogs, such that one can range widely through *The Hidden Life of Dogs* to *Famous Dogs of Famous People*.[8] Interest in real dogs can be fulsome and diverse, as attested by the success of both the science-oriented best-seller *Inside of a Dog: What Dogs See, Smell, and Know* and the connection-oriented *Dog Love* by the respected humanities scholar Marjorie Garber.[9] Such books in a sense tell history from below because they regularly speak of individual dogs with histories, and there are some especially nonhuman-focused accounts in best-selling biographies such as *Merle's Door: Lessons from a Freethinking Dog*.[10] Our connection to individual dogs is a natural outgrowth of our ability to care about other individuals in the larger community.

Maureen Adams's 2007 *Shaggy Muses: The Dogs Who Inspired Virginia Woolf, Emily Dickinson, Edith Wharton, Elizabeth Barrett Browning, and Emily Brontë* is a series of stories about famous humans associated with certain dogs.[11] These stories are told because of the famous women involved, but one passage by Edith Wharton reveals nicely how histories are intertwined: "The owning of my first dog made me into a conscious, sentient person fiercely possessive, anxiously watchful, and woke in me that long ache of pity for animals and for all inarticulate beings which nothing has ever stilled."[12]

While there are many other well-known stories of individual dogs (such as Byron's dog Boatswain), with diligence one can also discover now unknown individual dogs who walked with both obscure and famous humans, such as Descartes's canine Monsieur Grat. The ironies here are layered—Descartes notoriously claimed that nonhuman animals were more like machines, but his choices regarding Monsieur Grat, including bestowal of a name, were not at all like his actions toward mere machines.

Contemporary media continue to be fascinated by dogs' ability to communicate, often expressed as admiration for dogs' evident intelligence when picking up skills in our language.[13] Other media reports give attention to nonhuman beings for reasons suggested by a 2010 science-based story, "Pets Vital to Human Evolution": "Dogs, cats, cows and other domesticated animals played a key role in human evolution, according to a theory.... The uniquely human habit of taking in and employing animals—even competitors like wolves— spurred on human tool-making and language."[14]

Media reports of this kind still reflect a preoccupation with human-centered issues, but, equally, they reveal how many humans today choose to live alongside a range of other-than-human animals, often as family members thought of as individuals with an important history. Such choices go considerably beyond recognition of the biological reality that our world is obviously a more-than-human community. Humans connect, remember, and tell stories about their nonhuman family members.

Beyond the widespread stories about dogs, one easily finds geographically diverse reports of humans and dolphins interacting cooperatively. Cetaceans, as many people know though cultural lore, are highly intelligent and live in richly social contexts. Science has confirmed in many ways the brain size, intelligence, and communication abilities of these mammals, as well as their ability to learn certain features of human communication. But stories of humans who were rescued by dolphins, as well as stories of cooperative fishing, abound.[15] Some free-living cetaceans have associated with humans often enough to be given names, such as Pelorus Jack in New Zealand, Fungi in Dingle, Ireland, and the orca Luna in the US Pacific northwest. In such cases, the history of the nonhuman involved (and the name) is based solely on the interactions between individual cetaceans and specific human communities.

Elephant individuals, too, have remembered stories—the monastic codes known as the Vinaya found in the Buddhist scriptures tell separate stories of the matriarch Bhaddavatika and the bull Nālāgiri.[16] Modern scientists who study elephants have given more detailed accounts, including Cynthia Moss's stories about different matriarchs and their family groups' communications and interactions.[17]

Stories of other large-brained social mammals, such as the nonhuman great apes (gorillas, bonobos, chimpanzees, and orangutans), could easily be added. Though often completely ignored by modern education, religion, and policy or lawmaking circles, stories of other animals are part of humanity's common heritage.

Ancient Help in Seeing Other-Than-Human Communities and Cultures

As we contemplate the challenges of giving an honest version of history, two deep reservoirs can be tapped—the rich past of human observation, and present-day experience with nonhuman individuals. The former is possible only through interdisciplinary scholarship regarding our forebears who recognized that humans can tell a history that includes other animals as meaningful participants. In their essay "History from Below: Animals as Historical Subjects,"

Georgina Montgomery and Linda Kalof provide many examples of diverse scholars around the world who now tell animal histories and many others who did so in the past.[18]

A resurgence in such work has been developing for decades. In 1984, Robert Delort published his groundbreaking *Les animaux ont une histoire*. Four years later, Thomas Berry opened his groundbreaking and still influential *The Dream of the Earth* with this set of observations:[19]

> Paul Winter is responding to the cry of the wolf and the song of the whale. Roger Tory Peterson has brought us intimately into the world of the birds. Joy Adamson has entered into the world of the lions of Africa; Dian Fossey the social world of the gentle gorilla. John Lilly has been profoundly absorbed into the consciousness of the dolphin. Farley Mowat and Barry Lopez have come to an intimate understanding of the gray wolf of North America. Others have learned the dance language of the bees and the songs of the crickets.

Berry then called out with remarkable prescience that "individual wild animals are entering into history" which he illustrated with these words:

What is fascinating about these intimate associations is that we are establishing not only an acquaintance with the general life and emotions of the various species, but also an intimate rapport, even an affective relationship, with individual animals within their wilderness context. Personal names are given to individual whales. Indeed, individual wild animals are entering into history, as in the burial of Digit, the special gorilla friend of Dian Fossey.

The personal implications of such an intimate rapport can only be known by those who are open-minded enough to explore this possibility. Such personal connection can be expressed in many ways, but whether realistic or fanciful, all risk being perceived as heresy in various human-centered circles that recognize only human-focused history.

When facing such risks, it helps to tap the reservoirs of past or present observation. When one recognizes that such mind-opening questions have been asked before and are being asked now, it is easier to recognize how ideological are the denials of the mere possibility of telling the history of a nonhuman individual.

Accomplishing the Tasks of Animal Studies

Scientific projects and political efforts of animal protectionists, environmentalists, and educators continue to develop the debate over how human individuals can help accomplish Animal Studies' first and incomplete task of telling the larger story. The second task of Animal Studies is important to seeing how histories of other animals must be anchored by realities-based information. Commitments to meet this task, in turn, grow through work on the third task of exploring future possibilities. Such work invites humans to use their capacious ethical abilities to explore perspectives on other-than-human animals' lives. This kind of widespread communal work is essential for our own species' imagining and exploring of our shared, more-than-human world.

From the fourth task of Animal Studies—identifying limits to what humans might know about other living beings—yet one more history will emerge. By accomplishing this task, each of us can fully recognize that our personal story is told from our own point of view. This task creates the interesting dynamic of self-actualization through self-transcendence. We can recognize that the limits of our point of view give us reasons to maintain a fundamental humility toward our world. Because we grow as we attempt to tell the self-transcending larger story, we achieve the status of special animals when we are at our most humble.

It is not only through the collection of past stories that we can grow. We glimpse much more by seeing the variety and limited features of human stories—those of affirmation and dismissal from the past, those reflecting the great variety of our own species' cultures, those connecting dismissed humans to even more radically dismissed nonhumans. It is in this manner that any human glimpses the importance of seeing different stories. Through such variety, we stand a chance of creating a holistic account.

Trying to see the outlines of the larger story not only helps one get beyond the myopias of any one culture's narrative about animals. By helping humans see our own heritages and thinking about the larger story, we benefit through seeing the present versions of our own and our children's education, just as we see our philosophy and its limits, the narrow features of our social sciences, and, especially, how narrow has been the story we claim as our history. Pursuing the larger story helps us, for example, move beyond the domination themes that impoverish industrialized cultures where most people mindlessly accept the exceptionalist tradition that puts humans in the role of, to use Aldo Leopold's classic phrase, "conqueror of the land-community" rather than "plain member and citizen of it."[20]

"Has Anyone, Then, Ever Asked Which Calf?"

To the theologian's dismissive query, then, we can definitely say, "Yes, again and again." In 1995, a cow escaped from a slaughterhouse and became the subject of national and international media attention until her death in 2003. In her 2007 *The Story of Emily the Cow*, Meg Randa tells the story of this individual who found sanctuary in the Peace Abbey in Sherborn, Massachusetts, where there is now a statute.[21] Seeing cows and other animals as real individuals with their own history has deep roots traceable as far back as the Rig Veda, the most ancient of scriptures in the Hindu tradition. "She is like the mother of the cosmic Forces, the daughter of the cosmic Matter, the sister of cosmic Energy, the centre of the ambrosia. I address to men of wisdom—kill not her, the sinless inviolate cow."[22]

Gandhi often spoke of cows in personal ways that expanded on the basic values so evident in the Hindu scriptures: "Cow protection to me is one of the most wonderful phenomena in human evolution. It takes the human being beyond this species. The cow means the entire sub-human world. Man through the cow is enjoined to realize his identity with all that lives."[23]

Gandhi also challenged human exceptionalism, still startling us today: "I would not kill a human being to protect a cow, as I will not kill a cow to save a human life, be it ever so precious. My religion teaches me that I should by personal conduct instill into the minds of those who might hold different views the conviction that cow-killing is a sin and that, therefore, it ought to be abandoned."

It is easy to find parallels to such views in many small-scale societies and in today's industrialized societies, for citizens in the latter often develop comparable commitments to their dog and cat family members.

So think again of the mother of the calf in the theologian's example. Might not that calf's own mother have wondered about her calf after it was taken from her? This is a straightforward empirical question that will be explored more realistically and fairly only if one has not already dismissed cows and other nonhuman animals before the inquiry has begun. Honest empirical exploration will confirm, of course, that cows are social mammals with noteworthy curiosity. One scholar, referring to Aristotle's claim, "All men desire to know," suggested the same "is true of cows, as anyone who has walked down a country lane must know."[24]

Simply said, we can, if we wish, explore a great deal about cows that helps us recognize that even if no one who reads the scriptural story about Abraham slaughtering that calf at Mamre has ever asked, Which calf?, Animal Studies can and does imagine the question and more about real events like pain and loss from such an act. Each human can transcend inherited habits that cause so many people to refuse to explore, and thereby erase, nonhuman animals' histories.

Central Issues of the Larger Story

Creating an account that could meaningfully be called the entire larger-than-human story will require an intriguing complex of character traits, skills, and educational opportunities. Among these will be a pronounced willingness to inquire humbly, to promote academic freedom in learning centers, and to be patient with forms of imagination that do not automatically defer to education-based traditions of human-centeredness. One will also need robust exploration of many cultures, religious traditions, and secular dialogues. Such an encompassing account must be scientifically informed, ethically deep and balanced, and self-conscious of the breadth of the task being undertaken. With such challenges in mind, consider fundamental features of a full history of humans' long and complex interactions with other living beings.

The larger story is a more-than-human story. The very attempt to craft this multifaceted, multispecies chronicle prompts us to see better the forms of human-centeredness that dominate us. Through such work, we begin to understand and grapple with the implications of the narrow-mindedness of our past. As this story is learned, we more easily recognize that we live on a multicultural, more-than-human planet.

The larger story arises out of necessity and is as much art as science. The ubiquity of animal life makes the larger story more than important—it underscores how telling history so frankly is communal. The ubiquity of life is our community, which makes the larger story a necessity to our self-understanding. As Berry suggests, "the larger community" populated by "all our companion beings throughout the earth" "constitutes our greater self."

Telling the larger story is an art in several important senses. As with all narratives, perfecting it requires skill, craft, and rehearsal. Of course, telling the larger story goes beyond some definitions of art, such as Picasso's famous claim that art is "the lie that helps us to see the truth." The larger story aspires to be the truth, and in the spirit of Rilke's observation, "A work of art is good if it has arisen out of necessity," it will take the healthiest of forms if it arises out of a deep need to name our community and roots.[25]

Recognizing, crafting, and telling this story is an ongoing, collective enterprise. Each human knows only a part of the larger world, the larger community, and the larger story. Even as we recognize our own limited place, we also recognize that we have the imagination to tell the larger story—that, of course, takes the very best of human skills and cooperation.

The story makes it clear that the world was not made primarily for one species. Despite the claim of some humans that the world was designed for us, every one of us knows viscerally that the world was clearly not made for human purposes alone. One must get beyond the language and ethics of the exceptionalist tradition to recognize how destructive it is to reduce other-than-human animals to mere resources. Such domination requires forms of intentionality that, in the end, harm humans and nonhumans alike. Different cultural and religious traditions make it clear that the domination and dismissal so characteristic of industrialized societies is but one choice among many available to us.

As the larger story's narrators, humans inevitably play an integral part. Taking the role of narrator requires us to construct the story, but moral and scientific criteria arising out of an ethic of inquiry mandate that we carry out this role in ways that avoid the bias, harms, and disingenuousness nurtured by the exceptionalist tradition. In fact, employing critical thinking in the service of community can be a celebration of humans' unique imagination, caring faculties, and abilities to seek and honor the truth. Humans' uniqueness in taking this role confers no superiority and privilege. As a species that is imaginative, disciplined, and humble enough to identify the larger story, humans self-transcend in ways that create their richest form of self-actualization. Telling the larger story, then, has the prospect of creating health, communal maturity, and ever-greater creativity for our own species even as it affirms the larger community of life that is our greater self.

The larger story will undoubtedly have multiple versions. Different versions of the larger story are possible even if humans agree on the general outlines—such is the nature of our human cultural and social realities, our thinking, our abilities to belong to an entire series of nested communities. Alongside science traditions, ethical reflection, common sense, and our diverse arts can sit myths, stories, and other narratives that provide us with orienting explanations of our special human abilities and origins.

The larger story will counter the impoverishment of the carpentered world. Animal Studies prompts versions of the larger story anchored in the fact that some nonhuman animals have their own realities that need not be measured against human realities. The larger story also prompts richer reflection on cognition, intelligence, sentience, and much more. Further, humans return as animals to our greater community. Culture-specific stories of how humans developed our special abilities need not underwrite notions of superiority overtly or surreptitiously. By pushing us to explore what human animals can know of our fellow animals' lives, Animal Studies travels beyond carpentered worlds and their inevitable one-dimensionality. Similarly, the field can invite us to wonder how to get beyond what we now see as the limits of human imagination.

The larger story is capacious. As both the larger story and Animal Studies pull humans beyond human-centered precincts, they enlarge us. This occurs even as they make us "plain member and citizen" rather than an isolated species that works out its insecurities by naming itself, in Leopold's words, "conqueror of the land-community."

Animal Studies provides the conceptual space in which to balance science-based approaches alongside ethics-based approaches. It also provides room for a balance between recognizing other-than-human realities and due consideration to the important role of creative animal-based symbols in human arts and history.

Finally, the story is now alive. Some crucial elements of the story exist already. Even though many humans alive today have been raised, nurtured, and educated within human-centered circles that have produced extremely impoverished and dysfunctional visions of our own animality, some have broken through to a commitment to become more knowledgeable about the larger story. As the modern world reengages the living beings beyond our own species, many more people will be enabled to ask, Why should an integrated human know the whole story? Answers to such questions belong to each individual but also to future generations who will add their chapters to the larger story. The story, then, is alive and in process.

Frank Voices about Current Realities

Contemporary developments reveal that truly fundamental work in Animal Studies is only beginning. The virulence of the harms caused by human domination is reflected in Scully's term "massive punishments." Given Scully's indictment that we inflict them "with such complete disregard," we have reason to wonder whether our species really deserves the description "moral species." This question begs consideration of the extent to which humans perpetuate human-on-human harms even as we, with such complete disregard, harm so many nonhumans. Both problems prompt some to see the familiar claim that humans are a moral species as a facile, self-serving rationalization.

Given that so many modern educational institutions offer impoverished and misleading information about other-than-human animals, it is not surprising that many mainline leaders and thinkers continue to ignore nonhuman animals and can barely imagine why one might ask about the history of any nonhuman individual, let alone wonder about the larger story.

Ethical Blind Spots

The exceptionalist tradition has produced paradigmatic examples of conditioned ethical blindness. Humans may become so familiar with a situation that they no longer notice that the facts before them pose profound ethical dilemmas. Even when they are aware of the ethical dilemmas, some humans choose to continue to ignore the harms because they are characterized as "not so serious," "long-standing tradition," "necessary to make a profit," or some other facile rationalization.[26]

Conditioned ethical blindness can be found in the constellation of negative effects caused by the exceptionalist tradition, which can also perpetuate failure to engage the ways in which claims and practices are values-driven, self-interested, or otherwise problematic from the standpoint of critical thinking. The evident beauty of promoting human dignity should be honored even as critical thinking helps one see that some people use such claims merely to mask human-on-nonhuman harms. As animals, humans can be narrow. As animals who are humble enough to employ the best of human critical thinking skills, humans are capable of recognizing the shortcomings of the exclusivist tradition even as we construct beautiful edifices like ethics, encompassing narratives like the larger story, and a sense of community and identity that reaches well beyond the species line.

This is one reason that the ferment described in this book has produced many reactions to existing harms. Many contemporary citizens advance insights that parallel ancient insights. Many also recognize that extreme harms are now common, as well as that other animals' realities deserve respect when deliberating about whether such harms raise moral issues. Informed by science and cultural studies that confirm how widely human cultures have affirmed the importance of noncruelty as a cultural and personal achievement, modern citizens now openly propose that it is important to nurture rather than dominate the more-than-human world and its diverse nonhuman citizens.

Modern citizens have available to them a tapestry of perspectives that feature threads of traditional wisdom, intuitive experience, knowledge of one's local world, and scientific insights. Given how these threads can complement and mutually enrich each other, many humans advocate animal protection and environmental awareness as seamlessly presenting a picture of a complicated but interwoven world.

Scientists have the utmost importance in contributing to humans' vision of the world—recall that Stephen Jay Gould called Jane Goodall "one of the intellectual heroes of this century." Other scientists such as Donald Griffin pioneered the scientific world's willingness to ask questions and pursue investigations that contribute to Animal Studies. Said in the simplest terms, the realities of other animals, as discerned through science, daily encounters, and a host of other human efforts, are a basis for noticing other animals and taking them seriously. Doing so prompts us to lose our conditioned ethical blindnesses and then to use our large primate brains and surpassingly ethical hearts to notice an abundance of reasons that some other-than-human animals command the attention of humans' moral sensibilities.

We have reasons to ask whether many forms of intentional, economics-rationalized harms to other-than-human animals might be abandoned altogether. Questions in this vein are often, of course, muted out of respect for political realities (such as the complexity of immediate abolition) or because of entrenched opposition from religious or cultural authorities. Sorting through such considerations is complicated, for disentangling unproductive human-centerednesses from those which are healthy and promote life will require skills that only a few educational, religious, and governmental institutions promote in any degree. As thinking, meaning-making animals, however, we are capable of identifying bias, fantasy, wishful thinking, and the conditioned ethical blindnesses that keep us from seeing ourselves and other animals well.

Wildlife and the Larger Story

The free-living nonhumans we know as wildlife offer a paradigm for the other-than-human. By definition, how we understand this paradigm will exclude the control and coercion of both domestication and zoo-based captivity—it is, accordingly, hard for many humans to fathom what wild animals' lives might be like. It may seem counterintuitive, but as Abram suggests, not knowing stimulates us in important ways:

> When we choose to be fully aware of even the simplest features of our inevitable encounters with other living beings, we notice that our attention is drawn by the open and uncertain character of this being.... This is precisely what enables our senses to really engage and participate in this encounter. Another being, and indeed *any* entity that captures the gazer, is never revealed in its totality—there are *always* facets of the being that do not present to our human eyes or noses or general awareness. This is our finitude—and the result is a fundamental withholding of aspects of the encountered being from our direct apprehension. Think of the multiple ways this is true of any encountered other—a red maple, a bat, a wasp, a deer walking on "our property." ... We see *none* of these in their *entirety*. We may ignore this—but in that case, our ignorance takes on self-inflicted features that are truly debilitating for ourselves *and the beings we ignore*.[27]

For most humans, wild or free-living nonhumans can be inviting—Abram adds, "This tension between the apparent and the hidden dimensions of each being beckons steadily to my perceiving body, provoking the exploratory curiosity of my senses."[28] Further, we can see a number of specific aspects of a macroanimal's life and, possibly, personality—we thus are not all ignorance, as some handling of epistemological arguments would seem to imply. Thus even though the fourth task of Animal Studies is to recognize limits, this task does not prompt one to despair of *any* awareness. The question is often one of limits, not a complete inability to perceive, imagine, and guess in responsible ways. In the midst of the animal's invitation to us and our awareness of our limits, we can recognize, as Abram suggests, "No matter how long I linger with any being, I cannot exhaust the dynamic enigma of its presence."

Other animals, then, have a dynamic presence for us, disclosing a shared animality even as they bear some essentially inaccessible features that may appear in the guise of mystery, ignorance, or simple silence. While more familiar domesticated animals regularly offer invitations, too, wildlife and animal communities free of human distortion also contend for many people's fullest attention. We can see them alternatively as free-living nonhuman animals to be prized as bringers of blessings, or competitors, antagonists, and pests. They can be symbols of many things, including our own aspirations, just as they can be understood as fellow citizens in the larger community. They can ask us, in what ways are humans unique, or even a paradigm of possibility? They prompt us to wonder about their place, as well as our own, in the world we share.

So Animal Studies has at its heart important questions regarding any kind of nonhuman living being from companion animals to wildlife, from research subject to food resource. Such questions open us up, as do questions about other humans and how they relate to the many different kinds of beings with which we share the earth.

Marginalized Humans and Other Animals

Although governments and institutions in modern, industrialized societies often contend that every human has importance, realities belie such claims. In this chapter we explore the interrelation between (1) marginalizing human animals and (2) harms to nonhuman animals in order to call out as fully as possible the multiple, inevitable links between harms to humans and harms to other animals.

An ancient insight, today sometimes called "interlocking oppressions" or "the Link," is that forms of violence often connect to each other.[1] Harms to one group of living beings can foster, even facilitate, other forms of oppression against the same beings or others. Sometimes one form of violence is so interwoven with other forms that any occurrence of one makes the other not only more likely, but even more virulent and more resistant to critique or change.[2] There are two key insights here—first, forms of oppression are linked, even interlocked, and, second, abilities to oppress others are in some respects like a muscle that is strengthened by use but which can atrophy if left unused for long periods.

The British historian Keith Thomas mentions views advanced in ancient Athens, the Hebrew Bible, the writings of the medieval theologian Thomas Aquinas, and William Hogarth's famous eighteenth-century sketches depicting those who harm nonhuman animals as moving inexorably to harms of humans, concluding, "But this view did not originally reflect any particular concern for animals; on the contrary, moralists normally condemned the ill-treatment of beasts because they thought it had a brutalizing effect on human character and made men cruel to each other."[3]

Thomas's generalization here can be misleading unless it is confined to the Western cultural tradition. By no means, however, have all previous voices addressing interlocking oppressions argued their case in this fashion. The ancient Axial Age sages recognized interlocking oppressions, and they condemned harms to nonhuman animals in their own right, not merely because they portended problems for humans. "[Ahimsa] originally applied not to the relationship between humans but to the relationship between humans and animals. Ahimsa means 'the absence of the desire to injure or kill,' a disinclination to do harm, rather than an active desire to be gentle; it is a double negative, perhaps best translated by the negative 'nonviolence,' which suggests both mental and physical concern for others."[4]

Recall that Immanuel Kant claimed that "we have no direct duties" to other animals. To justify this odd claim, this immensely influential Enlightenment philosopher offered an exceptionalist tradition explanation that cited Hogarth's widely discussed engravings depicting four stages of cruelty—the first stage admittedly involved cruelty to nonhumans, but this stage was only significant for Kant in that it prompted humans to move on to harming humans.[5] Kant's equally famous philosophical predecessor, John Locke, in 1705 framed the question of interlocking harms as a sort of common sense, proclaiming, "they who delight in the suffering and destruction of inferior creatures, will not be apt to be very compassionate or benign to those of their own kind."[6]

While Locke's point may often be true, many commentators, ancient and modern, have gone much further, noting unequivocally that harms to nonhuman animals are problems in and of themselves, that is, even when they do not lead to harms to humans. Human-centeredness is, as we have seen, a particular preoccupation of Western mainline ethics. But such an exclusive preoccupation with humans alone in no way represents fairly or well all ethical traditions. Nor can it be said that all circles in Western culture lacked voices decrying gratuitous violence against living beings outside the human community. Some voices claimed prohibitions on cruelty were for the sake of nonhumans, as well as humans.

Nonetheless, the most influential explanations had features that consistently put humans in the foreground and nonhumans in the background of ethical reasoning. Thus, as nineteenth-century citizens inaugurated the organized efforts that became today's Society for the Prevention of Cruelty to Animals, recognition of the interconnectedness of abuses characteristically relied on reasoning that distanced humans from other animals—it was often argued, for example, that cruelty to other-than-human animals "degraded" humans "to the level of the brutes."[7] Such reasoning involves more than a refusal to recognize that humans are animals—it is deceptive to suggest that cruel humans are like "the brutes" because human cruelties are often uniquely harsh in the animal kingdom. Claiming that humans are degraded "to the level of the brutes" even though human-like forms of cruelty are exceptionally rare in the other-than-human world is, then, to use an ironic caricature, not facts, to dismiss nonhuman animals.

In the late twentieth century, those concerned about links among oppressions began to foreground the suffering of the victims whether they were humans or not. Other similarities were also called out—just as the abuse of human victims is often hidden behind closed doors, so too abuse of nonhuman victims is often perpetrated in private settings so that enforcement of anticruelty laws is impossible. Women and children were in earlier times held to be "property" within a legal system, just as nonhuman animals can be reduced to property in contemporary legal systems. The harsh implications of this form of dominion can be seen in the fact that it was standard law in many places that the rape of a young girl or wife was defined as a property crime not against the woman herself but, respectively, the girl's father or the wife's husband.[8] Since nonhuman animals remain property today, owners of nonhuman animals who abuse them often hide, as did "owners" of women and children, behind legal protections afforded property owners.

Discussion of the ethical dimensions of interlocked oppressions has often been straitjacketed by refusals to countenance the importance of harms to nonhumans except as harbingers of possible harms to humans. But using the interdisciplinary connections fostered by Animal Studies, it is now possible to recognize accumulated wisdom about the interlocking features of oppressions. Animal Studies must also engage human-on-human domination because this problem so often impacts nonhuman individuals and populations.

Legal, Conceptual, and Psychological Realities

Because of increasing recognition that apathy regarding human cruelty to nonhumans can result in an increase in the overall level of cruelty, there has been a surge in recent legislation, research, and discussion surrounding these links. Conceptually, what is at issue is the claim that violence begets more violence, such that a society that tolerates one form of violence is unintentionally courting additional forms of violence. Psychologically, what is at issue is increased use of violence to dominate "others" who happen to be nearby. Cruelty at home not only begets more cruelty within the family but also reaches beyond the home and out into society generally. Such oppression easily reaches across the species line as well. A 1997 study of women's shelters in the United States found that 85 percent of women in the shelter disclosed that there had also been pet abuse in the home while 63 percent of children talked about animal abuse at home.[9] A 1983 study showed that abused animals were found in 88 percent of homes of families where child abuse occurred.[10]

Such harms also reach across generations, since children of batterers all too often go on to commit the same kind of violence. If, however, the cycle of abuse can be interrupted, it is not only existing people and nonhuman animals who benefit but future generations as well. These links suggest how integrated humans and nonhumans are in matters of oppression—some researchers note, for example, that there will be no reduction in family violence until all victims, human and nonhuman alike, receive satisfactory legal protection.[11]

Practical Steps: Changed Definitions in Criminology

The following example suggests individual fields can, through studying humans' relationships to other-than-human animals, implement important changes. Prior to 1988, the widely used Diagnostic and Statistical Manual of Mental Disorders of the American Psychiatric Association contained no suggestion that cruelty to animals might be an indicator of any recognized disorder. In 1988, however, a revised edition of the manual (known as *DSM-IIIR*) finally listed cruelty to nonhuman animals in the section "Destruction of Property" as one of the indicators of conduct disorder.[12] In 1994, a further revision of the manual (*DSM-IV*) placed the problem of cruelty to nonhuman animals in the section titled "Aggression against Animals and People." The change linked human animals ("people") and nonhuman animals ("animals") and thereby recognized interlocking oppressions. The upshot of these changes has been that, as a practical matter in specific cases, conduct disorder and "aggression to people and animals" can be measured in part by the occurrence of cruelty to nonhuman animals. These practical changes offer prospects of protecting numerous others.

Women and Other-Than-Human Animals

Another practical approach developed recently in American law reveals a direct connection between protection of women and protection of companion animals. In 2001, the Yale Journal of Law and Feminism published an article titled "Including Companion Animals in Protective Orders: Curtailing the Reach of Domestic Violence."[13] Within five years (April 2006), the state of Maine, according to the *New York Times*, "spurred by growing evidence of a link between domestic violence and animal abuse,... enacted a first-in-the-nation law that allows judges to include pets in protection orders for spouses and partners leaving abusive relationships."[14] The opening paragraph of the same article reveals well the practical importance of recognizing how oppressions can be interlocked:

> Susan Walsh told Maine legislators a chilling tale in January. She said she had wanted many times to take her two children and leave her husband, ending a relationship she found frightening and controlling. Ms. Walsh says that her former husband would harm and even kill their animals as a means of keeping her under his control.... "It wasn't just the cats and the dogs I had, it was the sheep and the chickens—I was terrified for their welfare," Ms. Walsh, 50, said. "I knew if I were to leave, he wouldn't hesitate to kill them. He had done it before." Experts on domestic violence say accounts like that of Ms. Walsh, who is now divorced, are not unusual. They say many men who abuse wives or girlfriends threaten or harm their animals to coerce or control the women.

Maine had to pass new legislation to address the underlying problem because, under existing law, judges were not permitted to include nonhuman animals in protective orders in domestic violence cases. With the new law, however, judges in Maine were permitted to issue official court orders that specifically prohibited harms to nonhuman animals as well. Importantly, within only a few years, many other states quickly followed Maine's example.

"The Cause of Our Time"

Of all human animals, it is still women who, as a group, suffer the most oppression. Even though women are now recognized as equals in many countries, there remains not only profound, extraordinarily harsh subordination of women in many cultures around the world today, but worse. A tragic measure reveals how complicated it can be to unwind traditional oppressions even when they protect the most remarkable animals of all—although killing of citizens in war, genocide, and civil society remains rampant among our species, the largest form of human-on-human killing involves women.

In 2009, the New York Times Magazine published a cover story, "Why Women's Rights Are the Cause of Our Time."[15] The authors repeated a 1990 observation of the Nobel Prize–winning economist Amartya Sen that our human population is missing 100 million women: "The global statistics on the abuse of girls are numbing. It appears that more girls have been killed in the last fifty years, precisely because they were girls, than men were killed in all the wars of the twentieth century. More girls are killed in this routine 'gendercide' in any one decade than people were slaughtered in all the genocides of the twentieth century."

The continued marginalization of many women around the world means that fields known as feminist studies and women's studies must confront problems as important as any facing our species. Such problems are illuminated by work on interlocking oppressions. But in important other ways, Animal Studies is linked to the different social movements around the world that seek equality, protection, and justice for women. It has become axiomatic that liberations already achieved by feminist thinkers have opened up many minds to the vast array of oppressions suffered by women. Feminist critiques of many practices in different cultures make obvious how deep and long the history of human-on-human oppression has been.

Insights about humans' capacity to oppress others carry over to nonhuman issues, which is sometimes, but by no means always, called out by feminist thinkers—thus, despite prominent examples in the work of seminal figures like Carol Adams, Mary Daly, Wangari Maathai, Catharine MacKinnon, and Martha Nussbaum, many advocates of equal rights for women do not affirm either humans' harms to other living beings or to the possibilities of rich connections beyond the species line.[16] A reviewer of Adams's influential *The Sexual Politics of Meat* in 1992 unpacked some of the connections between women and other animals:

> Metaphor is a particularly powerful way of consuming, annihilating, another's reality. For example, woman's actual experiences of rape are made absent, appropriated and exploited in the metaphors, "the rape of nature" or "the rape of the wild," which some environmentalists and ecofeminists are fond of using. Feminists annihilate the reality of concrete animals' lives and deaths when they complain that patriarchy treats women like "meat." Such metaphors negate the reality of a specific form of violence— being raped, becoming "meat"—and make it difficult, if not impossible, to recognize the connections between such multiple forms of violence as racism and sexism, racism and speciesism, sexism and speciesism.[17]

Those who wish to root out interlocking oppressions must cross from one field to another, from one set of oppressions (as in gender or race) to others (across the species line). This is no easy task in modern circles where human-centered preoccupations hold sway.

Deeper Leadership

Today women play central roles in education, law, veterinary medicine, environmental protection, and a wide array of related challenges to injustice. Of central importance to Animal Studies are both women's dominance in rank-and-file roles in the animal protection movement and the increase in women in senior leadership in traditional animal protection organizations. In contemporary Animal Studies, many fields at the cutting edge feature a more evenly balanced mix of women and men. Women's work in Animal Studies, in animal protection around the world, and in untangling interlocking oppressions is so diverse as to defy simple description. Today the leadership skills of more and more women are recognized, thereby giving everyone ever greater access to a deep reservoir of feminist insights in many disciplines.

Of particular note are the contributions to ethical reflection by Vandana Shiva, Carol Gilligan, Nel Noddings, Martha Nussbaum, and ecofeminists like Carol Adams that open up discussion of problems beyond the species line. Ethics has often been extremely tradition laden when made a part of formal education and governance. Importantly, though, every generation must contribute to the constant renewal of ethical inquiry as a central human preoccupation. Important contributions have been made by an astonishingly talented group of women leaders such as Rachel Carson in environmental advocacy and Jane Goodall, Dian Fossey, and Birute Galdikas in primatology and science more generally. Joyce Tischler founded and stands as moving spirit of the Animal Legal Defense Fund, which has promoted the field of animal law in education and practice, and Ingrid Newkirk of People for the Ethical Treatment of Animals has done controversial but effective work in nonprofit activism. The successes and differences among these leaders model how diverse visions have animated hundreds of millions of people involved in modern animal protection.

Women's studies has also opened up assessment of our science traditions, which are susceptible to gender bias. Such bias is found in preferred methods, actual practices, and sociological realities, as well as general concepts and philosophies that have prevailed.[18]

Linked, Historical Dismissals

Women have often been compared to and linked with nonhuman animals for purposes of denigrating both groups—for example, the formidable philosopher Aristotle was, like so many other humans, a prisoner of his own culture in the sense of following blindly certain inherited assumptions about both women and nonhuman animals.

Later centuries provide examples of similarly narrow ethics. Immediately upon the 1792 publication by Mary Wollstonecraft of *A Vindication of the Rights of Women*, now a classic discussion of both liberty and equality, Thomas Taylor published a parody titled *A Vindication of the Rights of Brutes*. Taylor hoped to convince others of the absurdity of Wollstonecraft's advocacy for women by pretending to demonstrate that "beasts" also have "intrinsic and real dignity and worth."[19] While denigrating women by connecting them with other-than-human animals had negative effects, some scholars have used that connection to recognize traces of more favorable attitudes toward women in the work of ancients like Philo and Plutarch who were known to be animal sensitive at times. Perhaps such connections help explain in part why women artists of all kinds have pioneered creative uses of animal-related themes in their work.

Some contemporary approaches that connect women and other-than-human animals feature an altogether constructive, mind-opening quality, such as the list of "alternative people" mentioned in a recent book by a preeminent scholar of Hinduism: This book "tells a story that incorporates narratives of and about alternative people—people who, from the standpoint of most high-caste Hindu males, are alternative in the sense of otherness, people of other religions, or cultures, or castes, or species (animals), or gender (women)."[20]

Other-than-human animals are at the heart of the list, and women take the final position—one of strength. Significantly, the author is female and remarkably articulate about the central importance of females in all features of human and more-than-human life. By virtue of their inclusion in this list, other-than-human living beings are accorded a central place in the Hindu tradition's answer to the heartbeat question of all ethics, namely, Who are the others?

Anecdotal Evidence

More women than men are involved in Animal Studies, anthrozoology, the animal humanities, or research into human-animal relationships. While individual anecdotes fall short of scientific data, they can still be worthwhile.

Every group of students enrolled in the Animal Law course I have been teaching at Harvard Law School since 2002 has been overwhelmingly female (80-plus percent on the average). My summer-term Religion and Animals course (also at Harvard in 2009–2011) has had similar percentages. Whenever I have attended a religion and animals session at different conferences, the overwhelming majority of people in the room have been female. During one six-year period beginning in 1999, the graduate program in animals and public policy at Tufts' Cummings School of Veterinary Medicine admitted only female graduate students (the class ranged in size from eight to thirteen)—what's more, for a number of these years there were no male applicants at all in an applicant pool that numbered from thirty-five to fifty. Even in years when some males applied and were accepted, there were only one or two. Pools of both applicants and matriculating students were always over 90 percent female during the decade I was involved with that program.

Similar numbers appeared in the veterinary school classes I taught during that period. In the United States for the last decade, more than 75 percent of the 2,800–3,000 students entering veterinary school annually have been female. Some schools have had entering classes with only females. This dominance is a relatively new development, for it was only in the 1970s that women began to outnumber men in the entering classes of American veterinary schools. As a result, women became a majority of the veterinary profession in the United States by about 2005–2006.[21]

Such a major shift is not new. The vast majority of teaching profession in the United States, for example, was male until 1830, but by 1860 women outnumbered men in some states. By 1870, 60 percent of teachers were female, and by 1900 70 percent of all American teachers were women. The trend reached a high in 1925 of 83 percent.[22]

Illuminating Animal Studies

The topic of women and nonhuman animals is, then, for many reasons one of the most productive and wide-ranging inquiries pursued by Animal Studies. Women's studies has through its depth, breadth, and creativity modeled traditions of open inquiry, critiques of received concepts and values, and general scholarship. In sum, Animal Studies relies on—indeed, often stands on the shoulders of—a plethora of approaches, insights, and forms of creativity developed in women's studies.

Children and Other Animals

Even though our species' discussion of children's relationship to other living beings remains in a rudimentary stage, this topic has already emerged as one of the fastest-developing subjects in Animal Studies. Further, exploration of children and other-than-human animals involves multiple dimensions and questions that require interdisciplinary approaches. Indeed, even combining all forms of present-day science with the most encompassing forms of contemporary humanities still would not have good prospects of exhausting this fecund topic— in short, inquiries about children and other animals require one to examine a fundamental intersection that adult humans are only starting to explore in any detail.

Consider the Past

As the child development researcher and scholar Gail Melson has noted, "children's ties to animals seem to have slipped below the radar screens of almost all scholars of child development."[23] It is only now, through sensitivity to children's multifaceted fascination with other-than-human animals, that insightful researchers have discovered that children develop early in their lives what Melson calls "a core domain of knowledge about living things."[24] Such research is needed to grasp not only the dimensions of children's connections beyond the species line, but also the radical failure of some societies to recognize and honor these connections.

Importantly, recognition of the connections leads to important benefits in humane education. On the exclusively human side, such recognition helps one develop therapies based on the beneficial impact some nonhuman animals have on emotionally troubled children. And as many readers will already know, research has for decades confirmed that the mere presence of nonhuman animals has physiological benefits for adult humans, too.

Consider the Present

Children have demonstrable interest in, and often special relationships with, other animals. While any number of survey techniques honed in sociology circles can show how astonishingly fascinated children are with living animals, the simpler method of merely walking into any modern bookstore will show that images of animals move children dramatically.

On websites designed for children, elementary school classrooms, or advertising media intended for children, images of real and imagined nonhuman animals abound in the worlds that adults create for children. There is a darker side to adults' perception of such interests, for certain harms are done to nonhuman animals in the name of children's interests. Zoos explain animal captivity in terms of children's education, but holding many exotic nonhuman animals captive creates serious problems and even trauma. Foundations that lobby governments for fewer restrictions on use of live animals as research tools also cite children's interests regularly.[25] Proponents of dissection exercises using live animals in science courses have also argued that the interests of both society and children are such that even unwilling children should be required to perform such exercises. The claim, which is hotly disputed, is that requiring all children to go through such exercises prompts learning about the fundamental features of science. There is not, of course, much talk about how forcing unwilling students to participate in these exercises can cause negative attitudes toward science, or why parents should be able to choose the best option for their child (most legislation creating the possibility of opting out of this kind of education requires parental consent as well). Finally, when live animals are used in dissection, there is rarely, if ever, any discussion of the risk that such exercises teach children that humans have sufficient power to dominate, as well as dissect, animals in the laboratory setting.

But Especially Consider the Future

Any substantial engagement with Animal Studies puts one squarely amid questions about the future of children, education regarding other-than-human animals' actual abilities, and the harms and risks created by present human practices. Choices today in law, education, and public policy project an imagined future onto our children. Even as we wonder what sort of world we will leave for our children and their children, we answer this question implicitly by the kinds of education, laws, and social values we create and under which our children will mature. Some incongruities are evident when adults take charge of children's lives in these ways. For example, in choosing an imagined future for children, adults may have radically subordinated, dismissed, and otherwise made disappear the very animals in which children so naturally and fully delight. We can ask if present educational practices nurture or blunt the prospects of upcoming generations being open to the fact that humans live in a mixed community shared with many other forms of life.

The younger generations are impacting the future in a grassroots manner. Many children in industrialized countries push their parents on questions regarding nonhuman animals—the upshot is that any number of people in business, law, education, and government have become more attuned to the moral dimensions of modern societies' treatment of certain nonhuman animals.

Some children are taught by their parents or others in ways that are altogether open-minded. They may be schooled in patient observation, or in a tradition of speaking of other animals as kin or members of their own communities. They may be taught to recognize, tolerate, and even welcome competing claims. Some may have teachers who acknowledge that images frequently are less about the actual animals and more about cultural tradition, personal construction, or economic advantage. Some may be encouraged to use their own talents regarding other living beings, and even urged to wonder if inherited notions are driven by the realities themselves.

Some parallels to human-on-human problems are worth examining. What is to be done when, for example, a child has been taught negative caricatures about certain human groups? Some human groups have developed mechanisms by which negative stereotypes are repeatedly identified for children, with the expectation that maturity involves thinking critically about images that can be determined to be factually inaccurate. With nonhuman animals, of course, such open-minded training is only rarely the case (recall Midgley's comments about philosophers' failure to detect the problems with folk images of wolves), but from one vantage point, Animal Studies is an extended attempt to develop deeper and wider skills at seeing how our inherited views of other animals are often so superficial as to be fairly identified as willful, self-inflicted ignorance.

Adolescent Aspirations

Contemporary polls reveal something astonishing about connections felt by teenagers. While even a mild passing reference to the controversial group People for the Ethical

Treatment of Animals puts off many (animal consumers and protectionists alike), at least one study suggests that it is precisely this animal protection group that most interests the younger generation in industrialized countries. In 2006, Label Networks, Inc., a marketing company which described itself as "the leading global youth culture marketing intelligence + research company authentically measuring the most trendsetting and mainstream subcultures in the world," published its Humanitarian Youth Culture Study. According to the study, "PETA is the #1 overall non-profit organization that 13–24-year-olds in North America would volunteer for … peaking among 13–14-year-olds at 29.1 percent of this age group."[26] What is significant is that PETA held nearly a two-to-one margin over the runner-up (the Red Cross, which had just received great publicity after the 2005 devastation of Hurricane Katrina).

Such figures might shed light on why some children urge their parents to be concerned with "animals." But children who are far younger and have never heard of PETA often question their parents about diet, harms to local wildlife, and a great variety of other animal issues. Given the enormous harms inflicted on animals today, it may be inevitable that children lobby adults in this manner. Most children, however, are sheltered from the details of the harms that befall nonhuman animals in our food systems, zoos, and laboratories precisely because children so often are tenderhearted about injuries and captivity for nonhuman animals.

There are comparisons between children and other living beings meant in negative ways. But today there are special positive comparisons between children and certain nonhuman animals, such as reference to a family's companion animals as "our children." The implications and problems of such uses are complex enough to make this a separate issue in Animal Studies.

Choosing Children's Education

Making careful, examined choices about education requires reflection about one's own thinking, especially in settings where many different points of view, including both ethics-focused and nonanalytical approaches, are welcomed at the discussion table. Such habits of mind also help one see poignant features in the ways we educate children about language choices. Most adults train their children relentlessly to use the word "animals" to mean "all animals other than humans," not in the alternative sense "man is a political animal."

Some suggest that religious authorities, government officials, school board authorities, or parents should have the final decision as to what schools do in this regard. By virtue of its interdisciplinary potential and commitments, Animal Studies is capable of marshaling not only the most relevant information and perspectives about other animals, but also why language choices make a difference, as well as how specific or general cultural and sociopolitical factors play out in such discussions.

Dealing with the Exceptionalist Tradition

Animal Studies will be enriched if it keeps in constant dialogue with very diverse, confident voices from different human endeavors, including those whose voices are particularly strident on the issue of what children might learn about other-than-human animals. The exceptionalist tradition has supporters in many different circles. Some science advocates claim that their approach and their skills at critical thinking give them the definitive word on what counts as human knowledge of other animals. Such advocates will undoubtedly assert that science "supplies a lot of factual information, and puts all our experience in magnificently coherent order" even as they fail to add that the thrust of Schrödinger's famous comment is that science "keeps terribly silent about everything close to our hearts, everything that really counts."

Voices from different religious traditions, of course, are quite confident that religious believers have special abilities to discuss "everything that really counts." Philosophers and ethicists, too, have strong opinions about such matters. Members of each of these human endeavors might also feel, even publicly assert, that it is not science-intensive work but their own traditional forms of valuing and thinking that should be accorded the final word regarding what children should learn about the more-than-human world. Humanists, too, may make exclusive claims, for such advocates have insisted that only human beings should be members of the moral circle.

Whenever a vested interest of any kind claims that its advocates alone have the key to the form of knowledge most relevant to other-than-human animals, one can be confident that some key insights are being ignored. Whenever someone asserts that it is inappropriate (perhaps even irreverent) to give respect to a range of views, Animal Studies would be wise to convene an interdisciplinary gathering to probe such claims as reasonably and patiently as possible.

When it comes to children's education, though, the claims often become more shrill. Questions about what to learn or unlearn about the living beings in our larger community touch on some of the most sacred issues in humans' lives. Challenging the exceptionalist tradition means, in many cases, challenging not only long-standing harms, but also privileges that are part of certain worldviews, cultural heritages, religious beliefs, or researchers' preference to pursue science-based inquiries without any ethics-based constraints. Since challenges to the exceptionalist tradition are often brought by active citizens concerned that children be allowed to choose for themselves whether to respect and protect other animals, such claims can be framed in very powerful ways. Some suggest, for example, that when the principal way children encounter other animals is on their lunch plate at school, some alternatives are needed to be fair to the children. But conflict is inevitable since who decides what children eat and whether they might be afforded a choice is a deeply personal issue to parents, even as it is a moral issue to many nonparents. This debate will no doubt continue into the foreseeable future.

Animal Studies suggests that while it is disappointing that self-inflicted ignorance often dominates what adults count as their own knowledge, this problem rises to the level of tragedy with children's education since shortcomings need not dominate young humans' education about other living beings. There is now, thankfully, a well-developed discussion about the need for better study of children's relationships with other animals. Such discussions need to be as inclusive as they are interdisciplinary, for this is likely the only way that our species can learn how best to study why and how animal issues are so important at the earliest levels of formal education.

Imprisonment as Education?

It has been argued that zoos imprison us by locking us and our children into a mentality that requires coercion and domination of nonhumans (some zoo advocates openly admit that coercion and domination are part of zoo-based captivity, while others refuse to acknowledge this obvious point).[27]

Domination is, as a mentality, in the end detrimental to humans even as it creates far worse harms for the animals that we use for entertainment, profit, and captivity-based "education." Many children easily recognize such realities, thus piercing through zoos' claims that the presentation of captive animals is richly educational. At what age do children visiting a zoo learn the important, ethics-fraught message that our society approves of the captivity of nonhuman animals for human animals' entertainment? Note the power in this kind of question—it explains why it is fair to ask if zoos imprison our imaginations. The question does not require a denial of zoos' popularity, but it does ask about the quality and implication of zoo-based education about other animals. In fact, zoos' own conservation materials in many ways implicitly acknowledge what everyone knows, namely, the best place for some animals is not in our human society but in their natural habitat living a free life amid healthy social groupings and opportunities.

Learning Well

As remarkable young animals, children learn about both their conspecifics and other animals in a great variety of ways. They are socialized into their birth culture's ideas and practices, and thus as traditional animals they characteristically begin to think the views and claims they have learned are "knowledge." Attending rituals, meals, or other ceremonies and events, a child can become invested with an appreciative or dismissive attitude toward particular nonhumans.

When children have not learned through encounters with other living beings themselves, but only through social constructions and less-than-accurate generalizations about "animals," some unlearning will be needed if the goal is a responsible view of other living beings. Animal Studies must often go forward in societies where most children, juveniles, adults, and the elderly live their lives convinced that humans are far more than a distinct species. They often are certain that humans are a separate category of life, entirely unlike all other animals, uniquely possessing abstractions like "soul" or "intelligence." As an affirmation of humans' extraordinary abilities, such claims have their place, but when they are used to dismiss nonhuman animals, the issue of self-inflicted ignorance and the need for unlearning will likely be in many other people's hearts and minds.

An Instructive Comparison

In the cultural context of contemporary education of children regarding nonhuman animals, there are interesting lessons to be learned. For example, by studying the way children are taught about animals in the Jain culture or in a small-scale indigenous culture, researchers can see all the better certain features of contemporary education of children in industrialized societies.

When children are raised in a Jain religious community, primary importance is given to ahimsa or nonviolence. Jains have from time immemorial held that all life is sacred, and, like Buddhists and some Hindus, they start the day with a number of affirmations, one of which is a commitment not to harm any living being. Similarly, children raised in a small-scale society like that of the Rock Cree in eastern Canada receive lessons replete with themes that outsiders would likely deem religious in nature. Game animals, while hunted, are nonetheless deemed "reactive social others, alternately collaborating in and obstructing the designs of men and women who kill them with guns and traps."[28] Members of the tribe learn to relate to (and certainly to talk and think about) other animals as "social others" with "human and animal categories [that] are themselves continuous rather than discrete." The upshot is that Rock Cree hunters have "moral commitments" to the nonhumans they pursue.

Those who were not raised in either of these communities can easily notice that what the Jain and Rock Cree parents teach their children about other animals is an account at least as fully dominated by ethics and morality as it is by practical knowledge of the animals' lives. The lessons are positive and produce powerful psychological anchors for a lifetime of constant awareness of other animals. In effect, family and social conditions in these communities create and then sustain a learning environment that foregrounds ethical awareness.

In a widely read 2005 study, Richard Louv addressed a complex of problems occasioned by children's removal from the natural world in modern industrialized societies, creating the risk of nature deficit disorder.[29] Louv argued that among the specific disadvantages of limited exposure to the more-than-human world is impaired development of cognitive and ethical abilities. Much literature underscores both educational and therapeutic benefits to adults and children from the presence of other animals. Other advantages include increased environmental awareness. Such benefits can spur forms of responsible citizenship by prompting learning about which consumer practices are particularly harmful. While such benefits are only occasionally mentioned in debates about protecting other living beings, they have additional relevance to debates about how any individual human can lead a meaningful, integrated life.

Louv's book is not specifically about children's experiences with individual nonhuman animals, although the general trend of pulling children away from natural places has obvious relevance to their opportunities to notice how ubiquitous life is in natural spaces. Children still have some opportunities to learn about other-than-human animals, of course—they encounter companion animals, backyard wildlife, and elusive figures in the urban, suburban, and rural interstices of their home community. While such experiences may be common, it is nonetheless obvious that children who grow up in an environment where these opportunities are circumscribed miss something important about the natural world.

These children's experiences may be further limited by their cultural environment. When children's encounters with other animals are expressed, and thus understood, by the dualism "humans and animals," then there is a double problem. There are not only risks of nature-deficit disorder, but also a dearth of opportunities to think again and again, and thus critically, of the relationship of human animals to other-than-human animals. If their curricular content is relentlessly geared to human superiority, the result is myopias and crass oversimplifications, and thereby conditions that allow the exceptionalist tradition to prevail and grow.

Imagine the difference between the sensitivities of children educated in societies that notice and pay attention to other living beings and the insensitivities encouraged by dismissive attitudes. Children in the latter societies are not well equipped, and certainly not encouraged, to ask frank questions about the actual realities of the nonhuman living beings that share the earth with our species. Such children are encouraged to ignore harms done to other animals subjected to domination and coercion—for example, they often learn to see animals used in agribusiness as mere commodities. Similarly, they are not encouraged to inquire about the abilities or suffering of the individual animals science uses as mere experimental tools. Further, they are not taught to see the domination and coercion that zoos necessarily maintain in order to present captive animals for the children's "education."

If a society does not teach its children to notice other animals (including unrealistic caricatures), then many of its children will not notice other animals—their feel for animal-related issues will be empty unless they break through to such issues in some other way. While some might argue that such apathy benefits science (because unconcern implicitly supports laissez-faire policies), there is a powerful argument that lack of concern harms science. Apathy in no way promotes the intellectual freedom that is a precondition of fundamental questioning. In effect, schooling children to be unconcerned stops the very heartbeat of the scientific tradition, "an alliance of free spirits in all cultures rebelling against the local tyranny that each culture imposes on its children."[30] Said another way, such one-dimensional education hardly prompts children to "question everything." In the humanities as well, occasionally students will have the opportunity to ask questions about ethical blindnesses and other living beings. Yet, since the topic is the humanities (named after our own species), no one is surprised when the agenda is a relentless focus on humans as the Earth's lone special animals.

Such one-dimensionality leads to lost opportunities. Children's deep and natural fascination with other living beings has educational traction in matters of science, pursuit of personal interests, and development of responsible actions toward the community and environment. In light of this, one might plausibly ask, how does one open up education in ways that helps children freely explore and, if they wish, continue to embrace the more-than-human world?

Animal Studies provides key information and insights that help foster curricula that stimulate children's full range of inquiry along these lines. It also carries deep capacities to stimulate thinking skills and ethical development. The easily observed fascination and affinity that most young children have for other living beings has been only recently explored in any detail because pioneering child psychologists such as Freud and Piaget misgauged children's interactions with other animals. Like that of indigenous peoples, so, too, children's interest in other animals has often been seen as foolishness. Children are strongly encouraged to shed this interest when they enter adulthood, and failure to do so becomes a hallmark of immaturity.

Study of a wide range of human cultures, however, reveals that it has been common for adults in other cultures to "recognize the affinity of children for animals and build on...images that link children to animals."[31] It is telling that modern industrialized cultures have for more than a century worked to undo this affinity. "Children in Western cultures gradually absorb a worldview of humans as radically distinct from and superior to other species, the human as 'top dog' in the evolutionary chain of being. What [Myers] calls 'the categorically human self' emerges—a strict division between human attributes and often negatively valued animal characteristics."[32]

Animal Studies can help those educated away from a concern for nonhuman animals recognize that many human cultures choose to do just the opposite. Further, even when modern societies pull children out of the natural world, opportunities for exposure to other living beings remain. It is fortuitous that our cities provide more diversity of life than is normally imagined.

Throughout our learning of language and our journey through the different stages of formal, institution-based education, there are other ways we gain awareness of other living creatures. For example, we can learn a great deal from our own bodily experiences of "the overwhelmingly obvious similarities" between humans and some other animals.[33] They can help us understand certain things some nonhuman macroanimals do. Of course, we do not have much awareness of what a mosquito feels like when it is flying.[34] But when a chimpanzee or some other social mammal is bored and restless in a cage, our extraordinary biological similarity provides some basis for imagining why that individual is not thriving.

Wildlife as Teachers

A primary way in which we become aware of other living animals is that we encounter them. The domesticated dogs, cats, horses, rats, mice, birds, fish, and sundry other living beings we bring into our homes provide such encounters, but companion animals involve some complications that are often brushed aside. Domestication is a form of domination that can make these nonhuman animals conform to our lives in ways that hide what their lives apart from humans would be like. However, we learn otherwise inaccessible fundamentals from wild or free-living animals, for they are least affected by us. Their obvious complexities are played out in contexts and social realities unknown to us. No human really knows all that much about any oceanic or riverine dolphins in their social groupings. We have very little idea about what sperm whales do with their brains, the largest on earth—we simply do not know anything but the most obvious trivialities about sperm whale individuals. What we know of the actual realities, emotions, and intelligence of far more familiar animals, such as elephants or nonhuman great apes, also remains scant. What we can surmise about even less-well-known animals is humbling in an even greater degree.

Encountering one of the more complex nonhuman individuals in our shared world is, then, for virtually all of us a qualitatively unique learning experience. There are many testimonies about the awe-inspiring qualities of wild animals—the following comment by Thomas Berry inspired the book *A Communion of Subjects*:

> Even with all our technological accomplishments and urban sophistication we consider ourselves blessed, healed in some manner, forgiven and for a moment transported into some other world, when we catch a passing glimpse of an animal in the wild: a deer in some woodland, a fox crossing a field, a butterfly in its dancing flight southward to its wintering region, a hawk soaring in the distant sky, a hummingbird come into our garden, fireflies signaling to each other in the evening.[35]

The invitations to wonder, to curiosity, to exploration brought by the presence of free-living nonhuman animals are extremely diverse, of course. The English philosopher Mary Midgley has suggested that "animals, like song and dance, are an innate taste" for humans as members of a "mixed community." Children can teach us not only that Midgley's observation is good biology and ecology, but also that it is conducive to the open-mindedness needed for learning and unlearning regarding the nonhuman living beings in our larger community.

Other Marginalized Humans

It is not only women, children, and indigenous peoples who have been marginalized—so have people of a disfavored race or color, those viewed as holding "heretical" or nonorthodox positions in a religious tradition, political nonconformists, people whose lifestyle is not approved for some reason, and so many others. Two examples here illustrate the connection of these marginalized people to nonhuman animals. The first is a much-discussed comparison of the treatment of black slaves in the United States with nonhuman animals, and the second is a discussion of the way heretics in the Christian tradition have at times been compared to nonhuman animals.

The Dreaded Comparison

Marjorie Spiegel's detailed comparison of the treatment of black slaves in the nineteenth century and contemporary treatment of nonhuman animals is well known.[36] Perhaps because Spiegel knew that some would find any comparison of a human group to nonhumans offensive, she sought the endorsement of Alice Walker, a prominent black writer, whose preface addresses the issue of comparing practices across the species line: "The animals of the world...were not made for humans any more than black people were made for whites or women for men. This is the essence of Ms. Spiegel's cogent, humane, and astute argument, and it is sound."[37]

Critical inquiry will prompt any number of questions about such comparisons. Among the most obvious inquiries is whether the parallels drawn by Spiegel between the treatment of blacks and nonhuman animals are historically correct. The book uses eight points of comparison:

1. The recurring association of blacks and nonhuman animals in daily language, literature, and art
2. The use of branding, masks, collars, and other binding techniques
3. Similarity in transportation techniques
4. Similarity in attitudes toward production of these "workers"
5. Hunting and experimentation practices
6. Patterns of defense and rationalization by the establishment (including appeals to God and scriptural justifications, economics, and a natural order that places the oppressing group atop a hierarchy)
7. Secrecy, hiddenness, and propaganda regarding actual conditions; and conditioned ethical blindness of those involved in daily practices
8. Caricatures of the marginalized group to cover up the dominant group's faults (black men said to rape white women as a way of distracting from white men's rape of black women, and "animals" said to be vicious as a way of distracting from the viciousness of human domination over and cruelty to nonhuman animals)

Historically, comparison of humans to nonhumans was often meant to demean the human group, and this was surely the case with black slaves in the United States—this is, in fact, why Spiegel's title included "dreaded" as the adjective modifying "comparison." At a deep level, Animal Studies is capable of plumbing why any such comparison would offend. The answer, of course, lies in the psychological investment so many people have in humans being the earth's most remarkable species.

Alice Walker's endorsement also prompts questions. One might wonder, for example, if such an endorsement validates the analogy, as forceful as it might be, as politically acceptable. We can also inquire whether such analogies have been drawn previously without the intention to demean. The answer is that they have—the American Nobel Laureate Isaac Bashevis Singer, for example, used a comparison of the herding of Jews in Nazi Germany to the ways cattle were dominated in the pens he saw out of a window near the Chicago stockyards.

But, clearly, demeaning comparisons have occurred in many cultures and religious traditions. In "Moth and Wolf: Imaging Medieval Heresy with Insects and Animals," the Harvard Divinity School scholar Beverly Kienzle points out how establishment figures in the Christian tradition appropriated images of nonhuman animals (and some specifics of their behavior) and applied them in the late twelfth century to "heretics," that is, nonconforming believers.[38] Mere association of believers with nonhumans (both insects and more familiar macroanimals) was meant as a condemnation. Similar comparisons such as condemnation of Jews or Muslims by Christians, or condemnation of one Islamic group by another have been very common in history, and in many instances continue today.

The Expanding Circle Narrative

Even thinkers in fields with a developed tradition of critical thinking have often failed to alert us to the limits we face as ethical animals; and they sometimes fail to recognize the distorting features of our language, cultural heritage, widespread assumptions, and even what a majority of people take to be "common sense."

The image of an expanding circle can represent certain features of our own species' political and ethical history. But this image can beguile in a subtle but important way. If we fail to see the following problem, we miss an important feature of our relationship to other animals. The influential nineteenth-century historian of morals William Edward Hartpole Lecky famously observed, "At one time the benevolent affections embrace merely the family, soon the circle expanding includes first a class, then a nation, then a coalition of nations, then all of humanity, and finally, its influence is felt in the dealings of man with the animal world."[39]

The expanding circle appeals for several reasons—it is both simple and hopeful, and thus we relate to it easily. The image also helps us see important features of our moral lives— for example, humans' ideas about the reach of ethical abilities have not been static. In one era after another, societies have changed their notion of "the others" that deserve protecting.

The expanding circle idea is correct in one implication, namely, humans sometimes critique received views and replace them with other, more inclusive schemes. But most of us know that the history of ethics is more accurately described as pendulum-like, swinging back and forth, sometimes expanding, sometimes contracting. In the longer run, the issues are much more complicated than the expanding circle suggests.

The expanding circle misleads regarding a fundamental challenge to ethics that is inherent in humans' inevitable intersection with nonhuman animals. Lecky's sequence of expansions starts with the individual in a family and moves progressively through class, nation, coalition of nations, all of humanity, "and finally...the animal world." This sequence is by no means the actual trajectory in our own cultures—our ancient forebears often included some nonhuman animals at times when they very clearly did not include all humans. So the expanding circle image needs some tweaking and qualification if it is to be illuminating at all.

Lecky simply assumed that humans naturally and rightfully include all members of our own species before including any other beings. This is a radically misleading assumption, for many humans prioritize some nonhuman individuals—such as their companion animals—over the well-being of many other humans. One may not approve, but it is widespread today. In essence, Lecky's assumption that ethics broaden in the manner of an expanding circle was less description than prescription based on his culture's bias for humans first and foremost.

The geometric shape of a circle actually functions far better as an essentially ecological image, whereas Lecky's circle is primarily conceptual. Suppose we encounter someone who has always lived in one town. We can consider this person as the center point anchoring a local world of decision and action that, in a sense, encircles this person. Assume that this person is a morally capable being—as our individual wonders, "Who are the others about whom I might care?," family members will likely come first. Perhaps the person also rubs elbows with neighbors in this shared community.

Soon enough, though, our moral person will also encounter in his or her "circle" some nonhuman animals, perhaps dogs or cats or other-than-human neighbors. The simple point is that as our person moves further and further out into the ecological version of a widening circle that spreads out in all directions, she or he will encounter many local nonhuman animals long before other human beings, such as the faraway citizens of the same nation, let alone the people in other nations or even bigger units (the "coalition of nations" or "all of humanity" mentioned in Lecky's sequence).

Lecky's sequence is likely to mislead us into assuming that we must first care for all humans, and only then nonhuman animals. But such an approach lacks realism—people unavoidably live and act locally. Any set of prescriptions that makes living beings on the other side of the world more important than local beings will struggle amid day-to-day ethics.

In essence, Lecky takes for granted the exceptionalist tradition's one-dimensional portrayal of the moral human. Each human does care first about other humans who happen to be family members and perhaps nearby neighbors. But we do not have an instinct to protect all humans in day-to-day life. Lecky's sequence does not reflect the most important realm of real ethical lives (namely, the local world) but highlights the familiar biases of the exceptionalist tradition.

The idea of an expanding circle can work if an ecological sense of the moral circle replaces the human-centered conceptual circle. In a very real sense, each of us does inhabit the center of a real circle where we are constantly making choices. As individuals, we are citizens of the entire earth in a minor, mostly metaphorical sense, but we are certainly not the earth's center in any discernible way. True, today we make consumer choices in a globalized world, but this important phenomenon does not eclipse the central reality of our individuality in a truly local world where we make daily choices. Like all other animals, we live first and primarily in a specific local community. This is the ground-zero reality of the ethical life, and through it we have potential relationships with specific people and nonhuman animals who coinhabit our shared bioregion or econiche.

There is simply no sense in which one can say "we share the earth with all humans" that is not also equally true of the claim that "we share the earth with all other animals." We may think of ourselves as citizens of a humanized earth, but this is beautiful and true in only limited senses that hardly are more important than the fact that we remain citizens of a more-than-human world. It is this citizenship that gives us the chance to be responsible moral actors within our local world and moral circle.

Animal Studies and the Moral Core of Human Life

It is a central task of Animal Studies to be realistic about humans' ethical potential and actual realities. Further, Animal Studies has the corollary task of critiquing received ideas of ethics that do or do not pay attention to such potential and actual realities.

Arguably, what puts all other humans into each human's moral circle is not a biological urge but, instead, a commitment driven by the beautiful idea that each and every human being matters immensely. This idea-based commitment is far more than mere sleight of hand (or perhaps a sleight of mind). It is a decision, a matter of choice that grows out of a mature, developed character capable of the kind of moral imagination committed to treating all other humans as members of our community of "others." Yet, as beautiful as such a commitment is to us, it can become a source of problems if it is used as a justification for not caring about any other-than-human living beings. It is an important theme of this book that countless individuals in many cultures have chosen in their daily lives not to exclude nonhuman animals even though many high-profile modern cultures have allowed this foundational feature of human life to atrophy.

A problem emerges when the important notion that all humans count is surreptitiously converted into the claim that only humans count. When members of our species tout only humans as our real community, they miss the genius of so much that is the human moral animal.

In a spatial, practical, and truly ecological sense, then, we live our lives in a circle, our specific local world, in which some other animals are much closer coinhabitants than are remote humans. Because our true moral circle is, first and foremost, local, individual humans are not likely to find integrated, happy ways of living if they refuse to be cognizant of their truly local neighbors.

The Question of Leadership
Getting beyond Pioneers and Leaders to Individual Choices

What constitutes leadership in Animal Studies varies considerably as people work to engage the past, present, and future of the human-nonhuman intersection. In circles where the exceptionalist tradition prevails, leadership takes forms completely different from those one encounters in nonindustrialized societies that have long practiced a lifeway imbued with respect for living beings beyond the species line. Leadership takes yet other distinct and specialized roles in those communities and subcultures transitioning away from traditional harms and dismissals of other animals.

As young and old citizens in both industrialized and developing societies grow toward more open engagement with other-than-human animals, what reveals best the diversity of leadership is a feature of the change that often goes unremarked—the work of any well-known advocate for other-than-human animals is possible only because of the work of, literally, thousands upon thousands of anonymous women and men whose efforts make the achievements of high-profile leaders possible. It is this grassroots work that reveals the breadth and depth of the worldwide social movement discussed in previous chapters.

When the Real Difficulties Begin

"It's hard enough to start a revolution, even harder still to sustain it, and hardest of all to win it. But it is only afterwards, once we've won, that the real difficulties begin."[1] These insights suggest why leadership in any social movement must feature tremendous diversity—what counts as leadership evolves constantly. For example, some of the leaders that sparked the modern reemergence of animal protection came out of grassroots efforts that organized volunteers working on specific community problems (like euthanasia of unwanted companion animals), while other leaders worked to create movements in legal education, philosophy, and legislative circles. Still others founded nonprofit organizations that focused on specific problems.

In the decades following the mid-1970s reemergence of an animal protection movement in industrialized societies, some important changes have been accomplished even though many problems remain and some have worsened dramatically. Sustaining, consolidating, and building on the accomplishments requires leaders capable of expanding efforts at the grassroots level, and others with the skills to grow and fund organizations at the local, regional, national, and even international levels. Yet other leaders work in academic circles to expand description of the underlying problems and develop student interest. This work produces more opportunities, new leaders, and ever more sophisticated challenges through litigation and legislative lobbying on a wider range of problems. With volunteer efforts and organizations expanding, legislative and court-based possibilities growing, and academic work diversifying and becoming ever more sophisticated, new leadership opportunities emerge for change and even abolition of certain harms to nonhuman animals.

It is not easy to say whether and when ferment becomes a revolution, or when a revolution has been sustained, but it is clear that as tens of millions of people have become interested in animal protection issues, there are fundamental changes that will be "hardest of all to win" (for example, abolition of the most harmful practices, or adjusting the central legal concept of property so that it no longer produces such terrible harms to so many billions of nonhuman animals). Finally, even if such a high level of change is realized, leaders will continue to face major challenges because "it is only afterwards, once we've won, that the real difficulties begin."

What Matters Is the Quotidian

It is choices and day-to-day efforts of grassroots volunteers, consumers, factory workers and managers, artists and business owners, teachers and students at all levels of education, professionals and academics of many kinds, and countless others that drive change. Leaders may be the most visible, but it is everyone together who create or dull the insights and energies that drive Animal Studies. Because each human individual faces first and most often a local world, it is within that world that each person constantly chooses. Such daily choices are the obvious foundation of every major movement for social change. Individuals shape present options and future possibilities through daily opportunities, thereby choosing to lead lives that move toward or resist a particular choice or change.

Leadership of a most important kind, then, takes place in such personal and local worlds. The lives of many ordinary people feature actions and sacrifices that would clearly be thought of as achievements if known; such people have often existed at local levels even though their individual work did not become well known because "history from above" failed to record their achievements.

It is at this level of local worlds that student petitions for animal studies courses are imagined and then circulated, and it is at this level that people notice and take seriously the macroanimals that are neighbors. It is locally that voter initiatives on wildlife, food choices, dissection options for students, and other animal-related problems are won or lost. It is also locally that individuals engage or refuse to care about the actual realities of other animals because daily choices have ripple effects well beyond one's own life—because "every choice we make can be a celebration of the world we want."[2]

Listening as Leadership

Listening to others is another key but less-than-high-profile form of leadership. The work of Wangari Maathai, Vandana Shiva, Jane Goodall, and many others has shown that listening primes others for communal participation. The importance of listening to other humans is well demonstrated by certain problems in the subfield of religion and animals. I have on several occasions been told that members of the animal protection movement (who in many industrialized societies have characteristically been suspicious of religion) assume religious people need a new vocabulary. Religious community insiders, of course, recognize that something else is really needed, namely, for movement activists to learn to listen to each different religious group and then talk to that group in ways that the group already values, not with an entirely new vocabulary that implies the group's existing ways of talking, thinking, and living are morally deficient.

A particularly primal form of listening occurs when humans pay attention to other animals' individual and social realities. Similarly, a sensitive form of listening is needed to assess our own limitations as we attempt to explore other living beings' actual lives. Many people, such as Griffin, Goodall, and Bekoff, have pioneered attempts to listen to other animals in a fair and full way. Such listening has enabled countless others to explore local animals better, to identify caricatures, and to propose new ideas so that the human community may see issues more clearly.

Such work is also the foundation of much ethics-inspired work. For example, when the work of those who document harms to other animals can be joined to well-developed information about other animals' realities (such as their intelligence, or their need for freedom of movement or society with members of their own species), the prospects of the public supporting change (through their consumer choices, volunteer efforts, or calls to legislators) are much better.

Pioneers

The acknowledged pioneers who have challenged the exceptionalist tradition have had special, even if greatly varied, skills. Many of them recognize that the truest leadership is the work of the unknown people whose efforts comprise the worldwide movement.

What is commonly known about pioneers is often but a very small part of their contribution to awareness of life outside our species. Many people around the world know Rachel Carson's name because her *Silent Spring* was an important book in the history of environmental thinking—she also wrote best-selling books about the ocean.[3] Few people now know Ruth Harrison's name. In 1964 she published *Animal Machines: The Factory Farming Industry*, for which Carson wrote the preface. In that preface, Carson said, "Her theme affects practically every citizen, for it deals with the new methods of rearing animals destined to become human food. It is the story that ought to shock the complacency out of any reader." Harrison is not often talked about today, nor is Carson's support of her early indictment of harms to nonhuman animals. This shared obscurity shows how "history from above" fails to record many important forms of leadership even among pioneers in human thinking.

A Commonality: Our Indigenous Ancestors

Less obvious pioneers are our ancestors who lived in small-scale, indigenous societies—these peoples produced a vast array of observations and wisdom about the living beings outside our own species. Each of us is, through our own culture, heir to some portion of this lore, whether practical knowledge or story narratives that teach in a variety of ways. Importantly, Animal Studies has been developing quickly precisely because modern industrialized societies have handled this complex, sometimes opaque heritage in ways that minimize or eliminate altogether key insights such as humans' relatedness to other living beings.

Every human alive today is, of course, a descendant of ancient peoples who characteristically thought of other-than-human animals as bringers of blessings. But these common ancestors are now mostly unknown, and certainly their insights are no longer factors in daily choices given that many moderns have been trained to deride them as "savages." But if people conduct "personal archaeologies" (chapter 8) and then complement that with research regarding their own ethnic, religious, or cultural heritage, it is still possible to recognize how pieces of ancient wisdom influence one's own heritage. Ancient pioneers also include the Axial Age sages (chapter 2). By emphasizing the importance of respecting living beings beyond the species line, these pioneers developed humans' collective thinking about all living beings' place in what Berry called the larger community.

Aristotle: Pioneer and Humbling Model

Charles Darwin is probably the most acknowledged name today around the world when it comes to people who have exercised a profound influence on contemporary thinking about animal-related issues. Darwin's remarkable mid-nineteenth-century achievements moved human understanding of all animals into the scientific realm. In fact, by virtue of the Darwinian synthesis (chapter 3), Darwin may now have eclipsed Aristotle as the single most influential individual in the history of human thinking about other animals. But Aristotle, because of his special place in shaping many perspectives that continue to play central roles in Animal Studies, provides the opportunity to talk about both positive and negative aspects of leadership.

Aristotle and Everyone's Limits

For more than 1,500 years, this remarkable Greek philosopher was an astonishingly influential figure in many humans' understanding of other-than-human animals. He bequeathed us much, including his un-Platonic conviction that humans were in continuity with the rest of the physical world: "In applying his method to the study of animals, Aristotle says that we should investigate all of them, even those that are mean and insignificant, for when we study animals we know that 'in not one of them is Nature or Beauty lacking.'"[4]

A list of the words that Aristotle invented sits at the very heart of both common sense and many modern enterprises that still shape modern sciences today: category, energy, actuality, motive, end, principle, form, faculty, mean, maxim.[5] It was Aristotle's pervasive influence in areas as diverse as philosophy, early sciences, and our day-to-day understanding of the world that led Thomas Aquinas, himself among the most influential philosophers ever, to refer to Aristotle with the simple and respectful name "the Philosopher."

But Aristotle's penchant for explanations that favored humans, and then only some humans (particularly male Greeks of what are often termed the upper classes), caused him to make many statements that are disproportionately human-centered, a tendency that is integrally tied to "the Aristotelian spirit that reads the universe on an analogy to intelligent construction."[6] Aristotle subordinated two major classes of humans to civilized males: women and slaves. He followed Plato and other male Greeks in holding males naturally superior to females—perhaps his most famous statement in this regard comes from his influential Politics: "the male is by nature superior, and the female inferior; the one rules, and the other is ruled; this principle, of necessity, extends to all mankind."[7]

Aristotle is equally well known for his view of slaves expressed in Politics.[8] The influential moralist William Paley challenged Aristotle on this point in his 1785 Principles of Moral and Political Philosophy:

> Aristotle lays down, as a fundamental and self-evident maxim, that nature intended barbarians to be slaves and proceeds to deduce from this maxim a train of conclusions, calculated to justify the policy which then prevailed. And I question whether the same maxim be not still self-evident to the company of merchants trading to the coast of Africa.
>
> Nothing is so soon made as a maxim; and it appears from the example of Aristotle, that authority and convenience, education, prejudice, and general practice, have no small share in the making of them; and that the laws of custom are very apt to be mistaken for the order of nature.[9]

Aristotle also viewed nonhuman animals as inferior, advancing the questionable analogy "as plants are for animals, animals are for humans" (examined in chapter 6). Thus it is possible to notice that Aristotle's claims about nonhuman animals were influenced by "authority and convenience, education, prejudice, and general practice." We can wonder then with Paley if, in the matter of humans' intersection with nonhuman animals, this remarkably creative, obviously intelligent philosopher mistook Greek "laws of custom . . . for the order of nature."

Aristotle, then, provides a key lesson in humility. Though he was clearly one of the most formidable analytical thinkers of all time, he again and again offered arguments beset by shortcomings. These arguments were made at key ethical junctures dealing with justification of domination (men over women, Greeks over foreign slaves, humans over other animals).

Such observations about the most influential thinker on nonhuman animals in the Western cultural tradition are not made with disrespect, for Aristotle pursued research regarding all animals, human and nonhuman alike, with much passion. Yet Aristotle was, it turns out, only too human. Like the ordinary and extraordinary humans we all know, he was deeply biased by the privileges he held by virtue of his gender and social class and by his own cultural claim to be superior to others. Like so many other humans, then, Aristotle succumbed to self-affirming beliefs and the rationalization that his privileges were "the order of nature."

Such mistakes reveal that all humans are at risk when generalizing about "animals," "human nature," and what constitutes truth and justice. Since many familiar cultures have regularly promoted inequalities among different human groups as a divinely mandated natural order, we have become all too familiar with injustices for women, children, the males and females of marginalized ethnic or religious groups, and, of course, other macroanimals. Studying "the Philosopher" can help us understand the value of humility as we try to present a fair and honest version of the larger story.

Aristotle's errors are not merely ancient history—of great relevance to Animal Studies is that such reasoning has become a mainstay of both secular and religious circles that support the exceptionalist tradition. Aristotle's views were passed along by other influential, otherwise capable thinkers, such as Cicero and Thomas Aquinas.[10] By virtue of its association with such luminaries, this form of thinking has become a central tenet in the exceptionalist tradition, thereby anchoring many humans' sense of privilege over all other lives.

Animal Studies' Tasks

Accounts of Aristotle's creativity and power can still underscore why his work is an impressive achievement. He models well both leadership and connection—by forthrightly stating his conviction that our human group features substantial, existentially important continuity with the rest of the animal and physical world, Aristotle courageously disagreed with his mentor Plato and provided subsequent generations with a basis for exploring humans' place in the natural world. His multiple, extensive treatises on nonhuman animals more than increased knowledge—they displayed humans' peculiarly large capacity for inventorying the universe in a manner that connects all of us to each other and other animals. As importantly, Aristotle's observation in Parts of Animals that it is the whole living being that matters, not the component parts, matches ethical insights that appear again and again from the time of the Axial Age sages to the present. Humans clearly have the ability to notice whole organisms as the level of biological reality that matters most. This insight is today one of the most integral features of Animal Studies as it surveys whole animals in community, not the parts of living beings that are used as mere resources for human profit.[11]

Luminaries beyond Aristotle

One can sense when perusing Aristotle's works that Aristotle himself was often captivated by the living beings he studied. Fascination of this kind occurs in every culture. Such fascination, rather than the dismissiveness in some misleading and troubling generalizations found in Aristotle's works, was one reason Aristotle influenced many later thinkers. Further, Aristotle's range of thought in addressing so many different kinds of nonhuman animals established a theme of broad inquiry that facilitated the work of those cultural successors who prized humans' interest in and connections to other living beings. Some of these successors' scientific inquiries about the natural world led to Darwin's insights. Many of the contemporary scientists cited in this book (such as Griffin, Goodall, Tinbergen, Lorenz, Wrangham, and Bekoff) also model that deep fascination with living beings in context can be a driving passion. The best-known contemporary leaders include Jane Goodall from southern England and the Australian philosopher Peter Singer, who at different times has been referred to as "the most influential living philosopher" and "the most dangerous man in the world today."[12] Each has, through personal efforts and by spurring thousands of others to act, stimulated awareness of specific problems and thereby a renaissance in worldwide interest in nonhuman animals.

Those we call leaders most often emerge from shadows and margins, "from below," and as individuals. Examples can be found in scholarly traditions—for example, in Islamic circles, Al-Hafiz B. A. Masri's name is known because of his work on nonhuman animal issues, just as the name Andrew Linzey is known in Christian circles. Each is well known for books that focus discussion of nonhuman animal issues in their own traditions.[13] In India, the on-the-ground work of Maneka Gandhi and Raj Panjwani has made each well known in the world's most populous democracy. Such individuals, who have emerged not only as scholars or organizational leaders but also as moral authorities, present startlingly diverse approaches that correspond to their backgrounds and interests. Their variety corresponds with the diversity of nonhuman animal issues.

Organized Realities

At the beginning of the twenty-first century, more than 17,000 animal protection organizations could be identified in more than 170 countries.[14] A catalog of what is being done in these organizations is impossible to compile because animal protection and advocacy work are extremely diverse and growing. Nonprofit organizations also play key roles in protecting human welfare linked to other animals. The fact that it is not only groups focused on the connection of human and nonhuman harms that qualify for special tax exemptions but also, in a number of countries, those focused solely on nonhuman animals reflects how government policy can help promote animal protection in very basic ways.

It is at the local level that problems can be identified with sufficient specificity to make meaningful change possible. The approaches used to create locally effective remedies are as various as the organizations and the issues they address—they include the standard approaches of filing lawsuits and proposing legislation, as well as voter initiatives on companion animal and wildlife protections, disclosure of harms involved in food production, dissection-choice options for students, and other animal-related problems in educational systems, cities, states or provinces, and so on. Less familiar approaches like student initiatives to create Animal Studies courses also abound.

The net effect of such diverse and wide-ranging work is communal, that is, people working together both informally and formally to address a particular problem. But individuals play the most crucial role in such social ferment. This reflects Margaret Mead's wise observation that we should "never doubt that a small group of committed citizens can change the world" for, "it is the only thing that ever has."

Returning to the Individual in Daily Life

The key feature of organizations remains individuals. It is individual people who make the difference in any organization—the charismatic founder, the insightful administrator who makes others more effective, the tireless rank-and-file volunteer who grasps how to create lasting learning and protections. Thus, while individuals may often be involved in organizations, the issue of humans' relationship with other-than-human animals returns to the level of personal choice. It is at this level that both the startlingly commonsense features and the compelling internal logic of Frances Moore Lappé's observation "every choice can be a celebration" play out most fully.[15] In this single phrase, human individuals' formative ability to care is called out in its basics and in all its power.

Promise: What Might Be Done

Any account of pioneers and leaders in Animal Studies will reveal not only the central role played by individuals, but also two particulars. First, each individual, from the highest-ranking to the newest arrival in any organization, has obvious limits as to what they can attempt or accomplish. There are so many issues that attempting to address too many of them is, in effect, a covert way of avoiding any significant advance in a particular area. As one sensitive observer of the human condition suggested, "The rush and pressure of modern life are a form, perhaps the most common form of its innate violence. To allow oneself... to surrender to too many demands, to commit oneself to too many projects, to want to help everyone in everything, is to succumb to violence. More than that, it is cooperation in violence."[16]

Second, each individual helps the human community with the task of taking responsibility for our inevitable impacts upon other living beings. In this, each individual illustrates something true of every human and the very structure of the ethical life. It is in precisely each person's daily choices for or against harms that we choose or avoid caring about others. Choices at this level are, to use a biological image, the very heartbeat of ethics, the place where noticing and taking seriously our own acts make a difference in the life of another. Assessing the facts and then taking responsibility for our quotidian acts is how one shapes a life. It is precisely at this local, personal level where we contend with our immersion in constructed worlds of self-focus and human-centeredness.

This task is not simple. How could it be? We are not simple animals. Indeed, this particular task is, in a human-centered society, among the most complicated. We are, surely, animals with special moral capacities, but we realize these only when we work hard and together toward this rich human possibility. In a vain way, we could even dream that we are the most special animals, but ironically this is true only when we are at our most humble and willing to take our full but still limited place in the community of life.

So Animal Studies must focus on leadership and vision at the level of daily life in order to explore fully humans' relationship with other-than-human animals. This observation dovetails with humans' awareness of and pride in the rich individuality so evident in each human person. This feature of human existence prompts Animal Studies to inquire how individuals lead societies in general and our species as a whole as we accept or reject humans' evident animality and our ubiquitous, inevitable encounters with other living beings.

Arts and Possibilities: The Question of Nonhuman Leaders

Individuals practicing different arts offer virtues and visions to the human community. This is one of the reasons that myths, folktales, songs, paintings, sculpture, dance, and instrumental music play key roles in teaching about the human-nonhuman intersection in various cultures. More recently, literary creations like Anna Sewell's 1877 *Black Beauty* have prompted humans to imagine the vantage point of a nonhuman animal. Today, a startling variety of literary and visual arts portray, imagine, and otherwise lead people to explore how any human might imagine the life of other animals.

Such arts of imagination open up the question of whether leadership of a kind might be imagined as coming from members of another species. One author addresses special qualities of dogs:

> A single glance between dog and human companion can communicate subtle and complex emotion and meaning, proving without question that we have more in common than not. Friendships between humans and dogs have proven to be as strong as, or stronger than, those found between many humans.[17]

We can imagine that some of the dogs we meet invite, even teach us, in unique ways and thereby lead us to places we cannot arrive without them. Whether this is a kind of leadership or more a mutual phenomenon of each side leading the other, the experience described by this author is not unfamiliar to many humans. Thus while a number of humans remain skeptical, many others today easily settle into the riches described by this author.

Animal Studies is capable of exploring whether leadership comes from diverse sources, including some nonhumans. Further, it is capable of remaining open to undiscovered or merely unappreciated realities in other animals. A long-influential philosopher decades ago used questions about humans' relationship to nonhuman animals to explore the breadth and depths of our philosophical traditions:

> Aristotle, though in general he was much more convinced of man's continuity with the physical world than Plato, makes some equally odd uses of the contrast between man and beast. In the Nichomachean Ethics (1.7) he asks what the true function of man is, in order to see what his happiness consists in, and concludes that that function is the life of reason because that life only is peculiar to man. I do not quarrel for the moment with the conclusion but with the argument. If peculiarity to man is the point, why should one not say that the function of man is technology, or the sexual goings-on noted by Desmond Morris, or even exceptional ruthlessness to one's own species? In all these respects man seems to be unique. It must be shown separately that this differentia is itself the best human quality, that it is the point where humanity is excellent as well as exceptional. And it surely is possible a priori that the point on which humanity is excellent is one in which it is not wholly unique or that at least

some aspect of it might be shared with other beings. Animals are, I think, used in this argument to point up by contrast the value of reason, to give examples of irrational conduct whose badness will seem obvious to us.[18]

Through such careful, paced reasoning, this philosopher hopes to show the reader that even our species' most sophisticated thinkers err when they so avidly pursue human exceptionalism. We are most exceptional when we remain open to going beyond the species line. We know nonhumans' realities poorly if at all, but that is no excuse for turning away from these unknown others—to reason that ignorance justifies lack of concern is every bit as fallacious and morally questionable as turning away from exceptional human genius because the rest of us do not have such capacities.

Fostering such openness is the task of leaders in animal protection and Animal Studies. How such leaders move people into open, full engagement with other-than-human animals depends not only upon individuals' vision and creativity but on the underlying social situations. When women are subordinated, the nonhuman animal issue can be impoverished because that society is denied the full benefit of the extraordinary capacities and depth of understanding that women have. When children's natural connections to other-than-human lives are not nurtured, a society produces impoverished citizens and political discourse. Finally, when entire ethnic groups, indigenous communities, and religious traditions are marginalized, leadership talents are often directed so fiercely to demolishing these human-on-human exclusions that issues such as the animal question, are shunted aside.

Historical accident has played a role in which nonhumans any cultural tradition will encounter. In India, elephants played a major role long after they were removed from their native habitats in China, the Middle East, and northern Africa. The fact that "the bear is the most significant animal in the history of metaphysics in the northern hemisphere" caused Shepard to observe, "The great circumpolar 'bear cult' is the salient religious and ritual association of people and a wild animal."[19] The overwhelmingly anthropocentric European ethics tradition was birthed and thrived in urban and rural environments cleared of large-brained, long-lived, socially complex nonhumans—the unsurprising end result was a fundamental, self-inflicted ignorance about many other animals.

What counts as leadership and pioneering can be framed, then, in almost countless ways. This explains why scores of human individuals have been mentioned as the previous chapters have recounted various aspects of human efforts to fathom, explore, explain, hunt, control, and otherwise deal with the astonishing multitude of living beings outside our own species. While these individuals can be thought of as pioneers, leaders, and revolutionaries in the larger story, the reality remains that innumerable individual humans have impacted humans' past dealings with ubiquitous nonhuman life.

The Future of Animal Studies

The future remains to be chosen. Some people believe Animal Studies faces complex, debilitating problems. Nonhuman animals are being harmed in astonishing ways and many people feel deeply that this pattern of human activity must change. For others, however, Animal Studies is primarily the academic approach to studying a multispecies world.

For the former group, ideological and moral challenges of great complexity must be met immediately because they arise in the core activities of daily life and will, if unmet, mean that many nonhumans continue to suffer serious harms. How one perceives these challenges is shaped by powerful forces such as one's birth or chosen culture, religious tradition, or the civic and secular commitments of modern industrialized nation-states. For the latter group, these challenges are real, but there are also formidable philosophical and educational challenges to be met, such as questions about how much of other animals' lives we can know, and how human limitations can be acknowledged and surmounted.

The differences between these two approaches to Animal Studies by no means outweigh the common elements—both approaches reflect a deep, organizing commitment to live in a healthy, multispecies world, just as both foreground the goal of exploring animals' abilities (humans included). Both are committed to the key role of education and the importance of learning the larger story. There are differences in how much effort should be spent on learning about the past, but both agree that doing so is important because it provides one key to understanding present-day views and future prospects. Both groups also recognize clearly that science literacy is a key component of a vibrant Animal Studies field. Finally, both groups focus on interdisciplinary approaches and hone a variety of skills.

The Near Term: A Future We Can Glimpse

If options for the near term can be outlined, our prospects of glimpsing some feature of the future will grow more certain. Clearly, new options will eventually present themselves—because of political work, discoveries about other animals' realities, helpful theoretical work, and coordinated visions synthesized out of many different disciplines and key insights.

Identifying present options for the next few years involves extrapolating trends. Animal Studies presently touches an astonishing range of disciplines. The future of individual courses that are already popular—animal law, animals in literature, history-based courses, sociological and cultural approaches to the depths and breadth of human-nonhuman connections, religion and animals, critical animal studies, and on and on—is bright. That future includes expansion of these offerings. For example, law schools must offer more than a survey course that is supposed to cover all of animal law. A single course has no reasonable prospects of covering the topic well. Given that one American law school has attracted much attention and high enrollment by offering well over a dozen courses touching on animal law,[1] a trend to multiple animal law courses will likely develop soon.

Second, the near term will undoubtedly see more linking up of existing courses. Many colleges already offer courses in which animal-related issues bridge topics traditionally taught in separate courses (for example, literature, religion, and animal-related themes). Since existing disciplines are regularly adding animal-related discussions, the near future is likely to see novel combinations of courses and crossing of disciplinary approaches.

Third, interdisciplinary connections will undoubtedly be developed further. For example, religion and animals courses focusing on a single religious tradition can include other traditions or new findings about animals that have had significant roles in specific religious traditions. Literature and animals courses can be expanded beyond the literary traditions or forms now typically studied, and animal law courses can be enriched by cross-disciplinary exchanges.

Distance learning, social media, and Internet conferencing can add immensely to student opportunities to attend courses and seminars. Less expensive, even free education via recorded lectures offered by cooperating educational institutions has almost unlimited prospects for increasing general awareness of Animal Studies. Such efforts will surely create additional synergies.

The near future also will include program-level innovations aimed at coordinating groups of single-discipline courses. There are already, for example, burgeoning animal studies centers and degree-granting programs (chapter 1). Educators will group courses with different interdisciplinary emphases—for example, focusing on different claims and knowledge problems regarding other animals, or on different sorts of ethics or a variety of cultural achievements. Creativity in weaving themes together will surely open up even richer possibilities for degree-granting programs.

Continuing education opportunities and online offerings will allow students from different cultures and parts of the world to meet and work together. Such work offers the prospect of mature and younger students developing publications, such as journals and edited volumes that address highly specific topics, just as it will facilitate joint projects at local, national, and international levels. Social networks will also tap into the extraordinary resources of nonprofit groups, government agencies, and other collections of people working on animal-related problems.

As the Midcentury Arrives

For the last decade, I have put a diverse range of students through a classroom exercise that asks them to imagine the shape of ethics or public policy or law regarding nonhuman animals at the beginning of the twenty-second century. I suggest that they ask questions about the subject matter their own future children, then grandchildren, and then great-grandchildren might encounter in courses studying nonhuman animal issues for the rest of the twenty-first century.

Questions about such a distant future usually produce more silence than answers. Only a few are daring enough to make specific guesses. Some focus on how to sustain the renaissance of interest in other-than-human animals that emerged in the second half of the twentieth century. Some think we will overcome humans' limits in knowing the inner workings of nonhuman animals' lives. Some simply remain silent, but many humbly explore and then, listening to others' tentative answers, begin to elaborate about what may happen in the coming decades. At times, the communal exchange has been remarkable, at other times weighed down by uncertainty. Yet there are consistently positive signs—for example, everyone senses that the community effort is far more important than that of an individual. In this development lies an insight—humility, community exchanges, and open-mindedness are virtues that must prevail if we are to guess responsibly about the future. Interestingly, these three virtues are also mainstays of the best practice in science and ethics.

My own personal guess is that by the middle of the twenty-first century, communal work will likely prompt members of our species to explore which of our multiple intelligences could be the province of some nonhuman animals. We might glimpse something far more alien, namely, which intelligences different from ours can be found in certain nonhuman animals.

In guessing what might happen by the middle of the twentieth-first century, my sense has been that it will most likely be a small group of women, perhaps young to middle-aged, who will make breakthroughs that allow our species to understand key features of the communications of some other-than-human social mammal group. Such an achievement would help the citizens of our increasingly dominant industrialized societies recognize that the earth is populated by other intelligences. Such a breakthrough may or may not give us a truly deep understanding of these newly appreciated social mammals, but, at the very least, it will poignantly confirm that our species has never been alone in the universe. Such work will prompt ever deeper commitments to protect other animals whose lives will for many centuries continue to seem foreign to human minds.

Before the End of the Twenty-First Century

Thinking to the end of the century, all of my students—the graduate students at Tufts University's Cummings School of Veterinary Medicine, the Harvard Law School students, the anthrozoology graduate students at Canisius College—fell silent. For obvious reasons, thinking so far out is hard in any discipline, but particularly in law or policy or ethics. Silence of this kind produces teaching opportunities, though, and here the relevant fact is that those of us who are heirs to the exceptionalist tradition are not at all skilled in seeing the range of possibilities at the human-nonhuman intersection. This is one reason that the core concerns giving Animal Studies its heartbeat will do fundamental educational work. Engaging the realities of other animals and telling the larger story will become essential to the vitality of individual fields and the growth of interdisciplinary approaches. Perhaps most relevant on the human side of this inevitable intersection, Animal Studies will provide existentially meaningful education. It will foster personal connection, pursuit of the truth, scientific literacy, informed ethics, and affirmation of our abilities to care within and beyond the species line.

The Three-Part University

It is possible to frame Animal Studies as a megafield that will complement and bridge the different kinds of work being done in the natural sciences, the social sciences, and the humanities. Animal Studies is not strictly a science, though it requires intensive commitments to both natural and social sciences. Nor is Animal Studies well positioned as a mere subfield within the humanities even though it must be in constant conversation with the humanities to study the past, present, and future features of the human-nonhuman intersection. When seen as an integrated battery of inquiries, Animal Studies will enhance the descriptive work of the natural sciences, the exploration of humans' social dimensions that drives our social sciences, and the soaring imagination and meaning-based searches of the humanities.

Educating humans effectively will require more than substantial commitments to the science and humanities megafields. In addition, each level of education must foreground a passionate but unbiased pursuit of the truth about the whole universe. The exceptionalist tradition is driven by the view that education should justify human privilege with standard answers ("standard facts") to standard questions that lead inexorably to an affirmation of humans' superior intelligence and character. Education, however, is far better understood as a process of questioning through not only empirical investigation but also awareness of humans' deep and wide ethical capacities.

Good education of this kind cannot coexist with constant reassertion that some portion of the human race holds a privileged position. This claim by no means excludes claims that humans should, as a species, be gratified and confident. Clearly, it is a source of justifiable self-esteem when humans recognize that we can choose to be remarkable living beings, to surmount selfishness, to actualize our fullest potential by transcending mere self-interest. But Animal Studies will make it clear in both the near and longer-term future that we are remarkable animals only when we recognize that on earth we are not alone in the matter of impressive skills and perceptions lived out by individuals in a rich social context. When we slip into Thrasymachus-like "might is right" arguments (explained below) that humans merely by virtue of our physical and political power justly enjoy the privilege of dominating other-than-human animals, we fall far short of an ethically justifiable position.

Because Animal Studies foregrounds other species and their individuals in social context, education in this vein necessarily prompts humans to become inquiring, science-interested individuals even as they also become caring, compassion-concerned individuals who realize humans' potential to be a moral species. To achieve such education, Animal Studies necessarily will push well beyond any single field found today in the university. In this opening up and deepening of education, Animal Studies complements the native human-centeredness of the humanities and arts and the broad exploratory commitments of the sciences.

Continuing Risks

Animal Studies will, in this role, confront the exceptionalist tradition, which has predisposed many people to go well beyond certain mild human-centerednesses that now prevail. In such an environment, it has been easy to assume Aristotle-like views of our own privileges as the order of nature. For some humans, the widespread prevalence of the exceptionalist tradition already provides compelling evidence that we deserve to be dominant because humans have creative minds that foster our arts, humanities, and precise sciences. But the lure of the exceptionalist tradition in the end is self-defeating because it pulls us out of our larger community and home. Further, superiority-invested reasoning also fails critical thinking tests—syllogistic reasoning by which those seeking privileges over all other living beings move seamlessly from a premise asserting humans' superiority to a conclusion justifying privileges is subject to multiple challenges. One can, for example, dispute the facts used to construct the premise. (Which facts, after all, go toward determining factual superiority?) One can also question whether the reasoning is distorted by bias, ignorance, wishful thinking, and the like. Justifications of privilege may seem to have a basis in reason, but to the oppressed they often seem a facade behind which selfishness, small-mindedness, and self-aggrandizement lurk. Thus, it is questionable to assert that privileges are supported by carefully reasoned, fair arguments.

Instead, such arguments seem more often anchored in ethical blindnesses. Plato's account of Socrates's refutation of the arguments made by Thrasymachus in the first book of the *Republic* long ago made the shortcomings of power-based privilege notorious. Thrasymachus, asked to give a definition of justice, replies, "justice is nothing else than the interest of the stronger."[2] Socrates proceeds to show how inconsistent, unjust, and morally wanting such a definition is.

Critical thinking requires a fair and full answer to the question, just what sustains the moral right to assert privileges over allegedly inferior groups (or, in Thrasymachus's terms, to make justice serve only the interests of the stronger)? A creative analogy illustrates why power-based domination invokes suspect reasoning—"we may find it difficult to formulate a human right of tormenting beasts in terms which would not equally imply an angelic right of tormenting men."[3]

One enfeebles Animal Studies by approaching it through exceptionalist assumptions such as Protagoras's claim, "of all things the measure is man."[4] The possibility that Animal Studies can, if pursued narrowly and in terms of the exceptionalist tradition, become a human-centered enterprise prompts discussion of ways in which a more balanced, three-part university can create the best possible education for humans even as it contributes to seeing our multispecies world for what it is.

Future Realism in Academia

Animal Studies will expand what counts as academic—it will problematize compartmentalization and separation of thought into categories, the effect of which is to discourage holistic thinking and thereby deny other-than-human animals an integrated place in a world dominated by humans.

The enigmatic sixteenth-century figure Paracelsus said, "The universities do not teach all things, so a doctor must seek out old wives, gypsies, sorcerers, wandering tribes, and such outlaws and take lessons from them."[5] Animal Studies, by its very nature, systematically seeks out and takes lessons from both the people inside universities and those whom universities have historically ignored, as well as nonhuman animals themselves. Education thereby becomes "a place of daring" (chapter 2) that will push students to reckon with our citizenship in our ever-so-diverse, populous, and interwoven larger community. No longer will the vast majority of life disappear because an exceptionalist ethic has reduced them to mere resources for privileged humans.

Because of its educational impacts, Animal Studies will prompt recognition of the profound limits of today's two-part university as it is connected to the exceptionalist tradition. A three-part university will have far greater capacity to create non-anthropocentric science and ethics. It will also deepen, even as it draws from, the humanities' embrace of values, creative arts, and humans' other remarkable talents. Above all, a three-part university will put humans back into the world, underscoring the search for truth rather than self-serving affirmations of privilege premised on superiority to and separation from all other life.

Some will find it ironic that a new educational endeavor might help both existing megafields accomplish their tasks better. But today's universities have let higher education slip away from the search for the truth, which in no way automatically favors just one species. Scientific, ethical and pedagogical norms require that humans be free to pursue the ethic of inquiry (chapter 2).

A broad, more encompassing three-part university, then, nurtures realism because it prompts teacher, student, and society to move beyond the "standard facts, standard questions" form of exceptionalist education, replacing it with a full, natural exploration of the earth as a more-than-human community. Like the emergence of concerns for different kinds of social justice, Animal Studies will become a force making programs and institutions more relevant to our actual lives.

With such improvements, a three-part university will assist science educators in making science a pursuit full of curiosity, beauty, and play. A three-part university will also host humanities that are more creative and openly connected to the life of a citizen in a more-than-human world. Above all, a three-part university will renew the college environment because it will nurture a communal commitment that crosses the species boundary. In this way, Animal Studies can make college education as fully relevant to local environments and their nonhuman neighbors as it is to the whole human. In this way, a three-part university will help prepare students to enter and take responsibility in the more-than-human world we coinhabit with other-than-human animals.

Metaquestions: Pedagogy and Classroom Dynamics in a Three-Part University

My own experience based on courses in law, religion, ethics, public policy, anthrozoology, and veterinary ethics is that students are eager to discuss animal-related issues—in effect, a large majority of students "lean into," rather than away from, work that helps them learn about real animals.[6] Further, students welcome ways to create win-win outcomes (that is, where both nonhumans and humans benefit in some way).

Animal Studies will grow teaching techniques to take advantage of such dynamics, for whenever students share their own experiences and ideas, they learn easily to call out connections and problematize exclusions. They also will recognize sooner and more fully that science-based approaches have distinctive power and yet definite limits. The same is true of approaches that identify what values impact one's view of the ethical dimensions of their situation, as it also is of art-based approaches that tap into humans' creativity and imagination.

The potential of Animal Studies as a megafield for broadly enhancing classroom dynamics is illustrated by examples from legal education and the academic study of religion. Because legal education already features an entrenched tradition of open discussion known as the Socratic method, and because students typically opt to take an animal law course because of deep personal interest in the subject, this form of education features lively, wide-ranging exchanges. These discussions are, in fact, so typically wide open that anyone in the classroom (including instructors and mere observers) will often be prompted to think about the larger topic "law."[7] An example can be drawn from the first animal law course at Boston College Law School in 2001—one of the student evaluations claimed, "This is the course in which I learned the most about the legal system." Animal law courses, by virtue of the fact that they deal with the question of who is important enough to protect and who is excluded from legal protections, will in the future have a remarkable power to illuminate some of the broader features of legal systems generally.

Impressive classroom dynamics are also noticeable in religion and animals courses. While students typically enroll for personal reasons, some cite fascination with religion rather than concern for nonhuman animals as the motivating factor. These different kinds of personal motivation create opportunities for discussion of religious traditions' strikingly different claims and social constructions regarding other-than-human animals. For those motivated by either religious concerns or animal protection concerns (or both), such courses in the future will provide many instructive lessons in how views of other living beings are developed, passed along, and persist over time and in the face of science-based information. Such courses will continue to be popular in both secular and religion-oriented colleges—for example, at the American Academy of Religion it has been reported for a number of years that these courses fill up even faster than the very popular religion and ecology courses.[8]

Helping Educators in Human-Centered Domains

Apart from such beneficial classroom dynamics, other salient features of modern teaching complicate contemporary education about other-than-human animals. Complex, sometimes debilitating problems of polarization arise when a teacher attempts to teach about other animals in human-centered domains. Developing a three-part university will offer prospects for those educators who now frown upon open discussion of ethical issues. Many questions are raised by students aware of the robust debates going forward around the world regarding treatment of nonhuman animals (for example, many students participate in, even lead, animal rights clubs, which are common in secondary education). Given that so many educators recognize the many educational lessons available only through open discussion, the future availability of expertise in ethics, history, religious and cultural diversity, and social and legal developments would provide both educators and students meaningful opportunities to explore problems, critique existing practices or proposed alternatives, and experience a group dealing forthrightly with complex real-world problems. Through such opportunities, educators gain experience in assessing how they can best shape education to help students learn.

Further, fostering truly open discussions gives Animal Studies educators the opportunity to integrate local, national, and international perspectives and thereby nurture an informed view of much that is happening around the world, including war, climate disruption, habitat and ecosystem destruction, depletion of resources, political abuse and marginalization of humans, food production, and so much more. Such breadth is important for countless reasons, but particularly with regard to young humans who, as ethically inclined and capable beings, live in a complex world full of challenges for the human heart and mind. Though Animal Studies will require that students in the future continue to be immersed in discussions revealing that some members of our species act regularly in selfish ways, future discussions will also offer good news about humans' compassionate actions. Animal Studies, then, offers future options and connections that help students deal with our morally deep world as it presents numerous opportunities to exercise care for others.

A Robust Future

Looking at the near-term future, the possibilities at midcentury, and the prospects at the beginning of the twenty-first century provides reason to conclude that Animal Studies courses will offer one prospect after another for immersing students in real problems in all their complexity. Further, by dint of immersion in thinking about issues at the human-nonhuman intersection, students attending a three-part university will encounter possibilities that help them resolve the inherent limitations, biases, and weakness of the two-part university.

"Almosting It" through Collective Effort

Just as one learns by traveling the world that knowledge of but one human language puts one at a disadvantage relative to those who know multiple languages, so, too, in the academic world one learns that it weakens communication and learning to insist that discussions proceed solely in terms of one's own discipline. Communicating across disciplines takes patience, imagination, and humility—sometimes the right images or metaphors or theories can give a discussion group the agreeable feeling that they are coming close to a shared understanding on an issue. These moments are, of course, encouraging, and when they occur over several different discussion opportunities they increase each group member's confidence that the group can meet the challenges of communicating across disciplines. Groups that communicate this effectively may be able to take on the challenge of speculating about elusive realities such as what another animal's mental life may be like. With a combination of patient observation and sensitivity to this task's complexity, it may become possible at times to hope that, in the words of Stephen Dedalus in James Joyce's *Ulysses*, a group is "almosting it" as they guess about what another animal's life may be like.[9] Yet even when a group or an individual "almosts it," their conclusions will surely be fragile and open to revision.

Consider the 1984 discovery by scientists who first used sophisticated equipment to confirm that elephants communicate regularly with subsonic sounds.[10] The term "subsonic," of course, is pitched to human comprehension, for the sounds, though below the range of human hearing, are clearly heard by elephants and some other nonhuman animals. Modern societies use the word "discover" whenever science illuminates information new to educated modern citizens. But the occurrence of subsonic communications was long ago "discovered" by a number of indigenous peoples who recognized circumstantial evidence confirming sub-sonic communication. Close observation of individual elephants will reveal that elephants' temples vibrate when they make subsonic sounds. More indirectly, it is possible to deduce that the sounds exist from the fact that all elephants stop at the same time as if listening for such sounds, or from observations of coordinated action between elephant groups even when the groups are separated by enough distance to make visual signals impossible. Careful observers who notice these clues can imagine that communication is occurring even though humans do not hear what is being "said."

Confirming subsonic communication by recording merely opens a door—going through the door, beginning to grasp the why and what of the now-acknowledged com-munications, presents immense challenges. Elephant researchers today are uniform in their affirmations that we are only at a rudimentary stage of understanding elephants' multiple forms of communication because we are only "just beginning to scratch the surface of the language—all their body language communication."[11]

Given that there are multiple traditions of observing and describing elephants from Africa and Asia, and abundant observations and comments from ancient times to the pres-ent day, getting those with different views to talk to each other requires great skill, multiple vocabularies, and so much more—and when this is done, there is only then perhaps the chance of beginning to "almost it" in the matter of saying something illuminating about what the inner and social lives of these remarkable animals may be like.

Our speculations about elephants' lives is, because of their "glorious and…infamous association" with so many humans over so many millennia, quite likely to be more realistic than our speculations about, say, sperm whales, who have the largest brains on earth. While we can observe that sperm whales, like elephants and ourselves, are intensely social beings, as well as record their communications through an array of clicks and utterances, sperm whales move in places and to great depths (diving as deep as 3 kilometers) that are, for humans, unfathomable. So while we can in some senses accompany them, this really amounts to watching them from afar and gathering limited information about their activities through technology. But these social mammals remain elusive and mysterious for us—accordingly, what they are now, and have been for millions of years, doing with the largest brains on Earth is not known to us in any detail at all.

Even if members of our species eventually accompany elephants and sperm whales long enough to claim rudimentary knowledge about what they do in their intensely social lives with their large brains and rich communication skills, it will take far more than recording, quantifying, and discussing their sounds and other activities to understand them as complex individuals living intensely social lives in their communities. "Almosting" a truly informed view of their lives will require human imagination enabled by the best of our empiricism, the great sensitivities of our poetic and spiritual gifts, the cautions of our skeptical sides, and, no doubt, much more. But just how humans might begin to "almost it" about the actual lives of these and other nonhuman animals, of course, will be determined by future humans. One thing is clear, though—simply dismissing these animals without attempting in the best scientific and ethical traditions to take account of their actual realities is nothing short of a radical failure to be as human and as animal as we can be.

On Finishing the Scientific Revolution

Animal Studies is an extension of the scientific spirit. Even more so, the emergence of Animal Studies is a means of completing that revolution. This is not only true in the sense that scientists' forthright acknowledgment of humans as fully animal is mandated by the integrities of science. It is also true that frank repudiation of the antiscientific dualism advanced by phrases like "human and animals" is needed to give the scientific revolution the chance to win. Similarly, the scientific revolution will not be completed until scientists insist that fundamental openness prevail across the curriculum, that is, in science courses, in the humanities, in every subject. Such openness is needed to allow Animal Studies to become the full subject it is, such that humans as members of a remarkably gifted animal species take seriously the task of telling the whole, larger story.

The impacts of scientists and others permitting the scientific revolution to come to fruition are not easy to inventory. One result would be that our sciences would no longer in any way support in overt or covert ways the fantasies and myopias of the exceptionalist tradition. Scientists would no longer explore the world in ways radically biased by human-centerednesses. Scientists and others who accept the scientific revolution as a key development in human history would thereby readily acknowledge what each of us must learn again and again in our personal lives—human knowledge is fragile, constructed, and very heavily affected by inherited notions. The same people would acknowledge that the best of our science traditions is supplemented by valuable thinking in other major human traditions even though we know from the history of science and the history of our humanities that all too frequently the notions we inherit from our cultures can be biased, ill-informed, or, when informed, limited in fundamental respects.

If through such changes our species can help push the scientific revolution across the finish line, we will no doubt have to say that only then will "the real difficulties begin." But that is the nature of human life in a more-than-human world.

On Finishing the Humanist Revolution

Given this book's many-sided challenge to the exceptionalist tradition, it may at first seem odd to hear that we should finish the humanist revolution that began in the thirteenth century in western European circles. After all, that development in history perpetuated, even deepened Western culture's dismissal of nonhuman animals. But just as Animal Studies is a continuation of the scientific revolution's commitment to pursue the truth without subordinating it to biased dualisms of any kind, Animal Studies can continue one very prominent feature that emerged as late thirteenth-century figures rebelled against an existing worldview that controlled free inquiry. In effect, humanists argued that such freedom was needed for human excellence. Without question, this theme was expressed in ways that reinforced the exceptionalist tradition that was a key feature in the Latin and Greek sources that the humanists found to be so compelling. For example, for Cicero, the term *humanitas* (the word from which "humanities" derives) meant forms of education that concentrated on what we call the liberal arts because these achievements (rhetoric, poetry, history, ethics, and politics) differentiated humans from nonhuman animals.[12]

A key point, though—and the one that Animal Studies needs to develop further—is that the pioneers who came to be called humanists emphasized education in the form of the humanities in order to foster human freedom. Such freedom was needed to inquire about the humanists' conviction that human beings were part of the natural world and "grasping the way things really are" because this is "the crucial step toward the possibility of happiness."[13] The early humanists sought freedom because they wanted to turn education, discussion, and speculation away from "a preoccupation with angels and demons and immaterial causes" so that they could "focus instead on things in this world."[14] Such a this-worldly focus was characteristic of many of the ancient texts that the early humanists studied (and, often, rediscovered after they had been lost following the fall of the Roman Empire almost a thousand years previously). The *studia humanitas* (the inspiration for today's humanities) not only permitted, but even pushed, much greater freedom of inquiry. Notice how this move away from what the humanists considered an oppressive educational and political environment parallels Dyson's description of the scientific tradition as "an alliance of free spirits...rebelling against local tyranny" (chapter 11).

It is surely true that, from the vantage point of Animal Studies, this revolution did not go far enough, for although it advanced human freedom to inquire about the ways "things are," it did so in a way that continued and deepened the radical dismissal of nonhuman animals. In effect, the humanists simply elaborated yet another chapter in the exceptionalist tradition, substituting at times their own explanations of human superiority and why humans deserve absolute privileges over other living beings.

But since Animal Studies has affinities with the humanists' reaction to their thought being controlled, to the need for interdisciplinary inquiries, to the importance of the arts, and, above all, to the conviction that humans belong to the natural world, Animal Studies can be seen as pushing this dimension of the humanist revolution to its logical conclusion.

On Fostering More Ferment

This book has suggested repeatedly that individuals in the human community will benefit from today's ferment over nonhuman animal issues because such developments foster ever more freedom to think creatively, much like the scientific revolution and humanists' early push to be free of assumptions that prohibited one from exploring humans' obvious connections to the natural world. Benefits will also accrue to disciplines and institutions through Animal Studies' emphases on thinking and exploring freely. A key example is that in both religious communities and the academic study of religion, concern for other animals has again blossomed. Scholars and members of religious traditions perceive that concern for our fellow living beings will fit well with respect for creation, indigenous views, ancient insights, ecumenical movements that respect religious pluralism, and much more.

Another key example is how media respond to popular demand by featuring ever more diverse discussions about different kinds of nonhuman animals. In law and ethics, too, honoring human abilities to care across the species line will continue to have popular appeal with students, but will also viewed by many scholars and students as a natural elaboration of much else that these two important human domains have long been doing. Accordingly, both law and ethics will increasingly include discussions with grassroots features pushing local and national communities to repudiate harmful forms of human-centeredness. Sociology, cognitive sciences, geography, and many other disciplines will increasingly reflect even more ferment through which compassion-informed reforms are proposed for education, public policy, the practice of our sciences, our philosophical speculation about reality, and the practices of our human-centered professions and businesses. The cumulative effect will be that major challenges to the exceptionalist tradition will become more and more common.

As these changes arrive, the topic of nonhuman animals will have increasing educational value precisely because it falls outside the traditional, canonical subjects that currently limit learning in our educational systems. Thus, those students who aspire to study reality rather than pursue narrow, human-centered versions of disciplines that superficially focus on nonhuman animals will inevitably pursue multidisciplinary approaches. The upshot will be that any student who confines his or her work to but a single traditional discipline will risk learning, in the matter of nonhuman animals, views that are one-dimensional and otherwise inadequate for understanding the actual individual and social realities of other animals. These problems underlie the observations of an important historian of ecological ideas:

> Intellectual fashion is largely set in either a deconstructive postmodern mode (a prevalent trend in the academy), where it is claimed that "the real" is ultimately nothing more than texts or interpretations of texts, or a modernistic mode (especially popular in Washington, DC, and other centers of power), where it is claimed that "economic man" is the measure of all things, and that the good society (and thus human flourishing) depends simply on the continued advance of industrial culture.[15]

Those who exhort us to become multidisciplinary, interdisciplinary, or transdisciplinary are, in effect, exhorting us to work at integrating knowledge, to let awareness have a holistic, even living, quality rather than allow it to remain artificially fragmented along traditional disciplinary lines. Such risks of fragmentation are real. Problems develop, for example, when religions are studied only by those who are religious, or when science education is dominated by grant-based researchers, or when philosophy is taught only by those with philosophy degrees. All of these human endeavors—understanding religion, learning about science, and philosophizing—belong to everyone, not merely those trained in an academic style which can be, from time to time, myopic, biased, or simply too technical to be relevant.

Animal Studies in particular will need to be taught by teachers of many different kinds. Further, Animal Studies will need new scholars who are not afraid to get beyond the recognized canon or to be politically incorrect under the exceptionalist tradition standards that prevail in the academy. Recall the description of the Animal Studies pioneer Paul Shepard as "comfortable with his outsider status, comfortable in the sense that he knew firsthand (as an academic himself) that intellectual culture is insecure, isolated from the biophysical context of life" (chapter 3). That description by the respected environmental historian Max Oelschlaeger is followed by this powerful critique of certain forms of academic life: "The books and articles typically generated by the intellectual class have nothing to say about the interrelations between the human and the more-than-human, the rest of nature. Leaving aside *littérateurs*, historians, and the like, who studied humanity in its splendid isolation, even intellectuals who extensively studied nature itself isolate themselves from it."[16]

Meeting the challenge of making Animal Studies a form of education that is relevant to life will require more than foregrounding living members of other species. Veterinarians deal with real animals every day, but they only rarely, if at all, engage the powerful cultural dimensions of humans' understanding of other animals that are decisive factors shaping the human-nonhuman intersection—without substantial engagement with this important dimension of human life as it impacts other-than-human animals, veterinarians miss key features of the breadth and depth of the human-nonhuman interactions.

For similar reasons explained in previous chapters regarding history, science, law, religion, and so many other individual disciplines, single-discipline approaches will always have important limits. Realistic description of animal issues requires that the inquiring mind be multifaceted and, if academic in orientation, then multidisciplinary. This is the principal thrust of observing that Animal Studies is more than a discipline—it is in reality a megafield whose sweep is broad in response to the great diversity of ubiquitous nonhuman life and the astonishing complexities of the human-nonhuman intersection. As Animal Studies goes forward into the future, engaging such diversity and complexities will require a richer set of options than those offered in two-part, altogether too exceptionalist universities.

Specific Fields in the Animal Studies Megafield

Previous chapters both explicitly and implicitly suggest that in order to engage issues involving all animals, not merely humans, many disciplines need visions much broader than the human-centered visions that have long been the foundation of these disciplines when they are part of the two-part university. Among the disciplines mentioned already are history, ethics, the arts, literature and poetry, law and public policy, religion, cultural studies, philosophy, politics and other social sciences, and many natural sciences. While these are not the only disciplines that would thrive in the Animal Studies megafield,[17] these and additional fields already existing in the two-part university might change and grow in a three-part university. Each will very likely benefit from the advantageous learning dynamics found in today's animal-focused courses in legal education and the study of religion. Consider how the following fields can remain an integral part of the humanities and social sciences even as they benefit from a broader, animal-focused version to be developed in the Animal Studies megafield. No doubt there are other human-centered courses that, with broadening, would fit into the curricular offerings of an Animal Studies megafield as well, but in particular the following "sibling courses" complement traditional course offerings in the two-part university and thereby reveal much about what it can mean in the future for education to be interdisciplinary. These sibling courses also suggest the importance of distinguishing, on the one hand, forms of education that merely pass along inherited bias and ideology from, on the other hand, forms of education that foreground organic, communal, and cross-disciplinary features of the quest for the truth.

History

Telling the larger story (chapter 10) is telling a meaning-full story for humans as meaning-making animals. Developing minimally biased accounts of the human-nonhuman intersection, as well as what we can piece together of the histories of nonhumans, in no way requires abandonment of the humanities-style versions where history is a special account of humans' remarkable capacities and achievements. But in the Animal Studies megafield, there will exist a far broader notion of history that can help reshape the humanities-intensive field and even push it to move beyond the myopic "great men and great deeds" approach that still impacts historical studies even today. A humanities-based version of history will likely always be an important enterprise, but it is one that can surely be enriched by an Animal Studies–based version that aspires to tell the larger story.

Ethics

The fact that ethics' root question, Who are the others, becomes clearer as one reviews human-nonhuman relationships over the millennia suggests a healthy future for ethics. Narrow answers are inadequate in light of the recurring wisdom that humans' ethical capacities go well beyond the species line. Like race-based, gender-based, class-based, education-based, or many other exclusions, narrow answers commit the fallacy of misplaced community. Further, given the number and diversity of animal-friendly ethical traditions in humans' many different cultural and subcultural groups, courses in ethics will be totally inadequate if they are taught solely in terms of the highly anthropocentric ethical tradition of Western culture. In this regard, ethics courses in the Animal Studies megafield will have many parallels to the introductory animal-related courses focused on law or religion. Further, given that some aspects of the exceptionalist tradition are now yielding, with animal ethics having become a popular and powerful form of inquiry about humans' remarkably wide-ranging moral inclinations, it is likely that the sibling version of this course in the humanities megafield will often benefit greatly from the Animal Studies version.

Literature

The traditional discipline of literature would benefit greatly if a human-centered version remained an integral part of the humanities while a broader, animal-focused version was developed in Animal Studies. These sibling disciplines would cross-fertilize and thereby create new meanings, syntheses, and methods that would foster growth in each version of the discipline.

Other Arts

Given the mysteries of nonhuman animals for the human heart and mind, many different arts can provide insights about features of the human-nonhuman intersection and thereby increase awareness in unique ways. For this reason alone, the arts are likely to be central players in Animal Studies as it responds to scholars' and students' need for personal connection.

In addition, courses that explore such artistic "reach," as it were, across the species line offer the prospect of seeing features of art not as easily recognized in exclusively human-centered courses.

Law and Public Policy

As paradigmatic examples of human abilities to project imagined futures upon coming generations of human and nonhuman lives, law and public policy need to be assessed from vantage points far more encompassing than those promoted by the exceptionalist tradition. These two disciplines would necessarily expand and thus likely deepen when multispecies vantage points are pursued in Animal Studies–based courses that address the complex histories and dynamics of these fields. Although education in law has already begun to change through some engagement with "the animal question," degree-granting public policy programs continue to be bastions of the exceptionalist tradition's business-as-usual. Nonetheless, studies of public policy can open up simply by, first, observing their own narrow history and focus and, second, asking whether "public policy as if only humans matter" need always be the operative assumption of education, scholarship, and programs in this important field.

Economics

Discussions of both consumer issues and public policy, where economics-based analysis is a mainstay, push one to look at the so-called dismal science.[18] "Dismal" is a mild characterization of economics from the vantage point of nonhuman animals because modern legal systems reduce living beings outside our own species to the status of mere personal property of humans. Animal Studies is, interestingly, capable of finding life in economics, as in the subtitle of E. F. Schumacher's groundbreaking 1973 work, *Small Is Beautiful: Economics as if People Mattered*. One natural question for Animal Studies is whether economics can be done as if all humans and some nonhumans matter. While this has not been the case with mainline, exceptionalist tradition–dominated economics, the field today is in great turmoil due to many factors. An Animal Studies–based course could explore what benefits economics-based thinking might derive from an engagement with a broader swath of life.

Cultural and Indigenous Studies

Studying cultures and indigenous traditions makes clear that the vast majority of cultures have recognized that human curiosity, caring, and responsibility reach across the species line. Animal Studies can, as a megafield, bring the deepest perspectives to the many different ways that human societies have pursued this curiosity, caring, and responsibility over millennia. In addition, Animal Studies brings the resources to ask questions about how to honor human diversity by describing in a responsible manner the visions of small-scale societies about humans' place in the more-than-human world. In addition, the commitment to developing a frank, full historical record equips Animal Studies to present forthrightly the problems in the past and the present regarding harms done to indigenous peoples and the parts of the earth they deem integral and even sacred in their world. Such work will be essential as smaller human cultures continue to deal with the enormous pressures they still face from industrializing and industrialized nations.

Philosophy and Other Reflective Thinking

Already a leading edge of Animal Studies, philosophy presents many different traditional topics (for example, epistemology, justice, and reflection on the wide history of attempts to think as carefully as possible) that can be studied in sibling courses. It is likely that traditional humanities-based forms of philosophy instruction will continue, but they can be deeply enriched by the fundamental open-mindedness and enriched critical thinking skills prompted by Animal Studies–based versions of philosophy instruction.

Broader Environmental Studies and Ecology

Because many recognize the importance of scholars and students engaging the question of how the worldwide environmental movement relates to Animal Studies, there are many roles that environmentally focused courses could play in Animal Studies as a megafield. Further, since engagement with the animal dimensions of this question is sorely lacking in either animal law or public policy courses offered today, Animal Studies will develop courses in which environmental and ecological questions are integrated with questions formulated in the traditional animal protection manner of focusing on individuals rather than species-level issues or habitat protection. Such questions will open up many other inquiries and thereby distinguish versions of these courses in the Animal Studies megafield from sibling environmental studies courses in the other two megafields. Similarly, environmental law and animal law, which in contemporary education are separate courses, will in the future be combined to produce far more interdisciplinary and comprehensive learning opportunities.

Veterinary Education

Veterinary medicine has some internal tensions as it balances its long history of human-centeredness against the deeply ethical nature of the healing commitments of rank-and-file veterinarians. Historically, the field has faced constant ethical questions, a fact that results in considerable difficulties and tensions arising out of the different commitments characteristic of each of the following groups—veterinary students; practicing veterinarians; and administrators of veterinary schools, officers and staff of provincial and national organizations, and researchers based at veterinary schools but funded by government or private corporation sources.

In this complex profession to which modern societies delegate the important task of protecting nonhuman animals' health, teaching students about ethics-intensive issues has long been complicated and, in the opinion of many, wanting. Some students clamor for wide-ranging ethical instruction that addresses social, cultural, and philosophical issues of the kind so often mentioned in this introduction to Animal Studies. But the most typical form of teaching ethics in veterinary education is to avoid entirely such broad issues. It favors confining ethics discussions to a single course that focuses on a far narrower range of issues arising under various professional ethics codes promulgated by veterinary associations.[19] Such ethics courses are often outliers in the veterinary education curriculum because broader forms of animal ethics are rarely, if ever, raised in any of the core courses in the principal curriculum. The result is that ethics as a broad topic engaging students' basic values is often shunted aside as irrelevant and merely a matter of opinion.

In the Animal Studies megafield, veterinary ethics will be handled in a frank manner that combines the substantive ethical issues that sit at the heart of this key profession with the important practical problem faced by veterinary medicine pursued in modern societies dominated by the exceptionalist tradition.

Peace and Conflict Studies

As a social science field, this new discipline has grown rapidly at North American and western European universities since the middle of the twentieth century.[20] While the subject matter has typically been handled as a human-on-human problem, any work that seeks to describe both violent and nonviolent behavior by humans will inevitably engage problems involving nonhuman animals (even if they are, for example, being fought over as mere resources). But discussion of harms to nonhumans has much relevance to discussion of harms done by humans to humans alone. Accordingly, sibling courses on peace and nonviolence could easily be offered in an Animal Studies megafield and both of the megafields now found in the two-part university.

Beyond This Century: Which World Will We Leave to Our Children?

The question of choosing which world we will leave to our children invokes, but goes well beyond, our ethical natures—the question is also, like so many family-related issues, deeply personal. What will we leave them as possibilities in regard to our larger, more-than-human community?

Making community is at once an ethical, personal, and practical set of problems. To achieve it, we need to address the four tasks called out in chapter 1. We need to tell the entire story about our past with other living beings. We need to provide perspectives on other living beings' individual and communal lives. We need to explore future possibilities of living in a shared, more-than-human world. And we need to come to terms with the nature and limits of what humans might know about other living beings.

For many people, accomplishing these tasks will succeed only if they have the kind of commitment that flows from both personal interest and connection. Further, while accomplishing these tasks clearly involves important idea-based elements, the overall challenge is truly a practical matter. As ethicists are fond of saying, "ought implies can." If someone says to us, "You should do X," they are obviously implying that we can do X. Although "ought implies can" sounds technical, only a small amount of reflection is needed to recognize that it is, plain and simple, an eminently practical observation. Further, it is not meant to limit, but rather to enable, for we can do much even if we cannot do everything. In practice, that is to say, in everyday life, then, our ethical abilities bid us to address those problems we can impact with decisions made in our personal lives. One of the most fundamental of such personal issues is how each of us will answer the ever-present and ever-so-ethical question, Who are the others? Doing nothing is not really an option, since this choice effectively answers the question with the narrowness of an egotistical "I am going to care only about myself." We know, because we are communal animals, that such narrowness is an asocial prison. At the other end of the spectrum, pretending that humans can easily transcend our obvious limitations also imprisons us or, worse, does the kind of violence that Thomas Merton warned follows from "surrendering to too many demands, committing oneself to too many projects" (chapter 12).

So answering Who are the others? is a basic, truly practical problem we face each day. Animal Studies suggests that part of our answer must involve both human and nonhuman living beings, for we can surely take responsibility both within and beyond the species line. Of course, since we cannot solve all problems, we must look for those that we can solve. We will see our options, as well as act in ways that serve other animals (human and nonhuman alike) far better through, first, noticing and, second, taking seriously Animal Studies as it pursues its four basic issues—answering, "What is Animal Studies?," asking, "Who are animals?," discerning why Animal Studies is important, and recognizing the important personal features of meeting living beings outside our own species. If we do this well, we can tell the entire story, recognize many other animals' realities, see our own future possibilities in a mixed-species world, and even celebrate our limitations as we accept our role as one citizen among others in the shared, larger community of all life.

Conclusion

Animal Studies is a breakthrough in two senses—it pushes us to an open-minded, informed approach regarding who and what the living beings outside our species are. As fully, it pushes us to consider who we are.

This book has argued that, through Animal Studies and its commitment to tell the whole story, it is clear that our species has, as a whole, caused great harm to many other-than-human living beings. It is most relevant to our ethical nature, however, that just as much past harm was avoidable, many harms now done each day to nonhuman animals can be stopped if we so choose. The losses have been and remain astonishing, but the world we share with other living beings is fecund—it can grow healthy again.

If we look closely at the way of life we have created in many of the societies we are wont to call advanced, we will notice that, tragically, that way of life puts our children at risk. If we look even more closely, we notice that it causes great harm to ourselves as well.

Animal Studies in its modern version is a response to this impoverishment—it is also a plea to see the consequences of our choices clearly so we can see our future possibilities for living in a world that is better in two ways. First, we can help other animals fare better in this world, for as Animal Studies reveals through its commitments to thinking carefully and honestly and being scientifically informed and ethically astute, humans have often chosen to honor the fact that we live in a multispecies world that comprises a larger, mixed community.

Second, we also can fare better today and in the future.

Tragically, this book's win-win argument will fall on some deaf ears and be resisted because many humans will not be able to let go of dysfunctional and cruel forms of human self-importance that repudiate the insight that other animals also have an important place in the community of life.

Because the future must still be chosen, Animal Studies has prospects of sustaining a revolution that has started and may prevail. If it does, as the text suggests, "the real difficulties begin"—that is, we must decide how to be plain citizens living among our fellow animals in a multispecies, larger community that is our real home. This will require that we choose to regulate our own privileges to create the win-win realities that are possible.

In all this, Animal Studies offers us the possibility of not only thinking as fully integrated humans, but also reflecting on the world through the eyes and mind of an ethical, science-informed animal. In this regard, Animal Studies helps us see that caring beyond the species line is more than a present possibility. It has often been an implicit foundation of human societies and cultures—indeed, it was at times so widespread that one can easily believe it is connected to our biological realities. There is much that suggests, then, that interest in other living beings has deep roots in our own psyche and is thus vital to many different aspects of our human existence.

Notes

INTRODUCTION

1. William James, *The Principles of Psychology*, Great Books of the Western World Series, vol. 53 (London: Encyclopaedia Britannica, 1952; originally printed 1891), 318. James uses this famous phrase to describe a human baby's perception as part of his broad discussion of all humans' perception; James's next comment reveals that he recognizes this feature in even mature humans' engagement of the world.
2. Henry Beston, *The Outermost House: A Year of Life on the Great Beach of Cape Cod* (Garden City, NY: Doubleday Doran, 1928).

CHAPTER 1

1. Alexander Pope, "An Essay on Man," in *Poetical Works*, ed. H. F. Cary (London: Routledge, 1870), 225–226.
2. "The first step toward making a captive animal happy, says Dr. Hediger, is to study its natural life in the wild." "Science: The Happy Prisoners," *Time*, September 18, 1950, 74.
3. Viktor E. Frankl, *Man's Search for Meaning: An Introduction to Logotherapy*, 4th ed. (Boston: Beacon Press, 1992), 115.
4. Thomas Berry, "Loneliness and Presence," in *A Communion of Subjects: Animals in Religion, Science, and Ethics*, ed. Paul Waldau and Kimberley Patton (New York: Columbia University Press, 2006), 5–10, at 5.
5. James Rachels, *Created from Animals: The Moral Implications of Darwinism* (New York: Oxford University Press, 1991), 86.
6. Ibid.
7. For example, "human-companion animal bond" is used by A. H. Katcher and A. M. Beck, *New Perspectives on Our Lives with Companion Animals* (Philadelphia: University of Pennsylvania Press, 1983). A well-developed website offering deep resources and a digital newsletter regarding the breadth of subjects covered in academic fields and conferences using these and other names, as well as international interest in nonhuman animal issues, is mindinganimals.com
8. C. P. Flynn, *Social Creatures: A Human and Animal Studies Reader* (New York: Lantern Books, 2008), xvi.
9. Ibid.
10. It has been suggested that established naming conventions using the Greek roots *anthropos* and *zoon* should have produced the more awkward-sounding "anthropozoology."
11. Marianne Moore, in *The Nation* 163 (August 17, 1946): 192. This passage also appears in *The Complete Prose of Marianne Moore*, ed. Patricia C. Willis (New York: Viking, 1986), 406.
12. Bob Torres, *Making a Killing: The Political Economy of Animal Rights* (Oakland, CA: AK Press, 2007), 1. The comment about the author's personal life is on p. 3.
13. Dennis Overbye, "Gazing Afar for Other Earths, and Other Beings," *New York Times*, January 30, 2011. A version of this article appeared in print on January 31, 2011, New York edition, A1.
14. Elizabeth McKey and Karen Payne, "APPMA Study: Pet Ownership Soars," *Pet Business* 18 (1992): 22.

15. In this book, I use both "ideology" and "ideological" in a narrow sense to refer to a frame of mind that is idea-driven but less than open. This is the sense the word has in phrases like "a term of ideological abuse" or "shaped by an ideological squint"—the reference is, then, to the use of ideas as inflexible and dogmatic, even to "ideas as weapons" (Daniel Bell, *The End of Ideology* [Glencoe, IL: Free Press, 1960]). Note the critical thinking implications raised when Bell says that ideology is an "inability to distinguish possibilities from probabilities, converting the latter into certainties" (372).

16. I first used this description in Paul Waldau, *Animal Rights* (New York: Oxford University Press, 2011), 20–21.

17. E. O. Wilson, *The Diversity of Life* (Cambridge, MA: Belknap, 1992), 5.

18. The date at which such larger organisms first emerged on earth has constantly been moved back. A report published online cites an emergence date of 650 million years ago: Adam C. Maloof et al., "Possible Animal-Body Fossils in Pre-Marinoan Limestones from South Australia," *Nature Geoscience* 3 (2010): 653–659.

19. Brigitte Resl, "Introduction: Animals in Culture, ca. 1000–ca. 1400," in *A Cultural History of Animals in the Medieval Age* (London: Berg, 2009), 1.

20. Henri Poincaré, *The Value of Science* (New York: Dover, 1920; repr. 1958), 12.

21. Bernard Rollin has illuminated the relevant issues in a number of different books—see, for example, his early *The Unheeded Cry: Animal Consciousness, Animal Pain, and Science*, Studies in Bioethics (Oxford: Oxford University Press, 1989); and his more recent *Science and Ethics* (New York: Cambridge University Press, 2006).

22. Sandra Harding, *The Science Question in Feminism* (Ithaca, NY: Cornell University Press, 1986).

23. Richard Primack, *A Primer of Conservation Biology*, 4th ed. (Sunderland, MA: Sinauer, 2008), 65–68, 245.

24. Diana Reiss, *The Dolphin in the Mirror: Exploring Dolphin Minds and Saving Dolphin Lives* (Boston: Houghton Mifflin Harcourt, 2011).

25. W. Zimmer, *Passive Acoustic Monitoring of Cetaceans* (New York: Cambridge University Press, 2011), 1.

26. Claude Lévi-Strauss, *Totemism*, trans. Rodney Needham (Boston: Beacon, 1963), 89 (where Lévi-Strauss is discussing Radcliffe-Brown's views): "The animals in totemism cease to be solely or principally creatures which are feared, admired, or envied: their perceptible reality permits the embodiment of ideas and relations conceived by speculative thought on the basis of empirical observations. We can understand, too, that natural species are chosen not because they are 'good to eat' but because they are 'good to think.'"

27. The term "deconstruction" is used in some versions of modern philosophy for a very technical, complicated approach to the way meaning works in written texts, but here I use it only to mean "unpacking" or "pulling apart" how we have put this category together and thereby constructed it for our use.

28. There are many other explanations that go deep into our psyche. For example, some scientists have suggested that humans' connections to other animals are evolutionarily based—see the different explanations given by E. O. Wilson, *Biophilia* (Cambridge, MA: Harvard University Press, 1984); and Larry O'Hanlon, "Pets Vital to Human Evolution," Discovery News, August 10, 2010, http://news.discovery.com/animals/pets-humans-evolution.html.

29. Katherine Kete, "Introduction: Animals and Human Empire," in *A Cultural History of Animals in the Age of Empire*, ed. Kathleen Kete (Oxford: Berg, 2009), 15.

30. Ibid.

31. Gary Francione, *Animals as Persons* (New York: Columbia University Press, 2008), 117.

32. For the history of veterinary medicine, see Robert H. Dunlop and David J. Williams, *Veterinary Medicine: An Illustrated History* (St. Louis: Mosby, 1996).

33. Bernd Heinrich, *Mind of the Raven: Investigations and Adventures with Wolf-Birds* (New York: Cliff Street Books, 1999), 31.

34. Ibid., 32.

35. Stephen Jay Gould, "Animals and Us," *New York Review of Books*, August 19, 1995, 20–25, at 23.

36. There is now extensive scientific research that utilizes many criteria, including other animals' actions, to suggest conclusions about these animals' preferences—see, for example, Marian Stamp Dawkins, "A User's Guide to Animal Welfare," *TRENDS in Ecology and Evolution* 21, no. 2 (2006): 77–82.

CHAPTER 2

1. It was Helvetius who famously suggested that education makes people stupid—see *A Treatise on Man; His Intellectual Faculties and His Education*, 5–6, the beginning of section 1, chapter 3, and again at 49–50. The phrase in the text is Bertrand Russell's summary of Helvetius's views, in his witty but caricature-prone *History of Western Philosophy*. See Bertrand Russell, *The Basic Writings of Bertrand Russell: 1903–1959*, ed. Robert E. Egner and Lester E. Denonn (New York: Simon and Schuster, 1961), 294.

2. Various scholars such as the South African psychologist William Hudson, researcher Richard Langton Gregory, and the cultural geographer Yi-Fu Tuan have since the 1960s and 1970s used the adjective "carpentered" to describe the built environment of industrial societies, which is dominated by straight lines and right angles.

3. Immanuel Kant, "Duties to Animals and Spirits," in *Lectures on Ethics,* trans. Louis Infield (New York: Harper and Row, 1963), 239–241, at 239.

4. Joseph Rickaby, *Moral Philosophy* (London: Longmans, Green, 1888), 250. The emphasis is in the original.

5. Leon Walras, *Elements of Pure Economics or The Theory of Social Wealth,* trans. William Jaffe (London: George Allen and Unwin, 1954), 62.

6. Carl Cohen, "The Case for Biomedical Experimentation," *New England Journal of Medicine* 315 (1986): 865–870, at 867. Cohen's argument is that there are profound differences between human and nonhuman animals, such that the pain of other animals does not have as much moral weight as human pain (at 868).

7. Letter to J. G. Gmelin, February 14, 1747, quoted in George Seldes, *The Great Thoughts* (New York: Ballantine, 1985), 247.

8. Importantly, the critical thinking approach described at the end of this chapter requires acknowledgment that the conclusion that humans are, like all other animals, members of very ancient lineages of life is not necessary for animal studies, but, as the text suggests, the evidence for this conclusion is of such a high quality that rigor and common sense do require that animal studies engage an evolutionary point of view constantly.

9. Richard Dawkins, "Meet My Cousin, the Chimpanzee," *New Scientist,* June 5, 1993, 36–38; a similar essay appears in Paola Cavalieri and Peter Singer, eds., *The Great Ape Project: Equality beyond Humanity* (London: Fourth Estate, 1993).

10. Poll numbers differ somewhat on this issue, but they are invariably high—for example, a 1985 report revealed that 99 percent of 1,500 survey respondents thought of their own companion animals as family members. This study is cited in Sonia S. Waisman and Barbara R. Newell, "Recovery of 'Non-economic' Damages for Wrongful Killing or Injury of Companion Animals: A Judicial and Legislative Trend," *Animal Law* 7, no. 46 (2001): 59.

11. Roger Fouts, *Next of Kin: What Chimpanzees Have Taught Me about Who We Are* (New York: William Morrow, 1997), 5.

12. Cicero, *De Oratore,* trans. E. W. Sutton and H. Rackham, Loeb Classical Library Series 348 (Cambridge, MA: Harvard University Press, 1974), book 2, 14, section 62, 243, 245.

13. Droysen's quote is from his 1875 *Grundriss der Historik* and is cited in Frederick Jackson Turner's 1891 essay "The Significance of History."

14. A chronology of relevant events is included in Barry B. Powell, *Writing: Theory and History of the Technology of Civilization* (Oxford: Blackwell, 2009). There is also wide-ranging information on the history of writing at "The History of Writing," The Scriptorium, www.historian.net/hxwrite.htm, which has a link to Donald Ryan's helpful website Ancient Languages and Scripts (www.plu.edu/~ryandp/texts.html).

15. See, for example, Mircea Eliade, *A History of Religious Ideas,* vols. 1 and 2, trans. Willard R. Trask, and vol. 3 trans. Alf Hiltebeitel and Diane Apostolos-Cappadona (London: University of Chicago Press, 1978, 1982, 1985); H. W. Janson, *Apes and Ape Lore in the Middle Ages and Renaissance* (London: Warburg Institute, 1952); Kenneth Clark, *Animals and Men: Their Relationship as Reflected in Western Art from Prehistory to the Present Day* (London: Thames and Hudson, 1977). An example of Elizabeth Lawrence's interdisciplinary work is her *Hunting the Wren: Transformation of Bird to Symbol: A Study in Human-Animal Relationships* (Knoxville: University of Tennessee Press, 1997).

16. Gavin Van Horn, "Howling about the Land: Religion, Social Space, and Wolf Reintroduction in the Southwestern United States" (unpublished PhD diss., University of Florida, 2008), chapter 5.

17. Paul Shepard, *The Others: How Animals Made Us Human* (Washington, DC: Island Press, 1996), 6–7, emphasis added.

18. Robertson Davies, The Manticore (1972), for which Davies was awarded the Governor-General's Literary Award. The text is at http://www.e-reading.org.ua/bookreader.php/79762/Davies_-_The_Manticore.html.

19. Karen Armstrong, *The Great Transformation: The Beginning of Our Religious Traditions* (New York: Knopf, 2006), xiv, emphases added.

20. James Jasper, *The Animal Rights Crusade: The Growth of a Moral Protest* (New York: Free Press, 1992); Lawrence Finsen and Susan Finsen, *The Animal Rights Movement in America from Compassion to Respect* (New York: Twayne, 1994); Diane Beers, *For the Prevention of Cruelty: The History and Legacy of Animal Rights Activism in the United States* (Athens, OH: Swallow Press, 2006).

21. Northcote W. Thomas, "Animals," in *Encyclopedia of Religion and Ethics*, ed. J. Hastings (Edinburgh: T. and T. Clark, 1908), vol. 1, 483–535, at 483.

22. On the former issue, a good collection by an anthropologist is Tim Ingold, ed., *What Is an Animal?* (New York: Routledge, 1994).

23. Mary Midgley, *Beast and Man: The Roots of Human Nature*, rev. ed. (New York: Routledge, 1995). The additional quotes by Midgley are found on pp. 25–27.

24. Ludwig Wittgenstein, *Philosophical Investigations*, 3rd ed., trans. G. E. M. Anscombe (New York: Macmillan, 1958), para. 109, p. 47.

25. Upton Sinclair, *I, Candidate for Governor: And How I Got Licked* (Pasadena, CA: Author, 1935).

26. Quoted by Natalie Angier, *The Canon: A Whirligig Tour of the Beautiful Basics of Science* (Boston: Houghton Mifflin, 2007), 22.

27. Ibid.

28. Quoted in Anthony Lewis, "Dear Scoop Jackson," *New York Times*, March 15, 1971. Available at http://query.nytimes.com/mem/archive/pdf?res=F30A11F73454127B93C7A81788D85F458785F9.

29. This decision, whose technical legal citation is 347 U.S. 483, can be found easily at dozens of Internet sites, such as FindLaw, http://caselaw.lp.findlaw.com/scripts/getcase.pl?court=US&vol=347&invol=483.

30. M. L. Wax and R. H. Wax, "American Indian Education for What?" *Midcontinent American Studies Journal* 6, no. 2 (1965): 164–170, at 164.

31. Gail F. Melson, *Why the Wild Things Are: Animals in the Lives of Children* (Cambridge, MA: Harvard University Press, 2001), 13.

32. David Orr, *Earth in Mind: On Education, Environment, and the Human Prospect* (Washington, DC: Island Press, 1994), 5.

33. Theodore Roszak, "On Academic Delinquency," in *The Dissenting Academy*, ed. Theodore Roszak (New York: Vintage, 1968), 3–42, at 4.

34. Two recent examples are Martha Nussbaum, *Not for Profit: Why Democracy Needs the Humanities* (Princeton, NJ: Princeton University Press, 2010); and Andrew Hacker and Claudia Dreifus, *Higher Education? How Colleges Are Wasting Our Money and Failing Our Kids—and What We Can Do about It* (New York: Times Books, 2010).

35. Jacques Derrida, "The Animal That Therefore I Am (More to Follow)," trans. David Wills, *Critical Inquiry* 28, no. 2 (2002): 369–418.

36. Paul Shepard, *Thinking Animals: Animals and the Development of Human Intelligence* (New York: Viking, 1978), 2.

37. Ibid. Shepard's contributions and significance are called out nicely in Jim Mason, "Animals: From Souls and the Sacred in Prehistoric Times to Symbols and Slaves in Antiquity," in *A Cultural History of Animals in Antiquity*, ed. Linda Kalof (Oxford: Berg, 2009), ch. 1.

38. See, for example, Sharon Bertsch McGrayne, *The Theory That Would Not Die: How Bayes' Rule Cracked the Enigma Code, Hunted Down Russian Submarines and Emerged Triumphant from Two Centuries of Controversy* (New Haven, CT: Yale University Press, 2011).

39. Critical Studies signifies the academic field, while more generic senses are signaled by lowercase letters.

40. Michael Smith, "Virtuous Circles," *Southern Journal of Philosophy* 25 (1987): 207–220, n. 47.

41. Felicia R. Lee, "How Nobility of Purpose Can Square with Meanness and Lies," review of *Wild Bill: The Legend and Life of William O. Douglas*, New York Times, May 24, 2003.

42. "That is why we've deliberately avoided using terms like cognitive and analytic, or phrases like critical thinking and moral reasoning. There's nothing inherently wrong with these rubrics, it's just that they've been recast to force freshmen to view the world through professorial prisms." Hacker and Dreifus, *Higher Education?*, 7.

43. Angier, *The Canon*, 19.

44. Ibid., 39.

45. Matthew Lipman, *Thinking in Education*, 2nd ed. (New York: Cambridge University Press, 2003), 6.

46. American Library Association, *Final Report of the American Library Association Presidential Committee on Information Literacy* (Chicago: American Library Association, 1989).

47. Alfred North Whitehead, *The Concept of Nature*, Tarner Lectures Delivered in Trinity College November 1919 (Cambridge: Cambridge University Press, 1920, rpt. 1955), 163.

48. B. L. Krause, *The Great Animal Orchestra* (New York: Little, Brown, 2012), 104, 68, 16.

49. Richard Arum and Josipa Roksa, *Academically Adrift: Limited Learning on College Campuses* (Chicago: University of Chicago Press, 2011), 36.

CHAPTER 3

1. See, for example, Paul Waldau, *Animal Rights* (New York: Oxford University Press, 2011), passim.

2. Max Oelschlaeger, "Introduction," in Paul Shepard, *Thinking Animals: Animals and the Development of Human Intelligence* (New York: Viking Press, 1978), xii.

3. An example of how truly multidisciplinary an approach to learning and unlearning about just one species of nonhuman animal can be is Linda Kohanov, *The Tao of Equus: A Woman's Journey of Healing and Transformation through the Way of the Horse* (Novato, CA: New World Library, 2007).

4. See the website Corrupted Science, Tufts University (www.tufts.edu/~skrimsky/corrupted-science.htm), for additional data about the withholding of information in academically based research on various topics that involve humans. For example, in 1997 the *Boston Globe* ran the headline "Biomedical Results Often Are Withheld" over a story that described a survey of thousands of biomedical scientists that revealed astonishing numbers. More than 410 within the previous three years had held back results, with 28 percent of this group indicating that the delay in publication was because their research produced undesired results. A significant number (9 percent) of the survey respondents had also within the previous three years refused to share research results or scientific materials with science colleagues. Richard A. Knox, "Biomedical Results Often Are Withheld," *Boston Globe*, April 16, 1997.

5. Bernard Rollin, "Annual Meeting Keynote Address: Animal Agriculture and Emerging Social Ethics for Animals," *Journal of Animal Science* 82 (2004): 955–964, at 958.

6. Michael Pollan, *The Omnivore's Dilemma: A Natural History of Four Meals* (New York: Penguin, 2006), 218.

7. Matthew Scully, *Dominion: The Power of Man, the Suffering of Animals, and the Call to Mercy* (New York: St. Martin's, 2002), x.

8. Ibid., 268.

9. Arnold Arluke, "Trapped in a Guilt Cage," *New Scientist*, April 4, 1992. Additional research by Arluke can be found in "Sacrificial Symbolism in Animal Experimentation: Object or Pet?," *Anthrozoos* 2 (1988): 98–117; "The Ethical Socialization of Animal Researchers," *Lab Animal* 23, no. 6 (1994): 30–35; and Arnold Arluke and Clinton Sanders, *Regarding Animals: Animals, Culture, and Society* (Philadelphia: Temple University Press, 1996), examines how different laboratory cultures create different moral climates. See, for example, chapter 5, "Systems of Meaning in Primate Labs."

10. Adrian R. Morrison, *An Odyssey with Animals: A Veterinarian's Reflections on the Animal Rights and Welfare Debate* (New York: Oxford University Press, 2009), 126, 118.

11. Steven Shapin, *The Scientific Revolution* (Chicago: University of Chicago Press, 1996), 4.

12. "Question everything" is the principal logo and advertising tool used throughout the Science Channel's website, http://science.discovery.com/videos/science-promos-question-everything.html (accessed May 9, 2012). "That's how the revolution starts" is the Science Channel's description of its first promotional video, "Question Everything," and is also part of the text read during this video (available at the website above).

13. Carl Sagan, *Cosmos* (New York: Random House, 1980), 233.

14. Erwin Schrödinger, Nobel Prize winner in physics, 1933, as quoted by Jean François Revel and Matthieu Ricard, *The Monk and the Philosopher: A Father and Son Discuss the Meaning of Life* (New York: Schocken, 1999), 214. Below, I use Schrödinger's image to suggest that approaches that give us some order but remain "terribly silent" about matters "close to our hearts" fail to pass "the Schrödinger test."

15. George Gaylord Simpson, *This View of Life: The World of an Evolutionist* (New York: Harcourt Brace, 1964), 106–107.

16. Importantly, there are other sciences that deal with subatomic realms, the mysteries of relativity, and the even stranger world of quantum mechanics. But since these baffling levels of reality are not easily recognized in day-to-day life, Simpson's point continues to help us understand why most humans' attention is captured by animals rather than, say, rocks or even plants.

17. Ludwig Wittgenstein, *Tractatus Logico-Philosophicus* (London: Routledge and Kegan Paul, 1981), 27.

18. Personal communication, Steve Meyer, September 18, 2004.

19. For an annotated introduction to science in non-Western traditions in China, India, Africa, Latin America, Native America, Japan, Australia, and the Pacific islands, see Douglas Allchin and Robert DeKosky, eds., *An Introduction to History of Science in Non-Western Traditions* (Seattle: History of Science Society, 1998).

20. See, for example, W. Zimmer, *Passive Acoustic Monitoring of Cetaceans* (New York: Cambridge University Press, 2011), for a description of how this highly technical science works.

21. The history and various aspects of this approach are described in E. Mayr and W. B. Provine, *The Evolutionary Synthesis: Perspectives on the Unification of Biology* (Cambridge, MA: Harvard University Press, 1980).

22. There are several versions of Ernst Rutherford's quote. This one is taken from P. M. S. Blackett, "Memories of Rutherford," in *Rutherford at Manchester*, ed. J. B. Birks (London: Heywood, 1962), 102–113, at 108.

23. Stephen Greenblatt, *The Swerve: How the World became Modern* (New York: Norton, 2011), 191, summarized the views advanced by the Roman writer Lucretius in the century before Jesus was born—these views were part of the widely discussed Epicurean tradition so eloquently described in Lucretius's *De rerum natura* (On the Nature of Things).

24. Arthur O. Lovejoy, *The Great Chain of Being* (Cambridge, MA: Harvard University Press, 1936; rpt. New York: Harper and Row, 1960), 235.

25. See, for example, the opening lines of Stephen Jay Gould's introduction to C. Zimmer, *Evolution: The Triumph of an Idea* (New York: HarperCollins, 2001).

26. See Mayr and Provine, *The Evolutionary Synthesis*.

27. David Quammen, "Was Darwin Wrong?," *National Geographic* 206, no. 5 (2004): 4–35.

28. Lynn Margulis, Karlene V. Schwartz, and Michael Dolan, *The Illustrated Five Kingdoms: A Guide to the Diversity of Life on Earth* (New York: HarperCollins, 1994).

29. Jim Mason, "Animals: From Souls and the Sacred in Prehistoric Times to Symbols and Slaves in Antiquity," in *A Cultural History of Animals in Antiquity,* ed. Linda Kalof (Oxford: Berg, 2009), 20.

30. The second quote is Francis Bacon, "preface," *Novum Organum*, vol. 30, Great Books of the Western World Series (London: Encyclopaedia Britannica, 1952), 105. The first quote also comes from *Novum Organum* and is cited by James Gleick, *Isaac Newton* (New York: Pantheon, 2003), 66.

31. Voltaire, "Beasts," in *A Philosophical Dictionary*, vol. 3, part I, 222–225, at 223, in *The Works of Voltaire: A Contemporary Version with Notes* (New York: Dingwall-Rock, 1927).

32. Barry Lopez, *Arctic Dreams: Imagination and Desire in a Northern Landscape* (New York: Scribner's, 1986), 177.

33. Alan Cutler, *The Seashell on the Mountaintop: A Story of Science, Sainthood, and the Humble Genius Who Discovered a New History of the Earth* (New York: Dutton, 2003), 38–42.

34. See, for example, J. David Smith, "The Study of Animal Metacognition," *Trends in Cognitive Sciences* 13, no. 9 (2009): 389–396, summarizing growing evidence that various nonhuman animals share humans' capacity for metacognition, "that is, for monitoring or regulating their own cognitive states."

35. See, for example, G. A. Bradshaw, "Inside Looking Out: Neuroethological Compromise Effects in Elephants in Captivity," in Debra L. Forthman, Lisa F. Kane, David Hancocks, and Paul F. Waldau, eds., *An Elephant in the Room: The Science and Well-Being of Elephants in Captivity* (North Grafton, MA: Center for Animals and Public Policy, 2008). Some of the implications of this sort of study are addressed in Charles Siebert, "An Elephant Crackup?," *New York Times Magazine*, October 8, 2006.

36. Eileen Crist, *Images of Animals: Anthropomorphism and Animal Mind* (Philadelphia: Temple University Press, 1999), 89.

37. David Scofield Wilson, introduction to part 4, "Come into Animal Presence: Ethics, Ethology and Konrad Lorenz," in David Aftandilian, *What Are the Animals to Us? Approaches from Science, Religion, Folklore, Literature, and Art* (Knoxville: University of Tennessee Press, 2007), 259–265, at 260.

38. Collin Allen and Marc Bekoff, *Species of Mind: The Philosophy and Biology of Cognitive Ethology* (Cambridge, MA: Bradford/MIT Press, 1997), ix.

39. Ibid.

40. Paul Sears, "Ecology—a Subversive Subject," *BioScience* 14, no. 7 (1964): 11–13. This article led to (and also appears in) Paul Shepard and Daniel McKinley, eds., *The Subversive Science: Essays toward an Ecology of Man* Boston: Houghton Mifflin, 1969).

41. J. E. R. Staddon, "Animal Psychology: The Tyranny of Anthropocentrism," in *Whither Ethology? Perspectives in Ethology*, ed. P. P. G. Bateson and P. H. Klopfer (New York: Plenum, 1989), 123–135, at 133, emphasis in original. The first two quotes are, respectively, at 133 and 123.

42. This argument is made in Paul Waldau, "Pushing Environmental Justice to a Natural Limit," in *A Communion of Subjects: Animals in Religion, Science, and Ethics*, ed. Paul Waldau and Kimberly Patton (New York: Columbia University Press, 2006), 629–642.

43. Kenneth S. Norris, *Dolphin Days: The Life and Times of the Spinner Dolphin* (New York: Norton, 1991), 180.

44. Denise L. Herzing, "Dolphins: Focusing on an Understanding," *Ocean Realm: International Magazine of the Sea*, June 1995, 22–29, at 24.

45. Thomas White, *In Defense of Dolphins: The New Moral Frontier*, Blackwell Public Philosophy (Malden, MA: Blackwell, 2007).

46. R. Sukumar, *The Asian Elephant: Ecology and Management* (Cambridge: Cambridge University Press, 1989), 1.

47. Siebert, "An Elephant Crackup?"

48. Daphne Sheldrick, "The Rearing and Rehabilitation of Orphaned African Elephant Calves in Kenya," in *An Elephant in the Room: The Science and Well-Being of Elephants in Captivity*, ed. Debra L. Forthman, Lisa F. Kane, David Hancocks, and Paul F. Waldau, (North Grafton, MA: Center for Animals and Public Policy, 2008), 208–212, at 208.

49. Early books that have prompted much discussion include Frans de Waal, *Chimpanzee Politics* (London: Jonathan Cape, 1982); R. W. Byrne and A. Whiten, *Machiavellian Intelligence* (Oxford: Oxford University Press, 1988); and Andrew Whiten and Richard W. Byrne, *Machiavellian Intelligence II: Extensions and Evaluations* (New York: Cambridge University Press, 1997).

50. Whiten and Byrne, *Machiavellian Intelligence II*, 2.

51. Ibid., 1.

52. Quoted by Natalie Angier, "Confessions of a Lonely Atheist," *New York Times Magazine*, January 14, 2001. See as well de Waal's many books that expand on his *Chimpanzee Politics: Peacemaking among Primates* (Cambridge, MA: Harvard University Press, 1989); *Good Natured: The Origins of Right and Wrong in Humans and Other Animals* (Cambridge, MA: Harvard University Press, 1996); and *The Age of Empathy: Nature's Lessons for a Kinder Society* (New York: Harmony Books, 2009).

53. Harold Lasswell, *Politics: Who Gets What, When, How* (New York: McGraw-Hill, 1936; 2nd ed., Cleveland: World Publishing, 1958). Lasswell is the respected founder of the field of policy sciences.

54. The problems in slaughterhouses go far beyond the extraordinary harms suffered by the food animals—see, for example, Gail Eisnitz, *Slaughterhouse: The Shocking Story of Greed, Neglect, and Inhumane Treatment inside the U.S. Meat Industry* (Amherst, NY: Prometheus, 1997).

55. Pew Commission on Industrial Farm Animal Production, *Putting Meat on the Table: Industrial Farm Animal Production in America* (Pew Charitable Trusts and the Johns Hopkins School of Public Health, 2008); Food and Agriculture Organization, *Livestock's Long Shadow: Environmental Issues and Options*, 2006, www.fao.org/docrep/010/a0701e/a0701e00.HTM.

56. Based on the comment "Law is the projection of an imagined future upon reality," Robert M. Cover, "Violence and the Word," *Yale Law Journal* 95 (1986): 1601, 1604.

57. See, for example, William R. Jordan III and George M. Lubick, *Making Nature Whole: A History of Ecological Restoration* (Washington, DC: Island Press, 2011).

58. Rabindranath Tagore, *Sādhanā: The Realisation of Life* (New York: Macmillan, 1913), ch. 1, available at Project Gutenberg (www.gutenberg.org/).

59. Cited in Frank Graham Jr., *Since Silent Spring* (New York: Fawcett World Library, 1970), 173.

60. Personal communication, during a meeting with the faculty of Tufts Center for Animals and Public Policy, 2001. Professor Meyer died in 2006, the same year his *The End of the Wild* (Cambridge, MA: MIT Press) was published.

61. Diane Beers, *For the Prevention of Cruelty: The History and Legacy of Animal Rights Activism in the United States* (Athens, OH: Swallow Press, 2006), 7.

62. See, for example, any of the many editions of Thomas Dye, *Understanding Public Policy* (in 2011 this book's 13th edition was published). Another example of a definition in this vein is "public policy is how politicians make a difference," Peter Bridgman and Glyn Davis, *The Australian Policy Handbook*, 3rd ed. (Crows Nest, N.S.W., Australia: Allen and Unwin, 2004), 3.

63. Thomas Birkland, *An Introduction to the Policy Process: Theories, Concepts, and Models of Public Policy Making*, 3rd ed. (Armonk, NY: M. E. Sharpe, 2010), 9.

64. Frank Ackerman and Lisa Heinzerling, *Priceless: On Knowing the Price of Everything and the Value of Nothing* (New York: New Press, 2004), 150–151.

65. Deborah Stone, *Policy Paradox: The Art of Political Decision Making*, rev. ed. (New York: Norton, 2002), xi.

66. See, for example, the following regarding such views among Americans 1985–2000: Elizabeth McKey and Karen Payne, "APPMA Study: Pet Ownership Soars," *Pet Business* 18 (1993): 22.

67. A 1985 survey of 1,500 respondents found that 99 percent considered one or more nonhuman animals to be a family member—Victoria L. Voith, "Attachment of People to Companion Animals," *Veterinary Clinics of North America* 15 (1985): 289–290. Ten years later, a similar survey showed that 70 percent of those surveyed thought of their companion animals as children—see Carol Marie Cropper, "Strides in Pet Care Come at Price Owners Will Pay," *New York Times*, April 5, 1998, A16. A *USA Today* study in 2000 showed that 52 percent of people cook special foods for their companion animals and that 84 percent refer to themselves as

"mom" or "dad" in relation to their companion animals—see Cindy Hall and Elizabeth Wing, "Pets Are Part of the Family," *USA Today*, March 1, 2000, D9.

68. *Dille v St. Luke's Hospital*, 355 Mo. 436, 196 S. W. 2d 615, at 629.

69. Ibid.

70. Harold Lasswell, *A Pre-View of Policy Sciences* (New York: American Elsevier, 1971), 2.

71. Eisnitz, *Slaughterhouse*, 188–190.

72. Food and Agriculture Organization, *Livestock's Long Shadow*.

73. Pew Commission on Industrial Farm Animal Production, "Putting Meat on the Table."

74. Rollin has argued this in a number of places—see, for example, the introduction to Bernard E. Rollin, *An Introduction to Veterinary Medical Ethics: Theory and Cases* (Ames: Iowa State University Press, 1999), 48–50.

75. The Association of Zoos and Aquariums has been reporting this for more than a quarter of a century—in 2012, "over 175 million annual visitors—more visitors than NFL, NBA, NHL, and MLB annual attendance combined." See "Visitor Demographics," Association of Zoos and Aquariums, http://www.aza.org/visitor-demographics/.

CHAPTER 4

1. Steven M. Wise, author of the seminal *Rattling the Cage: Toward Legal Rights for Animals* (Cambridge, MA: Perseus Books, 2000).

2. The interesting story of law's relation to science, including past uses of science to buttress harms such as racial prejudice, is only now being told in detail—see, for example, David Faigman, *Laboratory of Justice: The Supreme Court's 200-Year Struggle to Integrate Science and the Law* (New York: Times Books/Henry Holt, 2004); and Angelo Ancheta, *Scientific Evidence and Equal Protection of the Law* (New Brunswick, NJ: Rutgers University Press, 2006).

3. Thomas White, *In Defense of Dolphins: The New Moral Frontier*, Blackwell Public Philosophy (Malden, MA: Blackwell, 2007), 8.

4. The concept of first-wave animal law was addressed in the 2010 conference at Harvard Law School convened by the Animal Legal Defense Fund and others—see, for example, Paul Waldau, "Catching the Second Wave of Animal Law," Paul Waldau's Work and Research, 2010, http://www.paulwaldau.com/publications-for-downloading.html.

5. Framing debates in terms of "rights versus welfare" is challenged by some animal protection advocates—see, for example, Paul Waldau, *Animal Rights* (New York: Oxford University Press, 2011), 95–99, regarding important differences in the way the word "welfare" is used.

6. For example, the widely quoted scientist Marc Bekoff (who also writes often about ethical issues and the importance of activism) suggested in 1998, "Narrow-minded primatocentrism and speciesism must be resisted in our studies of animal cognition and animal protection and rights. Line-drawing into 'lower' and 'higher' species is a misleading speciesist practice that should be vigorously resisted because not only is line-drawing bad biology but also because it can have disastrous consequences for how animals are viewed and treated." Marc Bekoff, "Deep Ethology, Animal Rights, and the Great Ape/Animal Project: Resisting Speciesism and Expanding the Community of Equals," *Journal of Agricultural and Environmental Ethics* 10 (1998): 269–296, abstract.

7. Jeremy Bentham, *An Introduction to the Principles of Morals and Legislation*, ed. J. H. Burns and H. L. A. Hart (London: Athlone Press, 1970), 283 n. 2.

8. See, for example, Waldau, *Animal Rights*.

9. Tom Regan, *The Case for Animal Rights* (Berkeley: University of California Press, 1983).

10. John Locke, "Epistle to the Reader," in *An Essay Concerning Human Understanding*, Great Books of the Western World Series, vol. 35 (London: Encyclopaedia Britannica, 1952), 89.

11. Martha Nussbaum, *Frontiers of Justice: Disability, Nationality, Species Membership* (Cambridge, MA: Belknap, 2006), 337, emphasis in the original.

12. Daniel Dennett, "Animal Consciousness: What Matters and Why," *Social Research* 62, no. 3 (1995): 691–710, at 693.

13. James Rachels, *Created from Animals: The Moral Implications of Darwinism* (New York: Oxford University Press, 1991), 220. A version of this dilemma also appears in Hugh Lafollette and Niall Shanks, "The Origin of Speciesism," *Philosophy* 71 (1996): 41–61.

14. Bernard E. Rollin, *Animal Rights and Human Morality*, rev. ed. (Buffalo, NY: Prometheus, 1992), 137.

15. Kwame Anthony Appiah, "What Will Future Generations Condemn Us For?," *Washington Post*, September 26, 2010.

16. Ernst Mayr, *The Growth of Biological Thought: Diversity, Evolution, and Inheritance* (Cambridge, MA: Belknap, 1982), 23.

CHAPTER 5

1. Sarah Boxer, "Animals Have Taken Over Art, and Art Wonders Why," *New York Times*, June 24, 2000, B9, B11.

2. Marcel Brion, *Animals in Art* (London: George C. Harrap, 1959), 15.

3. Diane Apostolos-Capadona, "On the Dynamis of Animals, or How Animalium became Anthropos," in *A Communion of Subjects: Animals in Religion, Science, and Ethics*, ed. Paul Waldau and Kimberly Patton (New York: Columbia University Press, 2006), 439–457, at 446.

4. Jim Mason, "Animals: From Souls and the Sacred in Prehistoric Times to Symbols and Slaves in Antiquity," in *A Cultural History of Animals in Antiquity*, ed. Linda Kalof (New York: Berg, 2009), 16–45, at 21 (Mason cites Kenneth Clark and Paul Shepard in support of this claim).

5. For an interesting perspective on changes that marginalized nonhuman animals in Western art traditions, as well as much on scholarship about the presence and absence of nonhuman animals in the arts generally, see Apostolos-Capadona, "On the Dynamis of Animals."

6. An illustrative example is the place given to dogs by Titian in his 1538 painting *Venus of Urbino*; on this subject more generally, see Simona Cohen, *Animals as Disguised Symbols in Renaissance Art* (Boston: Brill, 2009).

7. Linda Kalof, *Looking at Animals in Human History* (London: Reaktion Books, 2007), 1.

8. Ibid., 2–3.

9. Ibid., 3.

10. Ibid., 5 n. 21.

11. Ibid., 5.

12. Ibid., 7, referring to Juliet Clutton-Brock, *A Natural History of Domesticated Animals* (Cambridge: Cambridge University Press, 1999).

13. Kenneth Clark, *Animals and Men: Their Relationship as Reflected in Western Art from Prehistory to the Present Day* (London: Thames and Hudson, 1977), 14.

14. Multiple fields debate this issue—including history, anthropology, archaeology, comparative religion, and more.

15. See, for example, Steven Lonsdale, *Animals and the Origins of Dance* (New York: Thames and Hudson, 1982).

16. Aniela Jaffé, "Symbolism in the Visual Arts," in *Man and His Symbols*, ed. Carl Jung (Garden City, NY: Doubleday, 1964), part 4, 230–271, at 236.

17. Lonsdale, *Animals and the Origins of Dance*, 11, 18, 23.

18. Ibid., 12.

19. Ibid., 56–57, regarding the understanding of Blackfoot Indians about the origin of their Buffalo Dance.

20. Ibid., 18.

21. Ibid., 12.

22. Ibid., 9–11.

23. Ibid., 12.

24. See, for example, ibid., 18ff.

25. Ibid., 17.

26. Douglas H. Chadwick, *The Fate of the Elephant* (San Francisco: Sierra Club Books, 1994), 352–353.

27. Ibid., 348.

28. Ibid., 346.

29. More of this story appears in Paul Waldau, *The Specter of Speciesism: Buddhist and Christian Views of Animals* (New York: Oxford University Press, 2001), chapter 4.

30. Paul Shepard, *The Others: How Animals Made Us Human* (Washington, DC: Island Press, 1996), 6–7.

31. Richard Bauckham, "Jesus and the Wild Animals (Mark 1:13): A Christological Image for an Ecological Age," In *Jesus of Nazareth: Lord and Christ*, ed. Joel B. Green and Max Turner (Grand Rapids, MI: Erdmans, 1994), 3–4.

32. William B. Ashworth, "Natural History and the Emblematic World View," in *Reappraisals of the Scientific Revolution*, ed. David C. Lindberg and Robert S. Westman (New York: Cambridge University Press, 1990), 303–332, at 305–306.

33. Lionel Trilling, *The Liberal Imagination* (New York: New York Review of Books, 1950), xxi.

34. Alexander McCall Smith, "Big Cats," reviewing *A Tale of Two Lions* by Roberto Ransom, *New York Times Book Review*, January 28, 2007.

35. Percy Bysshe Shelley, "Defence of Poetry," Bartleby.com, http://www.bartleby.com/27/23.html.

36. Wallace Stevens, *Opus Posthumous* (New York: A. A. Knopf, 1957), 162.

37. Simon Critchley, *Things Merely Are: Philosophy in the Poetry of Wallace Stevens* (New York: Routledge, 2005). The first quote is from Critchley himself at p. 89, while the second quote is from the publisher's description of the book.

38. Mary Oliver, "The Summer Day," in *New and Selected Poems* (Boston: Beacon, 1992).

39. Mary Oliver, "The Swan," *Paris Review* no. 124 (Fall 1992).

40. Natachee Scott Momaday, *House Made of Dawn* (New York: Harper and Row, 1966).

41. Thomas Berry, *The Dream of the Earth* (San Francisco: Sierra Club Books, 1988), 4–5.

42. Daniel Quinn, *Ishmael* (New York: Bantam, 1992), 84; emphasis added.

43. Howard Gardner, *Frames of Mind: The Theory of Multiple Intelligences* (New York: Basic Books, 1985).

44. Martha Nussbaum, *Not for Profit: Why Democracy Needs the Humanities* (Princeton, NJ: Princeton University Press, 2010), 102.

45. There is much new, animal-informed scholarship available today about the complex Christian tradition and its many subtraditions—among the best book-length treatments are Laura Hobgood-Oster, *The Friends We Keep: Unleashing Christianity's Compassion for Animals* (Waco, TX: Baylor University Press, 2010); Laura Hobgood-Oster, *Holy Dogs and Asses: Animals in the Christian Tradition* (Urbana: University of Illinois Press, 2008).

CHAPTER 6

1. Michel Eyquem de Montaigne, "Apology for Raymond Sebond," in *The Complete Works of Montaigne: Essays, Travel Journal, Letters*, trans. Donald M. Frame (Stanford, CA: Stanford University Press, 1948), 318–457, 408.

2. Immanuel Kant, "Duties to Animals and Spirits," in *Lectures on Ethics*, trans. Louis Infield (New York: Harper and Row, 1963), 239–241, at 239; Immanuel Kant, "Duties towards Others," in *Lectures on Ethics*, 191–200, at 191.

3. See, for example, Descartes's famous "Letter to Marquess of Newcastle," written November 23, 1646, in *Descartes: Philosophical Letters*, ed. Anthony Kenny (Oxford: Oxford University Press, 1970), 205–208.

4. René Descartes, *Discourse on the Method of Rightly Conducting the Reason and Seeking Truth in the Sciences*, Great Books of the Western World Series, vol. 31 (London: Encyclopaedia Britannica, 1952), 41–67, opening paragraph of section 4.

5. Ibid., section 6, 4th paragraph.

6. Mary Midgley, *Animals and Why They Matter* (Athens: University of Georgia Press, 1984), 123.

7. This passage is from Descartes, "Letter to Henry More (Februay 5, 1649)," in *Descartes: Philosophical Letters*, 243.

8. Bernard le Bovier de Fontenelle, quoted in L. C. Rosenfield, *From Beast-Machine to Man-Machine*, enlarged ed. (New York: Octagon, 1968), 54.

9. Paul Shepard, *The Others: How Animals Made Us Human* (Washington, DC: Island Press, 1996), 325.

10. Bernard Rollin, *Putting the Horse before Descartes: A Memoir* (Philadelphia: Temple University Press, 2011).

11. Richard Sorabji, *Animal Minds and Human Morals: The Origins of the Western Debate* (London: Duckworth, 1993), 2.

12. See Augustine, *De Musica*, 1.4–6.

13. See Augustine, *De Trinitate*, 8.6.9.

14. Aristotle, *Politics* 1256b17–22, in *The Complete Works of Aristotle: The Revised Oxford Translation* (Princeton, NJ: Princeton University Press, 1984), 1993–1994.

15. Cicero, *De Natura Deorum*, II 14, 37.

16. Clarence J. Glacken, *Traces on the Rhodian Shore: Nature and Culture in Western Thought from Ancient Times to the End of the Eighteenth Century* (Berkeley: University of California Press, 1967), 47–48.

17. Perhaps the most famous passage advocating unrestrained experimentation on nonhuman animals comes from the late nineteenth century: "A Physiologist is not a man of fashion, he is a man of science, absorbed by the scientific idea which he pursues: he no longer hears the cry of animals, he no longer sees the blood that flows, he sees only his idea and perceives only organisms concealing problems which he intends to solve." Claude Bernard, *An Introduction to the Study of Experimental Medicine*, trans. H. C. Green (New York: Dover, 1957), 103 (originally published 1865).

18. Midgley, *Animals and Why They Matter.*

19. *Catechism of the Catholic Church* (London: Geoffrey Chapman, 1994), 516, para. 2415.

20. See, for example, Norman Kemp Smith, *New Studies in the Philosophy of Descartes* (London: Macmillan, 1952), 136, 140; and John Cottingham, "A Brute to the Brutes? Descartes' Treatment of Animals," *Philosophy* 53 (October 1978): 551–559. This claim is disputed by many; see, for example, Tom Regan, *The Case for Animal Rights* (Berkeley: University of California Press, 1983), 3–4.

21. I am indebted to the philosopher of religion John Hick for this quote, cited in Margaret Chatterjee, *Gandhi's Religious Thought* (Notre Dame, IN: Notre Dame University Press, 1983), 73. The original quote is from Mahadev Desai, *Day to Day with Gandhi*, vol. 7 (Varanasi: Navajivan, 1968–1972), 111–112.

22. Aristotle, *Metaphysics* 980b25.

23. Consider, for example, a list of words in use today that Aristotle coined—category, energy, actuality, motive, end, principle, form, faculty, mean, maxim—which still form a large part of our commonsense and scientific terminology. Will Durant, *The Story of Philosophy: The Lives and Opinions of the Great Philosophers of the Western World* (New York: Clarion/Simon and Schuster, 1967), 46.

24. Ibid., 54.

25. For example, see Christine Korsgaard, "Fellow Creatures: Kantian Ethics and Our Duties to Animals," Tanner Lecture, 2004, www.tannerlectures.utah.edu/lectures/documents/volume25/korsgaard_2005.pdf.

26. Steven Pinker, *The Better Angels of Our Nature: Why Violence Has Declined* (New York: Viking, 2011).

27. Daniel Robinson, *The Great Ideas in Philosophy* (Teaching Company, 1997, on CD), lecture 2.

28. Ibid.

29. Ibid., lecture 7.

30. Socrates, *The Dialogues of Plato*, trans. B. Jowett, vol. 2 (New York: Random House, 1937), 290. This form of argument, with variations, can be found in other classical Greek writers—see, for example, fragments 15 and 16 of Xenophon (regarding horses, oxen, and lions conceiving of divine bodies in terms of, respectively, horses, oxen, and lions), cited in Kathleen Freeman, *Ancilla to the Pre-Socratic Philosophers: A Complete Translation of the Fragments in Diels, Framente der Vorsokratiker* (Cambridge, MA: Harvard University Press, 1966), 22.

31. Donald R. Griffin, "From Cognition to Consciousness," *Animal Cognition* 1 (1998): 3–16, 13. This essay also appears in Waldau and Patton, *A Communion of Subjects.*

32. The "anthropomorphism" in the text should be distinguished from what is sometimes called "ethical anthropomorphism," a term that refers to traditions in ethical analysis that pay attention only to human issues and not at all to impacts of actions on living beings outside the human species.

33. Griffin, "From Cognition to Consciousness," 11.

34. Ibid.

35. See, for example, F. B. M. de Waal, *Good Natured: The Origins of Right and Wrong in Humans and Other Animals* (Cambridge, MA: Harvard University Press, 1996), 63–64.

36. F. B.M. de Waal, "Are We in Anthropodenial?," *Discover* 18, no. 7 (1997): 50–53.

37. John Henry Newman, *Parochial and Plain Sermons*, vol. 4 (London: Longmans, Green, 1909), 205–206.

38. John Muir, "Wild Wool," *Overland Monthly* 20 (April 1875): 364.

39. Thom White Wolf Fassett, "Where Do We Go from Here?," in *Defending Mother Earth: Native American Perspectives on Environmental Justice*, ed. Jace Weaver (New York: Orbis, 1996), 177–191, at 182.

40. The short quote comes from John Grim, "Knowing and Being Known by Animals: Indigenous Perspectives on Personhood," in Waldau and Patton, *A Communion of Subjects*, 373–390, at 380.

41. Vine Deloria Jr., *God Is Red: A Native View of Religion*, 2nd ed. (Golden, CO: North American Press, 1992), 66.

42. Daniel Jonah Goldhagen, *Worse Than War: Genocide, Eliminationism, and the Ongoing Assault on Humanity* (New York: PublicAffairs, 2009), 54.

43. Bernard Rollin, "Veterinary and Animal Ethics," in James F. Wilson, *Law and Ethics of the Veterinary Profession* (Yardley, PA: Priority Press, 1993), 24–49, at 30.

CHAPTER 7

1. See, for example, Paul Hawken, *Blessed Unrest: How the Largest Movement in the World Came into Being, and Why No One Saw It Coming* (New York: Viking, 2007); as well as Thomas Friedman, *The Lexus and the Olive Tree* (New York: Anchor Books, 1999), and *The World Is Flat: A Brief History of the Twenty-First Century* (New York: Farrar, Straus and Girous, 2005), both of which link the democratization of information to democratizations of technology and finance.

2. Mary Fairchild, "Christianity Today: General Statistics and Facts of Christianity," About.com, Christianity, christianity.about.com/od/denominations/p/christiantoday.htm, which lists the number as "approximately 41,000" as of October 2012. While the numbers given for the Islamic tradition's subdivisions or those of the many different Hindu traditions are considerably smaller, the same phenomenon of great internal diversity prevails in these and other religious traditions generally.

3. Researchers and scholars often use the encompassing term "lifeway" to describe how worldviews and practical actions are bound together. For example, the Yale scholar John Grim uses this term as he explores the views and lives of indigenous peoples: "Cosmologies, or oral narrative stories, transmit worldview values of the people, and describe the web of human activities within the powerful spirit world of the local bioregion. In this sense to analyze religion as a separate system of beliefs and ritual practices apart from subsistence, kinship, language, governance and landscape is to misunderstand indigenous religion." John Grim, "Indigenous Traditions and Ecology," Forum on Religion and Ecology, http://fore.research.yale.edu/religion/indigenous/.

4. There are many hybrids of the second and third comparative approaches—see, for example, Janos Jany, *Judging in the Islamic, Jewish and Zoroastrian Legal Traditions: A Comparison of Theory and Practice* (Burlington, VT: Ashgate, 2012).

5. See, for example, the constitutions of India, Switzerland, Germany, and Ecuador.

6. René David and John Brierley, *Major Legal Systems in the World Today: An Introduction to the Comparative Study of Law*, 2nd ed. (New York: Free Press, 1978), 2.

7. W. Bradley Wendel, "Jurisprudence and Judicial Ethics," in Cornell Law School, Legal Studies Research Paper Series, research paper no. 08-009, available from the Social Science Research Network Electronic Paper Collection: http://ssrn.com/abstract=1024316. Wendel cites Mitchel Lasser, *Judicial Deliberations: A Comparative Analysis of Judicial Transparency and Legitimacy* (New York: Oxford University Press, 2004), and Montesquieu's 1748 masterpiece The Spirit of the Laws.

8. See, for example, David Wolfson, "Beyond the Law: Agribusiness and the Systemic Abuse of Animals Raised for Food or Food Production," *Animal Law* 2 (1996): 123–151. Also in book form: *Beyond the Law: Agribusiness and the Systemic Abuse of Animals Raised for Food or Food Production* (New York: Farm Sanctuary, 1996).

9. Thoreau, "Slavery in Massachusetts" (1854), University of Pennsylvania African Studies Center, http://www.africa.upenn.edu/Articles_Gen/Slavery_Massachusetts.html.

10. Henry Mayer, *All on Fire: William Lloyd Garrison and the Abolition of Slavery* (New York: St. Martin's Press, 1998), 224.

11. The typology of four different forms of critical legal theory used in these paragraphs relies on the divisions called out in the "The Philosophy of Law," *Internet Encyclopedia of Philosophy*, April 19, 2009, http://www.iep.utm.edu/law-phil/#H3.

12. Andrew Altman, "Legal Realism, Critical Legal Studies, and Dworkin," *Philosophy and Public Affairs* 15, no. 2 (1986): 205–236, at 221.

13. Richard Posner, *Economic Analysis of Law*, 4th ed. (Boston: Little, Brown, (1922), 23.

14. Howard Zinn, *A People's History of the United States 1492–Present* (New York: Perennial Classics/Harper Collins, 2003), 288.

15. Cicero, "The Speech of M. T. Cicero in Defence of Aulus Cluentius Habitus," chapter 53, section 146. The original Latin is "legum denique idcirco omnes servi sumus ut liberi esse possimus."

16. Any use of slavery images for nonhuman animals requires two sensitive qualifications—first, such use has produced significant controversy; second, there are, in fact, terribly harsh realities not covered by a slavery image, such as what happens in slaughterhouses.

17. The legal citation for this opinion is 65 Mississippi 329, 331–332, 3 Southern 458, 458–459.

18. Susan D. Jones, *Valuing Animals* (Baltimore, MD: Johns Hopkins University Press, 2003), 4.

19. Wolfson, "Beyond the Law," offers the view of a sophisticated lawyer on the rationale and problems created by this historical development.

20. Jehuda Feliks, "Animals of the Bible and Talmud," *Encyclopaedia Judaica*, vol. 3 (Jerusalem: Keter, 1972), 7–19, at 10–16.

21. Lynn White Jr., "The Historic Roots of Our Ecologic Crisis," *Science* 155 (1967): 1203–1207, at 1205.

22. Jeremy Cohen, *"Be Fertile and Increase, File the Earth and Master It": The Ancient and Medieval Career of a Biblical Text* (Ithaca, NY: Cornell University Press, 1989), 309.

23. See, for example, Richard Foltz, *Animals in Islamic Tradition and Muslim Cultures* (Oxford: Oneworld, 2006); and Sara Tlili, *From an Ant's Perspective: The Status and Nature of Animals in the Qur'an* (New York: Columbia University Press, 2012).

24. For example, Paul Waldau, *The Specter of Speciesism: Buddhist and Christian Views of Animals* (New York: Oxford University Press, 2001), examines a commonly asserted contrast suggesting that while the Christian tradition is radically dismissive of nonhuman animals, the Buddhist tradition is more perceptive and accepting.

25. See, for example, the work of the Forum on Religion and Ecology at Yale, www.yale.edu/religionandecology.

26. Laura Hobgood-Oster, *Holy Dogs and Asses: Animals in the Christian Tradition* (Urbana: University of Illinois Press, 2008), 5.

27. On the number of cultures, see *The Encyclopedia of World Cultures* (New York: Macmillan Library Reference USA, 1999). On the number of languages, William J. Sutherland, "Parallel Extinction Risk and Global Distribution of Languages and Species," *Nature* 423 (2003): 276–279, refers to "6809 living languages."

28. For example, one scholar gives 1964 as the beginning of the field—see Carla Castricano, *Animal Subjects: An Ethical Reader in a Posthuman World* (Waterloo, ON: Wilfrid Laurier University Press, 2008), 3.

29. Cary Nelson et al., "Cultural Studies: An Introduction," *Cultural Studies* 1 (1992): 5; cited in Madhavi Sunder, "Cultural Dissent," *Stanford Law Review* (2001): 495–467, at 521 n. 138.

30. Max Horkheimer, *Critical Theory* (New York: Seabury Press, 1982), 244.

31. See, for example, "Critical Theory," *Stanford Encyclopedia of Philosophy*, http://plato.stanford.edu/entries/critical-theory/.

32. Ibid.

33. Both original writings and overviews of the continental tradition can be found in these collections: John Sanbonmatsu, *Critical Theory and Animal Liberation* (Lanham, MD: Rowman and Littlefield, 2011); M. Calarco, *Zoographies: The Question of the Animal from Heidegger to Derrida* (New York: Columbia University Press, 2008); G. Steiner, *Anthropocentrism and Its Discontents: The Moral Status of Animals in the History of Western Philosophy* (Pittsburgh: University of Pittsburgh Press, 2005); M. Calarco and P. Atterton, *Animal Philosophy: Essential Readings in Continental Thought* (New York: Continuum, 2004); and Peter H. Steeves, ed., *Animal Others: On Ethics, Ontology, and Animal Life* (Albany: State University of New York Press, 1999).

34. Martin Heidegger, "Existence and Being," in *Existentialism from Dostoevsky to Sartre,* ed. Walter Kaufmann, (New York: Meridian Books, 1956), 206–221, at 214.

35. Steiner, *Anthropocentrism and Its Discontents,* 204.

36. Descartes, *Discourse on Method* V, 11. Descartes's use of "flies and ants" as representatives of any nonhuman animals is a rhetorical device that commits what might be called "the fallacy of the poor representative." Descartes's use of this common technique is described in Waldau, *The Specter of Speciesism,* 92–93 (chapter 5 includes other examples as well). Such questionable reasoning was made possible by linguistic habits that collected all "animals" or "brutes" into one category that obscures the obvious, extraordinary differences among the beings outside our own species.

37. Emmanuel Lévinas, "The Name of a Dog, or Natural Rights" (1974), in Calarco and Atterton, *Animal Philosophy,* 47–50.

38. Ibid., 49.

39. See, for example, M. Calarco, *Zoographies: The Question of the Animal from Heidegger to Derrida* (New York: Columbia University Press, 2008), 55–59; and David L. Clark, "On Being 'the Last Kantian in Nazi Germany': Dwelling with Animals after Levinas," in *Animal Acts: Configuring the Human in Western History,* ed. Jennifer Ham and Matthew Senior (New York: Routledge, 1997), 165–198.

40. Particularly revealing comments in this regard are made in Clark, "On Being 'the Last Kantian in Nazi Germany'".

41. Jacques Derrida, "The Animal That Therefore I Am (More to Follow)," trans. David Wills, *Critical Inquiry* 28, no. 2 (2002): 369–418, at 397.

42. The first phrase is ibid., 400, and the second phrase occurs often—see, for example, 395, 396, 400, and 413.

43. This story is well told in Giorgio de Santillana, *The Crime of Galileo* (Chicago: University of Chicago Press, 1955).

44. Ibid.—the comment about learning is at 11, and "first victims" at 2.

45. Ibid., 20.

46. See the rich discussion of human rights in David Kennedy, *The Dark Sides of Virtue: Reassessing International Humanitarianism* (Princeton, NJ: Princeton University Press, 2004), 11. At p. 21, Kennedy asks whether "the human rights movement impoverishes local discourse," set up by this observation at 18: "That human rights claims to be universal but is really the product of a specific cultural and historical origin says nothing—unless that specificity exacts costs or renders human rights less useful than something else." For another challenge to the idea that "rights fits all circumstances," see Richard Ford, *Rights Gone Wrong: How Law Corrupts the Struggle for Equality* (New York: Farrar, Straus and Giroux, 2011).

47. See, for example, Jhan Hochman, *Green Cultural Studies: Nature in Film, Novel, and Theory* (Moscow: University of Idaho Press, 1998).

48. Terry Eagleton, *After Theory* (New York: Basic Books, 2003), 101–102, says, "Cultural theory as we have it promises to grapple with some fundamental problems, but on the whole fails to deliver. It has been shame-faced about morality and metaphysics, embarrassed about love, biology, religion and revolution, largely silent about evil, reticent about death and suffering, dogmatic about essences, universals and foundations, and superficial about truth, objectivity and disinterestedness. This, on any estimate, is rather a large slice of human existence to fall down on. It is also, as we have suggested before, rather an awkward moment in history to find oneself with little or nothing to say about such fundamental questions."

49. An illuminating use of theory can be found in Sue Donaldson and Will Kymlicka, *Zoopolis: A Political Theory of Animal Rights* (New York: Oxford University Press, 2011). Addressing only macroanimals, the authors propose a three-part division based on citizenship theory: domesticated animals (companion animals and those animals raised for food); wild animals; and "liminal" animals (squirrels, pigeons, and other macroanimals that have learned to live on their own in human communities). The division is used to discuss both legal and moral protections, tying the treatment of these macroanimals to basic principles used in discussions in liberal democracies about justice and human rights.

50. Brigitte Resl, ed., *A Cultural History of Animals in the Medieval Age* (New York: Berg, 2009), 2.

51. Ibid., 2–3.

52. John Grim, "Knowing and Being Known by Animals: Indigenous Perspectives on Personhood," in Paul Waldau and Kimberly Patton, *A Communion of Subjects* (New York: Columbia University Press, 2006), 388.

53. Ibid., 380.

54. Howard L. Harrod, *The Animals Came Dancing: Native American Sacred Ecology and Animal Kinship* (Tucson: University of Arizona Press, 2000), 138.

55. Robert A. Brightman, *Grateful Prey: Rock Cree Human-Animal Relationships* (Berkeley: University of California Press, 1993), 2.

56. Donald Worster, *Nature's Economy: A History of Ecological Ideas,* 2nd ed. (New York: Cambridge University Press, 1994), x.

57. Kathy Rudy, *Loving Animals: Toward a New Animal Advocacy* (Minneapolis: University of Minnesota Press, 2011).

58. Elizabeth Lawrence, "Love for Animals and the Veterinary Profession," *JAVMA* 205 (1994): 970–972.

CHAPTER 8

1. Various popular books have kept this idea before the public—David Brooks, *The Social Animal: The Hidden Sources of Love, Character, and Achievement* (New York: Random House, 2011), appeared on several American best-seller lists, and in 1972 the first edition of Elliot Aronson's extremely successful introduction to social psychology *The Social Animal* (New York: W. H. Freeman) was published—in 2012, the twelfth edition was published.

2. See, for example, Bert Hölldobler and Edward O. Wilson, *The Superorganism: The Beauty, Elegance, and Strangeness of Insect Societies* (New York: Norton, 2009); and E. O. Wilson, *The Social Conquest of Earth* (New York: Liveright, 2012).

3. For example, the *World Book Encyclopedia* states, "The first gorillas kept in zoos did not live long, and some people believe they died from loneliness." "World Books' Official Animal of the Day—the Gorilla," World Book, http://www.worldbook.com/world-books-official-animal-of-the-day-consumer/

item/973-world-books-official-animal-of-the-day-the-gorilla. The tentativeness of this suggestion reflects the great difficulty we have when attempting to identify a single reason for the death of a member of another species—such reports have circulated about gorillas because of the extremely close bond of a baby gorilla to his or her mother, which is described in Kristina Cawthon Lang, "Gorilla,"Primate Info Net, http://pin.primate.wisc.edu/factsheets/entry/gorilla/behav. For dolphins, similar stories have been widely discussed since the mid-1970s based on an account in Jacques Cousteau and P. Diolé, *Dolphins* (Garden City, NY: Doubleday, 1975). The authors describe the first attempts of a Cousteau-led group to capture dolphins in the late 1950s and write of the death of their first captive "perhaps out of loneliness" (53). They observe, "In our laboratory experiments at Monaco, we have seen dolphins who *allow* themselves to die. Dolphins are emotional, sensitive, vulnerable creatures—probably more so than we are" (24, emphasis in original). For a report that problematizes such reports, see "Do Dolphin Commit Suicide in Captivity?," Marine Animal Welfare, http://marineanimalwelfare.com/suicide.htm.

4. Many of de Waal's extensive publications provide insights of this kind—besides those cited in chapter 3, see also F. B. M. de Waal, *The Ape and the Sushi Master: Cultural Reflections by a Primatologist* (New York: Basic Books, 2001).

5. Aristotle, *Politics* I, 2.1253a, in *The Complete Works of Aristotle: The Revised Oxford Translation* (Princeton, NJ: Princeton University Press, 1984).

6. Peter L. Berger and Thomas Luckmann, *The Social Construction of Reality: A Treatise in the Sociology of Knowledge* (Garden City, NY: Doubleday, 1966), 1.

7. Ibid., 3.

8. Blaise Pascal, *Pensees [and] The Provincial Letters* (New York: Modern Library, 1955 [1669]), 294.

9. Berger and Luckman, *The Social Construction of Reality*, 3.

10. Elaine Pagels, *Adam, Eve, and the Serpent* (New York: Random House, 1988), xvii.

11. Stephen Clark, *The Moral Status of Animals* (Oxford: Clarendon, 1977), 7.

12. A modern treatment of the history and philosophical issues of this claim is Richard Rorty, *Philosophy and the Mirror of Nature* (Princeton, NJ: Princeton University Press, 1979).

13. R. Tagore, *Sanyasi, or the Ascetic*, in *Collected Poems and Plays of Rabindranath Tagore* (New York: Macmillan, 1942), 461–479, at 464.

14. The details of this effort are documented in David Nibert, "Origins of the ASA Section on Animals and Society—with a Bibliographic Appendix," *Sociological Origins: Journal of Research Documentation and Critique* 3 (Autumn 2003): 53–58.

15. David Nibert, "Humans and Other Animals: Sociology's Moral and Intellectual Challenge," in *Social Creatures: A Human and Animal Studies Reader*, ed. Clifton P. Flynn (New York: Lantern Books, 2008), 259–272, at 259–260.

16. For some idea of the diversity at the undergraduate and graduate level, see all six chapters on social sciences (anthropology, geography, law, psychology, social work, and sociology) in Margo DeMello, ed., *Teaching the Animal: Human/Animal Studies across the Disciplines* (New York: Lantern Books, 2010). In 2012, the first law school advanced degree program (an LLM degree in animal law through Lewis and Clark Law School in Portland, Oregon) opened.

17. G. W. Allport, "The Historical Background of Social Psychology," in *The Handbook of Social Psychology: Volume I, Theory and Method*, 3rd ed., ed. G. Lindzey and E. Aronson, (New York: McGraw-Hill, 1985), 1–46, at 3. Versions of this essay first appeared in the first edition (1954).

18. Note that many of the issues dealt with in this chapter might also have been raised in chapter 6 when the philosophical term "epistemology" was explored.

19. Philip Zimbardo, *The Lucifer Effect: How Good People Turn Evil* (London: Rider, 2007), 265.

20. For a concise summary of Milgram's and Asch's research, see ibid., 260–75. For Asch's own reports, see S. E. Asch, "Effects of Group Pressure upon the Modification and Distortion of Judgment," in H. Guetzkow, ed., *Groups, Leadership and Men* (Pittsburgh: Carnegie Press, 1951); "Opinions and Social Pressure," *Scientific American* 193 (1955): 31–35; and "Studies of Independence and Conformity: A Minority of One against a Unanimous Majority," *Psychological Monographs* 70 (1956). Milgram's experiment is described in Stanley Milgram, "The Behavioral Study of Obedience," *Journal of Abnormal and Social Psychology* (1963): 371–378.

21. See Linnaeus, *Systema Naturae*, 10th ed. (1758), 22—the Latin text is available at Biodiversity Heritage Library, http://www.biodiversitylibrary.org/item/10277#page/28/mode/1up. Although modern ideas of the peoples often referred to as "pygmies" can be much more sensitive (see, for example, Colin M. Turnbull, *The Forest People* [New York: Simon and Schuster, 1961]), there continues to be much mistreatment. Of

particular note is the exhibition of these humans in zoos as late as the early twentieth century—see Phillips Verner Bradford and Harvey Blume, *Ota Benga: The Pygmy in the Zoo* (New York: St. Martin's, 1992).

22. Berger and Luckman, *The Social Construction of Reality*, 1.

23. Katherine Kete, "Introduction: Animals and Human Empire," in *A Cultural History of Animals in the Age of Empire*, ed. Katherine Kete (Oxford: Berg, 2009), 15.

24. C. S. Lewis, *An Experiment in Criticism* (Cambridge: Cambridge University Press, 1961), 140.

25. Edward Armstrong, *The Way Birds Live*, 4th ed. (New York: Dover, 1967); Len Howard, *Birds as Individuals* (London: Collins, 1955); Bernd Heinrich, *Mind of the Raven: Investigations and Adventures with Wolf-Birds* (New York: Cliff Street Books, 1999); Bridget Stutchbury, *The Bird Detective: Investigating the Secret Life of Birds* (Toronto: HarperCollins, 2010).

26. Berger and Luckman, *The Social Construction of Reality*, 5.

27. This observation is taken out of context, for it comes from an analysis of how the very earliest Christians made meaning in the hostile Roman world in which they lived—see Wayne A. Meeks, *The Origins of Christian Morality: The First Two Centuries* (New Haven, CT: Yale University Press, 1993), 5.

28. E. O. Wilson, *Biophilia* (Cambridge, MA: Harvard University Press, 1984); Erich Fromm's use of term in 1973 is noted by Elizabeth Lawrence, "Love for Animals and the Veterinary Profession," *JAVMA* 205 (1994): 970 n. 2, where Fromm is quoted as defining biophilia as "the passionate love of life and all that is alive."

29. Nietzsche, *Beyond Good and Evil*, trans. Helen Zimmern, in *The Complete Works of Friedrich Nietzsche (1909–1913)*, chapter 4, 156, Project Gutenberg, http://www.gutenberg.org/files/4363/4363-h/4363-h.htm.

30. Steven M. Wise, in his breakthrough *Rattling the Cage: Toward Legal Rights for Animals* (Cambridge, MA: Perseus, 2000), cites this human-centered claim and other decidedly biased opinions that long prevailed among our forebears—see chapter 2.

31. Eugene Hargrove, "The Role of Zoos in the Twenty-First Century," in Bryan G. Norton et al., *Ethics on the Ark: Zoos, Animal Welfare and Wildlife Conservation* (Washington, DC: Smithsonian Institution Press, 1995), 13–19, at 14.

32. Ibid., 14.

33. C. S. Lewis, *The Four Loves* (New York: Harcourt, Brace and World, 1960), 53.

34. You can find this passage in A. I. Hallowell, "Ojibwa Ontology, Behaviour and World View," in *Culture in History: Essays in Honor of Paul Radin*, ed. S. Diamond (New York: Columbia University Press, 1960).

35. E. O. Wilson, *Sociobiology: The New Synthesis* (Cambridge, MA: Belknap, 1975).

36. See, for example, "Sociobology," *Stanford Encyclopedia of Philosophy*, http://plato.stanford.edu/entries/sociobiology/.

37. F. B. M. de Waal, *Good Natured: The Origins of Right and Wrong in Humans and Other Animals* (Cambridge, MA: Harvard University Press, 1996), 13. As noted below, de Waal also feels sociobiology has weaknesses.

38. Ibid., 18.

39. Mary Midgley, *Beast and Man: The Roots of Human Nature*, rev. ed. (New York: Routledge, 1995), xxii.

40. Ibid., xvi.

41. This approach was always applied more to nonhuman animals than to human animals—for an example of this kind of thinking, see B. F. Skinner, *Verbal Behavior* (New York: Appleton-Century-Crofts, 1957).

42. John Dunne, *The Way of All the Earth: Experiments in Truth and Religion* (New York: Macmillan, 1972), 17. I've replaced "man" with "human."

CHAPTER 9

1. See, for example, Michael Marshall, "Unique Life Form Is Half Plant, Half Animal," *New Scientist*, January 13, 2012, www.newscientist.com/article/dn21353-zoologger-unique-life-form-is-half-plant-half-animal.html, who explains several examples and includes this quote: "The division between plants and animals is collapsing completely."

2. W. C. McGrew, *Chimpanzee Material Culture: Implications for Human Evolution* (Cambridge: Cambridge University Press, 1992), 14, 197.

3. Richard W. Wrangham, Frans B. M. de Waal, and W. C. McGrew, "The Challenge of Behavioral Diversity," in *Chimpanzee Cultures*, ed. Richard W. Wrangham et al. (Cambridge, MA: Harvard University Press, 1994), 1–18, at 1. The book cited is Leslie A. White, *The Evolution of Culture: The Development of Civilization to the Fall of Rome* (New York: McGraw-Hill, 1959), 5.

4. Stephen Clark, *The Moral Status of Animals* (Oxford: Clarendon, 1977), 96.

5. McGrew, *Chimpanzee Material Culture*, 150.

6. Ruth Benedict, *The Chrysanthemum and the Sword: Patterns of Japanese Culture* (Boston: Houghton Mifflin, 1946).

7. Ibid., 14–15.

8. Michael Lambek, *A Reader in the Anthropology of Religion*, 2nd ed. (Malden, MA: Blackwell, 2008), 2.

9. Ibid., 3.

10. W. E. H. Stanner, "Religion, Totemism and Symbolism," in *A Reader in the Anthropology of Religion*, 2nd ed., ed. Michael Lambek (Malden, MA: Blackwell, 2008), 82–89, at 89. Stanner was speaking of Auguste Comte, Herbert Spencer, Edward Burnett Tylor, James Frazer, and Sigmund Freud.

11. Lambek, *A Reader in the Anthropology of Religion*, 3, quoting David Hoy.

12. Ibid., 2.

13. Ibid.

14. Ibid., 3.

15. Ibid.

16. Ibid., 5.

17. These issues, as well as the logical flaws mentioned below, are nicely set out in Bernard Rollin, "Veterinary and Animal Ethics," in *Law and Ethics of the Veterinary Profession*, ed. James F. Wilson (Yardley, PA: Priority, 1993), 24–49, at 29.

18. Ibid., 29.

19. Timothy J. Gorringe, *Capital and the Kingdom: Theological Ethics and Economic Order* (New York: SPCK/Orbis Books, 1994), 4. Gorringe supplies much more of the reasoning behind this approach, which is tied to a peace formula, *cuius regio, eius religio*, by which religious issues were resolved by deferring to different rulers' personal choice to follow either Catholic or Protestant Christianity.

20. Ibid., 5.

21. Ibid.

22. Ibid.

23. H. Skolimoski, "Reverence for Life," in *Ethics of Environment and Development: Global Challenge, International Response*, ed. J. Ronald Engel and Joan Gibb Engel (London: Bellhaven, 1990), 97–98, cited in Gorringe, *Capital and the Kingdom*, x n. 7, 5.

24. See, for example, John B. Cobb Jr., *The Earthist Challenge to Economism: A Theological Critique of the World Bank* (New York: St. Martin's, 1999).

25. J. Nollman, *Animal Dreaming: The Art and Science of Interspecies Communication* (New York: Bantam, 1987), 78–79.

26. See Mary Daly, *Gyn/Ecology: The Metaethics of Radical Feminism* (Boston: Beacon, 1990), for details and a powerful denunciation of these practices.

27. Tim Ingold, "From Trust to Domination: An Alternative History of Human-Animal Relations," in *Animals and Human Society: Changing Perspectives*, ed. Aubrey Manning and James Serpell (London: Routledge, 1994), 1–22, at 1.

28. Ibid.

29. Ibid., 2.

30. Ibid.

31. Ibid., 11.

32. This is a description of the views of the Koyukon people—see Richard K. Nelson, *Make Prayers to the Raven: A Koyukon View of the Northern Forest* (Chicago: University of Chicago Press, 1983), 241.

33. Ingold, "From Trust to Domination," 11.

34. Ibid., 12.

35. Ibid., 12, 14.

36. Ibid., 16.

37. See, for example, the views of the Ojibwa Sioux described by A. I. Hallowell, "Ojibwa Ontology, Behaviour and World View," in *Culture in History: Essays in Honor of Paul Radin*, ed. S. Diamond (New York: Columbia University Press, 1960), 19–52 (at p. 40 and following are references to "other-than-human persons").

38. Ingold, "From Trust to Domination," 16.

39. Tim Ingold, *The Appropriation of Nature: Essays on Human Ecology and Social Relations* (Manchester, UK: Manchester University Press, 1986), 272–273.

40. Ingold, "From Trust to Domination," 18.

41. Tim Ingold, *Hunters, Pastoralists and Ranchers: Reindeer Economies and Their Transformations* (Cambridge: Cambridge University Press, 1980), 96.

42. Jakob von Uexküll, *Theoretical Biology* (New York: Harcourt, Brace, 1926), 70.

43. The latter is from Kalevi Kull, "Umwelt," in *The Routledge Companion to Semiotics*, ed. Paul Cobley (London: Routledge, 2010), 348–349.

44. John Deely, "Umwelt," *Semiotica* 134, no. 1/4 (2001): 125–135.

45. Ingold, "From Trust to Domination," 19.

46. James Deetz, *Invitation to Archaeology* (Garden City, NY: Natural History Press, 1967), 3. Deetz elaborates, "Archaeologists are anthropologists who usually excavate the material remains of past cultures, and through the study of such evidence, attempt to reconstruct the history of man from his earliest past and to determine the nature of cultural systems at different times and places around the world." Deetz adds in a footnote that the distinct field of classical archaeology focuses on civilizations of the ancient Mediterranean world, which is usually taught as art history in university art departments. To identify the broader form of archaeology, Deetz uses the name "anthropological archaeology."

47. See Dave Aftandilian, "Archaeology and Animals," in *Encyclopedia of Human-Animal Relationships: A Global Exploration of Our Connections with Animals*, ed. Marc Bekoff (Westport, CT: Greenwood, 2007), 81–85, at 81.

48. Here and following, Hesiod, *Works and Days*, in *Hesiod: The Homeric Hymns and Homerica*, trans. Hugh G. Evelyn White (Cambridge, MA: Harvard University Press, 1974), 10–13, ll. 109–155.

49. See a summary of widespread views in Ivar Paulson, "The Animal Guardian: A Critical and Synthetic Review," *History of Religions* 3, no. 2 (1964): 202–219.

50. Barbara Williams, *Breakthrough: Women in Archaeology* (New York: Walker, 1981), x.

51. Tim Ingold, *What Is an Animal?* (Boston: Unwin Hyman, 1988), was the first book in the series that now has well over fifty volumes.

52. Deetz, *Invitation to Archaeology*, 73.

53. Ibid.

54. See, for example, Dave Aftandilian, "Archaelogy and Animals"; and Simon J. M. Davis, *The Archaeology of Animals* (New Haven, CT: Yale University Press, 1987).

55. Some of the information provided here comes from J. R. Hale, "Oracle Bones in Ancient China," lecture 14, *Exploring the Roots of Religion* (Chantilly, VA: Teaching Co., 2009).

56. Asko Parpola, "Religion Reflected in Iconic Signs in the Indus Script: Penetrating into the Long Forgotten Picto + Graphic Messages," *Visible Religion* 6 (1988): 114, quoted in Christopher Key Chapple, *Nonviolence to Animals, Earth, and Self in Asian Traditions* (Albany: State University of New York Press, 1993), 6 n. 14.

57. See, for example, Henry W. Henshaw, *Animal Carvings from Mounds of the Mississippi Valley* (Washington, DC: Government Printing Office, 1883).

58. Darcy F. Morey, "Burying Key Evidence: The Social Bond between Dogs and People," *Journal of Archaeological Science* 33, no. 2 (2006): 158–175, abstract.

59. D. F. Morey and M. Wiant, "Early Holocene Domestic Dog Burials from the North American Midwest," *Current Anthropology* 33, no. 2 (1992): 224–229, quoted in Aftandilian, "Archaelogy and Animals," 82.

60. Paul Waldau, *Animal Rights* (New York: Oxford University Press, 2011), 27.

61. Charles C. Mann, *1491: New Revelations of the Americas before Columbus* (New York: Alfred A. Knopf, 2005).

62. Jared Diamond, *Guns, Germs and Steel: The Fates of Human Societies* (New York: Norton, 1997).

63. William Cronon, *Changes in the Land: Indians, Colonists, and the Ecology of New England* (New York: Hill and Wang, 1983).

64. David Blackbourn, *The Conquest of Nature: Water, Landscape, and the Making of Modern Germany* (New York: Norton, 2006).

65. Cronon, *Changes in the Land*, 5.

66. Farley Mowat, *Sea of Slaughter* (Shelburne, VT: Chapters, 1996), 14.

67. Jehuda Feliks, "Animals of the Bible and Talmud," *Encyclopaedia Judaica*, vol. 3 (Jerusalem: Keter, 1972).

68. Such is the case, for example, with the respected environmental historians Alfred Crosby and Daniel Worster. See, for example, Alfred W. Crosby, *Ecological Imperialism: The Biological Expansion of Europe, 900–1900* (Cambridge: Cambridge University Press, 1986); and Donald Worster, *Nature's Economy: A History of Ecological Ideas*, 2nd ed. (New York: Cambridge University Press, 1994).

69. John Wright, *Human Nature in Geography: Fourteen Papers, 1925–1965* (Cambridge, MA: Harvard University Press, 1966), 250–285.

70. Jennifer R. Wolch, "Zoopolis," *Capitalism, Nature, Socialism* 7, no. 2 (1996): 21–47. Also in Jennifer R. Wolch and Jody Emel, *Animal Geographies: Place, Politics, and Identity in the Nature-Culture Borderlands* (New York: Verso, 1998), 119–138.Quotes in the following paragraphs come from the 1998 publication.

71. Wolch and Emel, *Animal Geographies*, 119–120.

72. Daniel B. Botkin, *Discordant Harmonies: A New Ecology for the Twenty-First Century* (New York: Oxford, 1990), 167.

73. Wolch and Emel, *Animal Geographies*, 121.

74. Ibid., 121 n. 7 (with text on 136).

75. Ibid.

76. Ibid., 121.

77. Ibid., 120–121.

78. More work of this kind appears in a second essay that sounds similar themes—Jennifer R. Wolch, Kathleen West, and Thomas E. Gaines, "Transspecies Urban Theory," *Environment and Planning D: Society and Space* 13, no. 6 (1995): 735–760. In this essay, Wolch, West, and Gaines put forth transspecies urban theory "that takes nonhumans seriously by critically assessing what we know about wild animals in the city" (736).

79. Ibid., 736–737.

80. Ibid., 755.

81. The website of the Animal Geography Specialty Group of the Association of American Geographers is http://www.animalgeography.org/

CHAPTER 10

1. Farley Mowat, *Sea of Slaughter* (Shelburne, VT: Chapters, 1996), 14.

2. W. Beebe, "The Evolution and Destruction of Life." *Scientific American* (Suppl.), no. 2234, October 26, 1918.

3. Jared Diamond, *Guns, Germs and Steel: The Fates of Human Societies* (New York: Norton, 1997).

4. Rod Preece, *Awe for the Tiger, Love for the Lamb: A Chronicle of Sensibility to Animals* (Vancouver: UBC Press, 2002).

5. Virginia DeJohn Anderson, *Creatures of Empire: How Domestic Animals Transformed Early America* (Oxford: Oxford University Press, 2004).

6. Oliver O'Donovan, *Begotten or Made* (Oxford: Clarendon, 1984), 51.

7. This point was made by the extraordinary Elizabeth Atwood Lawrence in a September 2000 lecture in the Human-Animal Relationships course, Cummings School of Veterinary Medicine, Tufts University, Grafton, MA.

8. Elizabeth Marshall Thomas, *The Hidden Life of Dogs* (New York: Houghton Mifflin, 1993); M. C. T. Smith, *Famous Dogs of Famous People* (New York: Dodd, Mead, 1943).

9. Alexandra Horowitz, *Inside of a Dog: What Dogs See, Smell, and Know* (New York: Scribner, 2009); Marjorie Garber, *Dog Love* (New York: Simon and Schuster, 1996).

10. Ted Kerasote, *Merle's Door: Lessons from a Freethinking Dog* (Orlando, FL: Harcourt, 2007).

11. Maureen B. Adams, *Shaggy Muses: The Dogs Who Inspired Virginia Woolf, Emily Dickinson, Edith Wharton, Elizabeth Barrett Browning, and Emily Brontë.* (New York: Ballantine, 2007).

12. Quoted only in part in the Wharton chapter of *Shaggy Muses*, the longer original is from Edith Wharton, *A Backward Glance*, Internet Archive, http://www.archive.org/stream/backwardglance030620mbp/backwardglance030620mbp_djvu.txt, 4.

13. This is the thrust of an article by the respected science writer Nicholas Wade, "Sit. Stay. Parse. Good Girl!," *New York Times*, January 18, 2011, D1. The story is about an individual named Chaser, a border collie living in South Carolina. Chaser "has the largest vocabulary of any known dog. She knows 1,022 nouns, a record that displays unexpected depths of the canine mind and may help explain how children acquire language."

14. Larry O'Hanlon, "Pets Vital to Human Evolution," Discovery News, August 10, 2010, http://news.discovery.com/animals/pets-humans-evolution.html.

15. Many books provide such information—see, for example, Peter Evans, *Dolphins* (London: Whittet Books, 1994), which offers examples from west Africa, Brazil, Australia, Burma, India, and China that involve six different types of dolphins.

16. This monastic code is six volumes long. *The Book of the Discipline (Vinaya-Pitaka)*, trans. I. B. Horner, Sacred Books of the Buddhists (London: Humphrey Milford, 1938–1966). These two elephants are mentioned in vol. 4, 392, and vol. 5, 273–274.

17. See, for example, the references to Teresias and Slit Ear in Cynthia Moss, *Elephant Memories: Thirteen Years in the Life of an Elephant Family* (New York: William Morrow, 1988), 124–125.

18. Georgina Montgomery and Linda Kalof, "History from Below: Animals as Historical Subjects," in *Teaching the Animal: Human/Animal Studies across the Disciplines*, ed. Margo DeMello (New York: Lantern Books, 2010), 35–47.

19. Robert Delort, *Les animaux ont une histoire* (Paris: Editions du Seuil, 1984); Thomas Berry, The Dream of the Earth (San Francisco: Sierra Club Books, 1988), 3–4.

20. Aldo Leopold, "The Land Ethic," in *A Sand County Almanac, with Essays on Conservation from Round River* (1948).

21. Meg Randa, *The Story of Emily the Cow: Bovine Bodhisattva, a Journey from Slaughterhouse to Sanctuary as Told through Newspaper and Magazine Articles* (Bloomington, IN: AuthorHouse, 2007).

22. Rig Veda viii, 102, 15–16; vi, 28, 1–8, trans. Swami Satya Prakash Sarasvati and Satyakam Vidyalanka.

23. Mahatma Gandhi, *The Gandhi Reader*, ed. Homer A. Jack (New York: Grove Press, 1956), 170.

24. Stephen Clark, *The Moral Status of Animals* (Oxford: Clarendon, 1977), 151.

25. Rainer Maria Rilke, Franz Xaver Kappus, and Stephen Mitchell, *Letters to a Young Poet* (New York: Random House, 1984), 9.

26. The oldest use I have found of this term appears in an essay by Don Barnes, who coined the term (he says his previous work "represented what I choose to call 'conditioned ethical blindness'")—see Donald J. Barnes, "A Matter of Change," in *In Defence of Animals*, ed. Peter Singer (New York: Basil Blackwell, 1985), 157–167, at 160.

27. David Abram, *Becoming Animal: An Essay on Wonder* (New York: Pantheon, 2010), 44.

28. Ibid., 45.

CHAPTER 11

1. See, for example, the developed work of the Humane Society of the United States on this topic, such as *First Strike: The Violence Connection*, Humane Society, 2008, http://www.humanesociety.org/assets/pdfs/abuse/first_strike.pdf.

2. Reports that confirm the ancient insight that oppressions are interlocked come from both science and more popular lore—an example of a science-based report is A. Arluke, J. Levin, C. Luke, and F. Ascione, "The Relationship of Animal Abuse to Violence and Other Forms of Antisocial Behavior," *Journal of Interpersonal Violence* 14, no. 9 (1999): 963–975. A journal article that harkens back to an 1851 speech by Sojourner Truth is Avtar Brah and Ann Phoenix, "Ain't I a Woman? Revisiting Intersectionality," *Journal of International Women's Studies* 5, no. 3 (2004): 75–86.

3. Keith Thomas, *Man and the Natural World* (New York: Pantheon, 1983), 150–151.

4. Wendy Doniger, *The Hindus: An Alternative History* (New York: Penguin, 2009), 9.

5. Immanuel Kant, "Duties to Animals and Spirits," in *Lectures on Ethics*, trans. Louis Infield (New York: Harper and Row, 1963), 239–241, at 240.

6. John Locke, *Some Thoughts Concerning Education*, quoted in Frank R. Ascione and Phil Arkow, eds., *Child Abuse, Domestic Violence, and Animal Abuse: Linking the Circles of Compassion for Prevention and Intervention* (West Lafayette, IN: Purdue University Press, 1999), at 114.

7. James Turner, *Reckoning with the Beast: Animals, Pain and Humanity in the Victorian Mind* (Baltimore, MD: Johns Hopkins University Press, 1980), 24.

8. The history of such laws can be found in Susan Brownmiller, *Against Our Will: Men, Women, and Rape* (New York: Simon and Schuster, 1975).

9. See F. R. Ascione, C. V. Weber, and D. S. Wood, "The Abuse of Animals and Domestic Violence: A National Survey of Shelters for Women Who Are Battered," *Society and Animals* 5, no. 3 (1997): 205–218.

10. E. DeViney, Jeffrey Dickert, and Randall Lockwood, "The Care of Pets within Child Abusing Families," *International Journal for the Study of Animal Problems* 4 (1983): 321–329 (reprinted in Ascione and Arkow, *Child Abuse, Domestic Violence, and Animal Abuse*, 305–313).

11. C. A. Lacroix, "Another Weapon for Combating Family Violence," in Ascione and Arkow, *Child Abuse, Domestic Violence, and Animal Abuse*.

12. "Conduct disorder" was defined as "a repetitive and persistent pattern of behavior in which the basic rights of others or major age-appropriate societal norms or rules are violated" as measured by the presence of a number of criteria over a specified time period. *American Psychiatric Association, Diagnostic and Statistical Manual of Mental Disorders,* 3rd ed. (Washington, DC: American Psychiatric Association, 1988).

13. Dianna J. Gentry, "Including Companion Animals in Protective Orders: Curtailing the Reach of Domestic Violence," *Yale Journal of Law and Feminism* 13 (2001): 97–116.

14. Pam Belluck, "New Maine Law Shields Animals in Domestic Violence Cases," *New York Times,* April 1, 2006.

15. Nicholas D. Kristoff and Sheryl WuDunn, "Why Women's Rights Are the Cause of Our Time," New York Times Magazine, August 23, 2009. There are numerous other articles in this magazine as well. The passages cited in the text come from the book-length treatment of the subject: Nicholas D. Kristof and Sheryl WuDunn, *Half the Sky: Turning Oppression into Opportunity for Women Worldwide* (New York: Alfred A. Knopf, 2009), xvii, xiv–xv.

16. Carol Adams is a feminist who has worked out in great detail the intersection of humans and nonhumans in our shared world—the titles of five of her books give one an idea of the range of her writing: *The Sexual Politics of Meat: A Feminist Vegetarian Critical Theory*; *The Pornography of Meat*; *Neither Beast nor Man: Feminism and the Defense of Animals*; *Animals and Women: Feminist Theoretical Explorations*; *Beyond Animal Rights: A Feminist Caring Ethic for the Treatment of Animals*.

17. Deborah Slicer, review of Carol Adams, *The Sexual Politics of Meat, Environment Ethics* 14, no. 4 (1992): 365–369.

18. See, for example, Sandra Harding, *The Science Question in Feminism* (Ithaca, NY: Cornell University Press, 1986); and Donna J. Haraway, *Primate Visions: Gender, Race, and Nature in the World of Modern Science* (New York: Routledge, 1989); and Donna J. Haraway, "Situated Knowledges: The Science Question in Feminism and the Privilege of Partial Perspective," *Feminist Studies* 14, no. 3 (1988): 575–599.

19. For an explanation of Taylor's general views on nonhuman animals, which were by no means generally negative, see Rob Boddice, *A History of Attitudes and Behaviours toward Animals in Eighteenth- and Nineteenth-Century Britain: Anthropocentrism and the Emergence of Animals* (Lewiston, ME: Edwin Mellen, 2008).

20. Doniger, *The Hindus,* 1.

21. See, for example, Yilu Zhao, "Women Soon to Be Majority of Veterinarians," *New York Times,* June 9, 2002.

22. Richard Hofstadter, *Anti-Intellectualism in American Life* (New York: Vintage, 1963), 316–317.

23. Gail F. Melson, *Why the Wild Things Are: Animals in the Lives of Children* (Cambridge, MA: Harvard University Press, 2001), 12.

24. Ibid., 13.

25. See, for example, a poster popularized over the last decades by the pro-research group Foundation for Biomedical Research (related to the National Association for Biomedical Research, a pro-research consortium that includes hundreds of universities and which is a powerful opponent of many efforts to limit use of nonhuman animals in research)—"Little Girl," www.fbresearch.org/store/. Critical thinking invites one to focus on a number of features of this poster's text explaining why it features a beautiful child: "It's the animals you don't see that really helped her recover."

26. Label Networks, "Humanitarian Youth Culture Study," 2006, Humane Research Council, http://www. humanespot.org/content/humanitarian-youth-culture-study-2006.

27. For the imprisonment argument, see Paul Waldau, "In the Case of Education, Captivity Imprisons Us," in *The Apes: Challenges for the 21st Century* (Chicago: Chicago Zoological Society, 2001), 282–285. On the necessity of coercion and domination, the respected zoo expert Jon Coe comments, "Traditional zoos, no matter how advanced, are founded on captivity and coercion." Jon Coe, "Design and Architecture: Third Generation Conservation, Post-Immersion and Beyond," Future of Zoos Symposium, February 10–11, 2012, Canisius College, Buffalo, NY, http://www.zoolex.org/publication/coe/design+architecture2012. pdf, 8.

28. See, for example, Robert A. Brightman, *Grateful Prey: Rock Cree Human-Animal Relationships* (Berkeley: University of California Press, 1993), 2, 3.

29. Richard Louv, *Last Child in the Woods: Saving Our Children from Nature-Deficit Disorder* (Chapel Hill, NC: Algonquin, 2005).

30. Freeman J. Dyson, *The Scientist as Rebel* (New York: New York Review Books, 2006), 3, 4.

31. Melson, *Why the Wild Things Are,* 20.

32. Ibid., 20, referring to Gene Myers, *Children and Animals: Social Development and Our Connection to Other Species* (Boulder, CO: Westview, 1998), 165.

33. Cora Diamond, "Eating Meat and Eating People," *Philosophy* 53 (1978): 465–479, at 470. This essay can also be found in her book, *The Realistic Spirit: Wittgenstein, Philosophy, and the Mind* (London: Bradford, 1991), 319–334.

34. One of the most famous essays in modern philosophy is Thomas Nagel, "What Is It Like to Be a Bat?," in *Mortal Questions* (Cambridge: Cambridge University Press, 1979), 165–180. This is an interesting question for humans to ponder because bats echolocate, which is unlike any of the sensory abilities that humans possess. Thus, asking "What is it like to be a bat?" provides a chance to pursue philosophical questions about limits in our ability to imagine what life is like for other animals.

35. Thomas Berry, "Prologue," in *A Communion of Subjects: Animals in Religion, Science and Ethics*, ed. Paul Waldau and Kimberly Patton (New York: Columbia University Press, 2006), 8.

36. Spiegel, Marjorie Spiegel, *The Dreaded Comparison: Human and Animal Slavery*, rev. and expanded ed. (New York: Mirror, 1996).

37. Ibid., 14.

38. Beverly Kienzle, "Moth and Wolf: Imaging Medieval Heresy with Insects and Animals," in *A Communion of Subjects: Animals in Religion, Science, and Ethics*, ed. Paul Waldau and Kimberly Patton (New York: Columbia University Press, 2006), 103–116.

39. William Edward Hartpole Lecky, *History of European Morals: From Augustus to Charlemagne*, 2 vols., 6th ed., rev. (London: Longmans Green, 1884), 100–101. In modern times, Peter Singer, *The Expanding Circle: Ethics and Sociobiology* (Oxford: Clarendon, 1981), re-sounded this important theme—Singer cites this quote at p. xiii.

CHAPTER 12

1. Larbi Ben M'Hidi to Ali la Pointe in *The Battle of Algiers*, dir. Gillo Pontecorvo (1966, documentary). This influential film in the history of political cinema focused on the 1957 revolution in Algeria.

2. Small Planet Institute, http://www.smallplanet.org/, described as "the online home of Frances Moore Lappé and Anna Lappé."

3. Carson's *The Sea Around Us* (1951) won the National Book Award in the United States. Both her *Under the Sea Wind: A Naturalist's Picture of Ocean Life* (1941) and *The Edge of the Sea* (1955) were best-sellers as well.

4. Clarence J. Glacken, *Traces on the Rhodian Shore: Nature and Culture in Western Thought from Ancient Times to the End of the Eighteenth Century* (Berkeley: University of California Press, 1967), 47. Glacken quotes Aristotle, *Parts of Animals* 645a 24–37.

5. Will Durant, *The Story of Philosophy: The Lives and Opinions of the Great Philosophers of the Western World* (New York: Clarion/Simon and Schuster, 1967), 46.

6. Bernard Williams, *Shame and Necessity* (Berkeley: University of California Press, 1993), 125–126.

7. Aristotle, *Politics* I, 5.1254b12–13, in *The Complete Works of Aristotle: The Revised Oxford Translation*, vol. 2, ed. Jonathan Barnes, Bollingen Series (Princeton, NJ: Princeton University Press, 1984), 1990.

8. Aristotle, *Politics* I, 5.1254b20–21.

9. Paley is quoted in J. B. Schneewind, ed., *Moral Philosophy from Montaigne to Kant: An Anthology*, vol. 2 (New York: Cambridge University Press, 1990), 448.

10. For example, Cicero passed these views along in *De Natura Deorum* II 14, 37.

11. Aristotle, *Parts of Animals* 645a24–37 in *The Complete Works of Aristotle*, vol. 1, 1004.

12. Both comments are made widely. The first appears, for example, in Ian Hacking, "Our Fellow Animals," *New York Times Book Review*, June 29, 2000; and the second in Kevin Toolis, "The Most Dangerous Man in the World," *Guardian*, November 5, 1999, http://www.guardian.co.uk/lifeandstyle/1999/nov/06/weekend.kevintoolis.

13. See, for example, B. A. Masri, *Islamic Concern for Animals* (Petersfield, UK: Athene Trust, 1987); and B. A. Masri, *Animals in Islam* (Petersfield, UK: Athene Trust, 1989); and Andrew Linzey, *Christianity and the Rights of Animals* (New York: Crossroad, 1987); Andrew Linzey, *Animal Theology* (Chicago: University of Illinois Press, 1994).

14. World Animal Net Directory, www.worldanimal.net/directory, is the source of the figures in the text. The number does not count the government-related or commercial efforts.

15. Small Planet Institute, http://www.smallplanet.org/.

16. Thomas Merton, *Conjectures of a Guilty Bystander* (New York: Image Books, 2009), 81.

17. Diane Jessup, *The Dog Who Spoke with Gods* (New York: St. Martin's, 2001), xvi.

18. Mary Midgley, *Beast and Man: The Roots of Human Nature*, rev. ed. (New York: Routledge, 1995), 44.

19. Paul Shepard, *The Others: How Animals Made Us Human* (Washington, DC: Island Press, 1996), 167.

CHAPTER 13

1. Lewis and Clark Law School in Portland, Oregon.

2. Plato, *The Republic*, book I, in *The Dialogues of Plato*, vol. 1, trans. B. Jowett (New York: Random House, 1937), 603.

3. C. S. Lewis, "Can Christians Support Vivisection?," *Anti-Vivisectionist* (March/April 1963): 154–155.

4. Kathleen Freeman, *Ancilla to the Pre-Socratic Philosophers: A Complete Translation of the Fragments in Diels, Framente der Vorsokratiker* (Cambridge, MA: Harvard University Press, 1966), 125.

5. Cited in Joseph F. Borzelleca, "Paracelsus: Herald of Modern Toxicology," *Toxicological Sciences* 53 (2000): 2–4.

6. Dynamics in the classroom are also the topic of several essays in Margo DeMello, ed., *Teaching the Animal: Human/Animal Studies across the Disciplines* (New York: Lantern Books, 2010).

7. For a fuller discussion, see Paul Waldau, "Law and Other Animals," in DeMello, *Teaching the Animal*, 218–253.

8. For a fuller discussion, see, Paul Waldau, "Religion and Other Animals," in DeMello, *Teaching the Animal*, 103–126.

9. James Joyce, Episode 3, "Proteus," in *Ulysses*, l. 365—the text is online and searchable at the Literature Network, www.online-literature.com/james_joyce/ulysses/.

10. Katy Payne, "A Throbbing in the Air," in *Silent Thunder: In the Presence of Elephants* (New York: Simon and Schuster, 1998).

11. Douglas H. Chadwick, *The Fate of the Elephant* (San Francisco: Sierra Club Books, 1994), 68.

12. This point and others, such as that the term *paideia* meant the same for Greeks, are made by Nicola Abbagnano, "Humanism," in *The Encyclopedia of Philosophy*, vol. 4, ed. Paul Edwards (London: Collier Macmillan, 1967), 69–72, at 70.

13. Stephen Greenblatt, *The Swerve: How the World became Modern* (New York: Norton, 2011), 199.

14. Ibid., 10.

15. Max Oelschlaeger, "Introduction," in Paul Shepard, *Thinking Animals: Animals and the Development of Human Intelligence* (New York: Viking, 1978), ix.

16. Ibid., xii.

17. Some disciplines will fall solely within this megafield—for example, anthrozoology, comparative animal studies, critical animal studies, history of other animals, animal law, and literature and other animals.

18. This description was coined in the nineteenth century by Thomas Carlyle, who had read T. R. Malthus's gloomy 1798 prediction in "An Essay on the Principle of Population" that the human population would always grow faster than food, thereby dooming many humans to unrelieved poverty and hardship.

19. See, for example, Paul Waldau, "Veterinary Education as Leader—Which Alternatives?," *Journal of Veterinary Medical Education* 34, no. 5 (2007): 605–614 (text based on keynote address at the Education Symposium on March 9, 2006, Washington, DC, at the annual meeting of Association of American Veterinary Medical Colleges).

20. See, for example, Ian M. Harris, Larry J. Fisk, and Carol Rank, "A Portrait of University Peace Studies in North America and Western Europe at the End of the Millennium," *International Journal of Peace Studies* 3, no. 1 (1998), www.gmu.edu/programs/icar/ijps/vol3_1/cover3_1.htm.

Select Bibliography

Abram, David. 2010. *Becoming Animal: An Essay on Wonder*. New York: Pantheon.

Adams, Carol J. 1991. *The Sexual Politics of Meat: A Feminist-Vegetarian Critical Theory*. New York: Continuum.

Adams, Carol. 1994. *Neither Beast nor Man: Feminism and the Defense of Animals*. New York: Continuum.

Adams, Maureen B. 2007. *Shaggy Muses: The Dogs Who Inspired Virginia Woolf, Emily Dickinson, Edith Wharton, Elizabeth Barrett Browning, and Emily Brontë*. New York: Ballantine.

Aftandilian, Dave. 2007. "Archaeology and Animals." In Marc Bekoff, ed., *Encyclopedia of Human-Animal Relationships: A Global Exploration of Our Connections with Animals*. Westport, CT: Greenwood, 81–85.

Aftandilian, David, ed. 2007. *What Are the Animals to Us? Approaches from Science, Religion, Folklore, Literature, and Art*. Knoxville: University of Tennessee Press.

Allen, Collin, and Marc Bekoff. 1997. *Species of Mind: The Philosophy and Biology of Cognitive Ethology*. Cambridge, MA: Bradford/MIT Press.

Angier, Natalie. 2007. *The Canon: A Whirligig Tour of the Beautiful Basics of Science*. Boston: Houghton Mifflin.

Apostolos-Capadona, Diane. 2006. "On the Dynamis of Animals, or How Animalium Became Anthropos." In Paul Waldau and Kimberly Patton, eds., *A Communion of Subjects: Animals in Religion, Science, and Ethics*. New York: Columbia University Press, 439–457.

Arluke, A., J. Levin, C. Luke, and F. Ascione. 1999. "The Relationship of Animal Abuse to Violence and Other Forms of Antisocial Behavior." *Journal of Interpersonal Violence* 14 (9): 963–975.

Armstrong, Karen. 2006. *The Great Transformation: The Beginning of Our Religious Traditions*. New York: Knopf.

Beers, Diane. 2006. *For the Prevention of Cruelty: The History and Legacy of Animal Rights Activism in the United States*. Athens: Swallow Press/Ohio University Press.

Bekoff, Marc. 2007. *The Emotional Lives of Animals: A Leading Scientist Explores Animal Joy, Sorrow, and Empathy, and Why They Matter*. Novato, CA: New World Library.

Bekoff, Marc, ed. 2007. *Encyclopedia of Human-Animal Relationships: A Global Exploration of Our Connections with Animals*. Westport, CT: Greenwood.

Berger, Peter L., and Thomas Luckmann. 1966. *The Social Construction of Reality: A Treatise in the Sociology of Knowledge*. Garden City, NY: Doubleday.

Berry, Thomas. 2006. "Loneliness and Presence." In Paul Waldau and Kimberly Patton, eds., *A Communion of Subjects: Animals in Religion, Science, and Ethics*. New York: Columbia University Press, 5–10.

Boddice, Rob. 2008. *A History of Attitudes and Behaviours toward Animals in Eighteenth- and Nineteenth-Century Britain: Anthropocentrism and the Emergence of Animals*. Lewiston, ME: Edwin Mellen.

Bradshaw, G. A. 2008. "Inside Looking Out: Neuroethological Compromise Effects in Elephants in Captivity." In Debra L. Forthman, Lisa F. Kane, David Hancocks, and Paul F. Waldau, eds., *An Elephant in the Room: The Science and Well-Being of Elephants in Captivity*. North Grafton, MA: Center for Animals and Public Policy.

Brion, Marcel. 1959. *Animals in Art*. London: George C. Harrap.

Byrne, R. W., and A. Whiten, eds. 1988. *Machiavellian Intelligence*. Oxford: Oxford University Press.

Calarco, M. 2008. *Zoographies: The Question of the Animal from Heidegger to Derrida*. New York: Columbia University Press.

Calarco, M., and P. Atterton. 2004. *Animal Philosophy: Essential Readings in Continental Thought*. New York: Continuum.

Castricano, Carla. 2008. *Animal Subjects: An Ethical Reader in a Posthuman World*. Waterloo, ON: Wilfrid Laurier University Press.

Chadwick, Douglas H. 1994. *The Fate of the Elephant*. San Francisco: Sierra Club Books.

Clark, David L. 1997. "On Being 'the Last Kantian in Nazi Germany': Dwelling with Animals after Levinas." In Jennifer Ham and Matthew Senior, eds., *Animal Acts: Configuring the Human in Western History*. New York: Routledge, 165–198.

Clark, Kenneth. 1977. *Animals and Men: Their Relationship as Reflected in Western Art from Prehistory to the Present Day*. London: Thames and Hudson.

Clark, Stephen. 1977. *The Moral Status of Animals*. Oxford: Clarendon.

Cohen, Carl. 1986. "The Case for Biomedical Experimentation." *New England Journal of Medicine* 315 (14): 865–870.

Crist, Eileen. 1999. *Images of Animals: Anthropomorphism and Animal Mind*. Philadelphia: Temple University Press.

Cronon, William. 1983. *Changes in the Land: Indians, Colonists, and the Ecology of New England*. New York: Hill and Wang.

Crosby, Alfred W. 1986. *Ecological Imperialism: The Biological Expansion of Europe, 900–1900*. Cambridge: Cambridge University Press.

Davis, Simon J. M. 1987. *The Archaeology of Animals*. New Haven, CT: Yale University Press.

Dawkins, Marian Stamp. 2006. "A User's Guide to Animal Welfare." *TRENDS in Ecology and Evolution* 21 (2): 77–82.

Dawkins, Richard. 1993. "Gaps in the Mind." In Paola Cavalieri and Peter Singer, eds., *The Great Ape Project: Equality beyond Humanity*. London: Fourth Estate, 80–87.

Dawkins, Richard. 1993. "Meet My Cousin, the Chimpanzee." *New Scientist*, June 5, 1993, 36–38.

DeMello, Margo, ed. 2010. *Teaching the Animal: Human/Animal Studies across the Disciplines*. New York: Lantern Books.

Dennett, Daniel. 1995. "Animal Consciousness: What Matters and Why." *Social Research* 62 (3): 691–710.

Derrida, Jacques. 2002. "The Animal That Therefore I Am (More to Follow)." Translated by David Wills. *Critical Inquiry* 28 (2): 369–418.

de Waal, Frans. 1982. *Chimpanzee Politics*. London: Jonathan Cape.

de Waal, F. B. M. 1996. *Good Natured: The Origins of Right and Wrong in Humans and Other Animals*. Cambridge, MA: Harvard University Press.

de Waal, F. B. M. 2001. *The Ape and the Sushi Master: Cultural Reflections by a Primatologist*. New York: Basic Books.

de Waal, F. B. M. 2009. *The Age of Empathy: Nature's Lessons for a Kinder Society*. New York: Harmony Books.

Donaldson, Sue, and Will Kymlicka. 2011. *Zoopolis: A Political Theory of Animal Rights*. New York: Oxford University Press.

Dunlop, Robert H., and David J. Williams. 1996. *Veterinary Medicine: An Illustrated History*. St. Louis: Mosby.

Eagleton, Terry. 2003. *After Theory*. New York: Basic Books.

Eisnitz, Gail. 1997. *Slaughterhouse: The Shocking Story of Greed, Neglect, and Inhumane Treatment inside the U.S. Meat Industry*. Amherst, NY: Prometheus Books.

Finsen, Lawrence, and Susan Finsen. 1994. *The Animal Rights Movement in America: From Compassion to Respect*. New York: Twayne.

Flynn, C. P., ed. 2008. *Social Creatures: A Human and Animal Studies Reader*. New York: Lantern Books.

Foltz, Richard. 2006. *Animals in Islamic Tradition and Muslim Cultures*. Oxford: Oneworld.

Forthman, Debra L., Lisa F. Kane, David Hancocks, and Paul F. Waldau, eds. 2008. *An Elephant in the Room: The Science and Well-Being of Elephants in Captivity*. North Grafton, MA: Center for Animals and Public Policy.

Fouts, Roger. 1997. *Next of Kin: What Chimpanzees Have Taught Me about Who We Are*. New York: William Morrow.

Francione, Gary L. 2008. *Animals as Persons*. New York: Columbia University Press.

Frankl, Viktor E. 1992. *Man's Search for Meaning: An Introduction to Logotherapy*. 4th ed. Boston: Beacon.

Glacken, Clarence J. 1967. *Traces on the Rhodian Shore: Nature and Culture in Western Thought from Ancient Times to the End of the Eighteenth Century*. Berkeley: University of California Press.

Gould, Stephen Jay. 1995. "Animals and Us." *New York Review of Books*, August 19, 20–25.

Greenblatt, Stephen. 2011. *The Swerve: How the World Became Modern*. New York: W. W. Norton.

Griffin, Donald R. 1998. "From Cognition to Consciousness." *Animal Cognition 1*: 3–16.

Ham, Jennifer, and Matthew Senior, eds. 1997. *Animal Acts: Configuring the Human in Western History*. New York: Routledge.

Haraway, Donna J. 1988. "Situated Knowledges: The Science Question in Feminism and the Privilege of Partial Perspective." *Feminist Studies 14* (3): 575–599.

Haraway, Donna J. 1989. *Primate Visions: Gender, Race, and Nature in the World of Modern Science*. New York: Routledge.

Harding, Sandra. 1986. *The Science Question in Feminism*. Ithaca, NY: Cornell University Press.

Harrod, Howard L. 2000. *The Animals Came Dancing: Native American Sacred Ecology and Animal Kinship*. Tucson: University of Arizona Press.

Hawken, Paul. 2007. *Blessed Unrest: How the Largest Movement in the World Came into Being, and Why No One Saw It Coming*. New York: Viking.

Heinrich, Bernd. 1999. *Mind of the Raven: Investigations and Adventures with Wolf-Birds*. New York: Cliff Street Books.

Henshaw, Henry W. 1883. *Animal Carvings from Mounds of the Mississippi Valley*. Washington, DC: Government Printing Office.

Hobgood-Oster, Laura. 2008. *Holy Dogs and Asses: Animals in the Christian Tradition*. Urbana: University of Illinois Press.

Hobgood-Oster, Laura. 2010. *The Friends We Keep: Unleashing Christianity's Compassion for Animals*. Waco, TX: Baylor University Press.

Horowitz, Alexandra. 2009. *Inside of a Dog: What Dogs See, Smell, and Know*. New York: Scribner's. Howard, Len. 1953. *Birds as Individuals*. Garden City, NY: Doubleday.

Ingold, Tim. 1980. *Hunters, Pastoralists and Ranchers: Reindeer Economies and Their Transformations*. Cambridge: Cambridge University Press.

Ingold, Tim. 1986. *The Appropriation of Nature: Essays on Human Ecology and Social Relations*. Manchester: Manchester University Press.

Ingold, Tim. 1994. "From Trust to Domination: An Alternative History of Human-Animal Relations." In Aubrey Manning and James Serpell, eds., *Animals and Human Society: Changing Perspectives*. New York: Routledge, 1–22.

Ingold, Tim, ed. 1994. *What Is an Animal?* New York: Routledge.

Janson, H. W. 1952. *Apes and Ape Lore in the Middle Ages and Renaissance*. London: Warburg Institute, University of London.

Jasper, James. 1992. *The Animal Rights Crusade: The Growth of a Moral Protest*. New York: Free Press.

Kalof, Linda. 2007. *Looking at Animals in Human History*. London: Reaktion Books.

Kalof, Linda, ed. 2009. *A Cultural History of Animals in Antiquity*. New York: Berg.

Kant, Immanuel. 1963. *Lectures on Ethics*. Translated by Louis Infield. New York: Harper and Row, 239–241.

Kete, Katherine. 2009. *A Cultural History of Animals in the Age of Empire*. New York: Berg.

Krause, B. L. 2012. *The Great Animal Orchestra*. New York: Little, Brown.

Lafollette, Hugh, and Niall Shanks. 1996. "The Origin of Speciesism." *Philosophy 71*: 41–61.

Lawrence, Elizabeth. 1994. "Love for Animals and the Veterinary Profession." *Journal of the American Veterinary Medical Association 205*: 970–972.

Lawrence, Elizabeth. 1997. *Hunting the Wren: Transformation of Bird to Symbol: A Study in Human-Animal Relationships*. Knoxville: University of Tennessee Press.

Lecky, William Edward Hartpole. 1884. *History of European Morals: From Augustus to Charlemagne*, 2 vols., 6th ed. rev. London: Longman's Green.

Lévi-Strauss, Claude. 1963. *Totemism*. Translated by Rodney Needham. Boston: Beacon.

Linzey, Andrew. 1987. *Christianity and the Rights of Animals*. New York: Crossroad.

Linzey, Andrew. 1994. *Animal Theology*. Chicago: University of Illinois Press.

Lonsdale, Steven. 1982. *Animals and the Origins of Dance*. New York: Thames and Hudson.

Lopez, Barry. 1986. *Arctic Dreams: Imagination and Desire in a Northern Landscape*. New York: Scribner's.

Louv, Richard. 2005. *Last Child in the Woods: Saving Our Children from Nature-Deficit Disorder*. Chapel Hill, NC: Algonquin.

Lovejoy, Arthur O. 1936. *The Great Chain of Being*. Cambridge, MA: Harvard University Press.

Manning, Aubrey, and Serpell, James, eds. 1994. *Animals and Human Society: Changing Perspectives*. New York: Routledge.

Margulis, Lynn, Karlene V. Schwartz, and Michael Dolan. 1994. *The Illustrated Five Kingdoms: A Guide to the Diversity of Life on Earth*. New York: HarperCollins.

Mason, Jim. 2009. "Animals: From Souls and the Sacred in Prehistoric Times to Symbols and Slaves in Antiquity." In Linda Kalof, ed., *A Cultural History of Animals in Antiquity*. New York: Berg, 16–45.

Masri, B. A. 1987. *Islamic Concern for Animals*. Petersfield, UK: Athene Trust.

Masri, B. A. 1989. *Animals in Islam*. Petersfield, UK: Athene Trust.

Mayr, Ernst. 1982. *The Growth of Biological Thought: Diversity, Evolution, and Inheritance*. Cambridge, MA: Belknap.

Melson, Gail F. 2001. *Why the Wild Things Are: Animals in the Lives of Children*. Cambridge, MA: Harvard University Press.

Midgley, Mary. 1984. *Animals and Why They Matter*. Athens: University of Georgia Press.

Midgley, Mary. 1995. *Beast and Man: The Roots of Human Nature*, rev. ed. New York: Routledge.

Morrison, Adrian R. 2009. *An Odyssey with Animals: A Veterinarian's Reflections on the Animal Rights and Welfare Debate*. New York: Oxford University Press.

Moss, Cynthia. 1988. *Elephant Memories: Thirteen Years in the Life of an Elephant Family*. New York: William Morrow.

Mowat, Farley. 1996. *Sea of Slaughter*. Shelburne, VT: Chapters.

Myers, Gene. 1998. *Children and Animals: Social Development and Our Connection to Other Species*. Boulder, CO: Westview.

Nagel, Thomas. 1979. "What Is It Like to Be a Bat?" In *Mortal Questions*. Cambridge: Cambridge University Press.

Nibert, David. 2008. "Humans and Other Animals: Sociology's Moral and Intellectual Challenge." In C. P. Flynn, ed., *Social Creatures: A Human and Animal Studies Reader*. New York: Lantern Books, 259–272.

Nollman, J. 1987. *Animal Dreaming: The Art and Science of Interspecies Communication*. New York: Bantam.

Norris, Kenneth S. 1991. *Dolphin Days: The Life and Times of the Spinner Dolphin*. New York: W. W. Norton.

Nussbaum, Martha. 2006. *Frontiers of Justice: Disability, Nationality, Species Membership*. Cambridge, MA: Belknap.

Nussbaum, Martha. 2010. *Not for Profit: Why Democracy Needs the Humanities*. Princeton, NJ: Princeton University Press.

Orr, David. 1994. *Earth in Mind: On Education, Environment, and the Human Prospect*. Washington, DC: Island Press.

Paulson, Ivar. 1964. "The Animal Guardian: A Critical and Synthetic Review." *History of Religions* 3 (2): 202–219.

Payne, Katy. 1998. *Silent Thunder: In the Presence of Elephants*. New York: Simon and Schuster.

Pollan, Michael. 2006. *The Omnivore's Dilemma: A Natural History of Four Meals*. New York: Penguin.

Primack, Richard. 2008. *A Primer of Conservation Biology*, 4th ed. Sunderland, MA: Sinauer.

Rachels, James. 1991. *Created from Animals: The Moral Implications of Darwinism*. New York: Oxford University Press.

Randa, Meg. 2007. *The Story of Emily the Cow: Bovine Bodhisattva, a Journey from Slaughterhouse to Sanctuary as Told through Newspaper and Magazine Articles*. Bloomington, IN: AuthorHouse.

Regan, Tom. 1983. *The Case for Animal Rights*. Berkeley: University of California Press.

Reiss, Diana. 2011. *The Dolphin in the Mirror: Exploring Dolphin Minds and Saving Dolphin Lives*. Boston: Houghton Mifflin Harcourt.

Resl, Brigitte, ed. 2009. *A Cultural History of Animals in the Medieval Age*. New York: Berg.

Rickaby, Joseph. 1888. *Moral Philosophy*. London: Longmans, Green.

Rollin, Bernard. 1989. *The Unheeded Cry: Animal Consciousness, Animal Pain, and Science*. New York: Oxford University Press.

Rollin, Bernard E. 1992. *Animal Rights and Human Morality*, rev. ed. Buffalo, NY: Prometheus.

Rollin, Bernard. 2006. *Science and Ethics*. New York: Cambridge University Press.

Rollin, Bernard. 2011. *Putting the Horse before Descartes: A Memoir*. Philadelphia: Temple University Press.

Rudy, Kathy. 2011. *Loving Animals: Toward a New Animal Advocacy*. Minneapolis: University of Minnesota Press.

Sanbonmatsu, John. 2011. *Critical Theory and Animal Liberation*. Lanham, MD: Rowman and Littlefield.

Scully, Matthew. 2002. *Dominion: The Power of Man, the Suffering of Animals, and the Call to Mercy*. New York: St. Martin's.

Sears, Paul. 1964. "Ecology—a Subversive Subject." *BioScience 14* (7): 11–13.

Shapin, Steven. 1996. *The Scientific Revolution*. Chicago: University of Chicago Press.

Shepard, Paul. 1978. *Thinking Animals: Animals and the Development of Human Intelligence*. New York: Viking.

Shepard, Paul. 1996. *The Others: How Animals Made Us Human*. Washington, DC: Island Press.

Shepard, Paul, and Daniel McKinley, eds. 1969. *The Subversive Science: Essays toward an Ecology of Man*. Boston: Houghton Mifflin.

Siebert, Charles. 2006. "An Elephant Crackup?" *New York Times Magazine*, October 8.

Simpson, George Gaylord. 1964. *This View of Life: The World of an Evolutionist*. New York: Harcourt Brace.

Sorabji, Richard. 1993. *Animal Minds and Human Morals: The Origins of the Western Debate*. London: Duckworth.

Spiegel, Marjorie. 1996. *The Dreaded Comparison: Human and Animal Slavery*, rev. and expanded ed. New York: Mirror.

Steeves, H. Peter, ed. 1999. *Animal Others: On Ethics, Ontology, and Animal Life*. Albany: State University of New York Press.

Steiner, G. 2005. *Anthropocentrism and Its Discontents: The Moral Status of Animals in the History of Western Philosophy*. Pittsburgh: University of Pittsburgh Press.

Stone, Deborah. 2002. *Policy Paradox: The Art of Political Decision Making*, rev. ed. New York: W. W. Norton.

Stutchbury, Bridget. 2010. *The Bird Detective: Investigating the Secret Life of Birds*. Toronto: HarperCollins.

Sukumar, R. 1989. *The Asian Elephant: Ecology and Management*. Cambridge: Cambridge University Press.

Thomas, Keith. 1983. *Man and the Natural World*. New York: Pantheon.

Thomas, Northcote W. 1908. "Animals." In J. Hastings, ed., *Encyclopedia of Religion and Ethics*. New York: Charles Scribner's Sons.

Tlili, Sara. 2012. *From an Ant's Perspective: The Status and Nature of Animals in the Qur'an*. New York: Columbia University Press.

Torres, Bob. 2007. *Making a Killing: The Political Economy of Animal Rights*. Oakland, CA: AK Press.

Turner, James. 1980. *Reckoning with the Beast: Animals, Pain and Humanity in the Victorian Mind*. Baltimore, MD: Johns Hopkins University Press.

Uexküll, Jakob von. 1926. *Theoretical Biology*. New York: Harcourt, Brace.

Waldau, Paul. 2001. *The Specter of Speciesism: Buddhist and Christian Views of Animals*. New York: Oxford University Press.

Waldau, Paul. 2010. "Law and Other Animals." In Margo DeMello, ed., *Teaching the Animal: Human/Animal Studies across the Disciplines*. New York: Lantern Books, 218–253.

Waldau, Paul. 2010. "Religion and Other Animals." In Margo DeMello, ed., *Teaching the Animal: Human/Animal Studies across the Disciplines*. New York: Lantern Books, 103–126.

Waldau, Paul. 2011. *Animal Rights*. New York: Oxford University Press.

Waldau, Paul, and Kimberly Patton, eds. 2006. *A Communion of Subjects: Animals in Religion, Science, and Ethics*. New York: Columbia University Press.

White, Thomas I. 2007. *In Defense of Dolphins: The New Moral Frontier*. Blackwell Public Philosophy. Malden, MA: Blackwell.

Wilson, David Scofield. 2007. "Come into Animal Presence: Ethics, Ethology and Konrad Lorenz." In David Aftandilian, ed., *What Are the Animals to Us? Approaches from Science, Religion, Folklore, Literature, and Art*. Knoxville: University of Tennessee Press, 259–265.

Wilson, E. O. 1984. *Biophilia*. Cambridge, MA: Harvard University Press.

Wilson, E. O. 1992. *The Diversity of Life*. Cambridge, MA: Belknap.

Wilson, E. O. 2012. *The Social Conquest of Earth*. New York: Liveright.

Wise, Steven M. 2000. *Rattling the Cage: Toward Legal Rights for Animals*. Cambridge, MA: Perseus.

Wolch, Jennifer R. 1996. "Zoopolis." *Capitalism, Nature, Socialism 7* (2): 21–47.

Wolch, Jennifer R., and Jody Emel. 1998. *Animal Geographies: Place, Politics, and Identity in the Nature-Culture Borderlands*. New York: Verso.

Wolch, Jennifer R., Kathleen West, and Thomas E. Gaines. 1995. "Transspecies Urban Theory." *Environment and Planning D: Society and Space 13* (6): 735–760.

Wolfson, David. 1996. *Beyond the Law: Agribusiness and the Systemic Abuse of Animals Raised for Food or Food Production*. New York: Farm Sanctuary.

Worster, Donald. 1994. *Nature's Economy: A History of Ecological Ideas*, 2nd ed. New York: Cambridge University Press.

Wrangham, Richard W., W. C. McGrew, Frans B. M. de Waal, and Paul G. Heltne, eds. 1994. *Chimpanzee Cultures*. Cambridge, MA: Harvard University Press.

Zimmer, C. 2001. *Evolution: The Triumph of an Idea*. New York: HarperCollins.

Index

gone wild. *See* feral animals
and narrative of bringing world under control,
235–237 (*see also* anthropology and dismissal
of cultures with positive views of nonhuman
animals; hunter-gatherers)
as representatives of nonhuman animals,
249–250
domestication of nonhuman animals, 131
Doniger, Wendy, 40, 260, 265
Droysen, Johan Gustav, 39
ducks, 243
Durant, Will, 150
Dyson, Freeman, *The Scientist as Rebel* (2006), 273,
299, 329

eagles, 139, 243
Eagleton, Terry, 189, 191
After Theory (2003), 189
ecofeminism, 77
ecological economics, 101
ecology, 22, 93–96
See also environment
economics, 232
Animal Studies as challenging basic assumptions,
104–105, 108
and claims about critical thinking, 60, 157
criticisms of, 104–108, 232, 303
as the dismal science, 303
exceptionalist tradition in, 34, 35, 104–105, 117
as pivotal in public policy circles, 104
See also consumers and consumerism; ecological
economics; economism; poverty
economism, 232
ecosystems. *See* environment
education, 32, 51–65, 78–79 161, 198
absence of nonhumans in, 26, 55
and children's learning about other animals,
140–141
as conservative, 128 (see also education, narrow
forms; Helvetius on ignorance and education;
Orr, David; Roszak, Theodore)
and critical thinking skills, 65 (*see also* critical
thinking)
culture-based. *See* culture
and dissent as important to learning, 205
through encounters, 274
and exceptionalist tradition. *See* exceptionalist
tradition
ferment today in, 55
formal, 33, 51–54, 123 (*see also* elementary education; legal education; universities; veterinary
medicine)
future prospects, 157, 161, 188–189,
211, 293

and human-centeredness. *See* human-centeredness
and ignorance, 32, 54, 198, 248, 310
informal, 33, 52–53, 55, 123, 210
narrow forms, 45, 53, 54 (*see also* education as
conservative)
as a place of daring. *See* Roszak, Theodore
as primary means by children are taught social
constructions, 211
and unlearning about animals, 50, 60, 62, 116 (*see
also* science, integrities of)
See also animal science; Animal Studies; educational
benefits of animal awareness; ethics; legal
education
eels, 243
Egypt, 40
and mummification of cats, 241
Einstein, Albert, 128
Eisnitz, Gail, 107–108
*The Shocking Story of Greed, Neglect, and Inhumane
Treatment inside the U.S. Meat Industry*
(1997), 107–108
Eliade, Mircea, 40
Elephant Studies, 93, 95
elephants, 31, 66, 94, 117, 118, 133, 134–135,
143, 144
in captivity, 95, 134–135, 207
as compassionate animals, 96
compared to humans, 96
and cognitive studies, 93
as dangerous animals, 31
"An Elephant Crackup?", 95–96
emotions in elephants, 96
and ferment, 95
history of humans with, 95
stress from captivity, 95
stress in wild, 96
and subsonic communications "discovered" in
1984, 296–297
and violence in the wild, 95
elk, 243
emotions of nonhuman animals, 61, 87, 90
denials of, 89–90, 187
difficulties in knowing. *See also* epistemology;
realities of nonhuman animals, difficulties in
identifying and studying)
emotion as basis for studying nonhuman animals.
See Rudy, Kathy
enforcement of laws, 102, 104, 107–108, 164, 168,
170, 171, 203, 261
reflecting operative public policy. *See* public policy
and enforcement of law
England. *See* Britain
entertainment animals, 121
environment, 34, 88–89, 100, 123, 157

MacKinnon, Catharine, 264
Machiavellian Intelligence (1998), 96–97
macroanimals, 21, 37, 38, 75, 76, 86, 157, 187, 211,
    218, 222, 231, 241
  defined, 21
  as familiar animals, 37
  as focal point of Animal Studies and animal protec-
    tion movement, 21, 121
  as social construction, 222
mammals, 37, 86
  as humans' ancestors and source of important
    human traits, 37, 195, 217
  as scientific classification, 85
marginalized humans, 260–278 (*see also* Animal
    Studies, harms to humans connected to harms
    to nonhumans; children; critical legal studies,
    concern with oppressions within the species
    line; ideology; indigenous peoples; women)
marine mammalogy, 25, 94, 95
  and ethics, 25
  and cognitive studies, 93
martens, 243
Marxist views, 45, 178, 227
Masai tribe, 250
Masri, Al-Hafiz B. A., 285
mathematics, 75, 76
Mayr, Ernst, 123, 186
Mead, Margaret, 285
media, 22, 23, 28, 33, 66. 113, 115, 161, 252
  See also social media
medieval views, 23, 173, 190–191, 214, 276
megafields in modern university
  animal studies as megafield, 292–306
  humanities and arts as one megafield, 22, 54–55
  science megafield, 54–55
  See also university
Melson, Gail, 273
  on child development psychology's discovery of
    nonhuman animal issues, 267
  *Why the Wild Things Are: Animals in the Lives of
    Children* (2001), 267, 312
Merton, Thomas, 286, 306, 331
Meyer, Steve, 78, 102–103
mice, 208, 243
microanimals, 20–21, 37, 75, 222, 231, 243
  defined, 20
  as social construction, 222
  See also microorganisms
microorganisms, 20, 85, 218
  scientific classification of, 84
Middle East, 144, 288
Midgley, Mary, 49, 63, 151, 268, 287–288
  and "absolute dismissal" of nonhuman animals,
    149, 184

Milgram, Stanley, 204, 205
Minding Animals International, 309
minds of animals (nonhuman), 90, 208 (*see also* cogni-
    tive revolution; Griffin, Donald)
minks, 243
Mohenjo Daro, 240
Momaday, Scott, 138–140
Monboddo, Lord, 82
monkeys, 206
Montaigne, Michel de, 143
Montgomery, Georgina. *See* history "from below"
moose, 243, 244
moral rights
  dismissal in western culture, 149
  See also ethics; Regan, Tom
morality and morals. *See* ethics
Moss, Cynthia, 252
mountain lions, 139
mourning doves, 243
Mowat, Farley, 242, 253, 326, 327
  and "horrendous diminishment" to North
    American wildlife, 242, 248
Muir, John, 156, 212
multiculturalism
  cross-species forms of, 232–234
  failures to appreciate cultures other than one's birth
    culture, 232
  principled multispecies forms of, 234
  and toleration of human-on-nonhuman oppres-
    sions, 234
multidisciplinary. *See* interdisciplinary approaches
music, 40, 129, 131, 132–133
  origin in imitation of nonhuman animals, 132
muskrats, 243
Myers, Gene, *Children and Animals: Social
    Development and Our Connection to Other
    Species* (1998), 273
myth, 41–43, 133, 135–136, 155, 158–159
  and Animal Studies, 42–43
  as controversial and difficult topic, 43
  and human limitations, 42
  as misleading at times about nonhumans, 42, 43
  modern versions of, 41, 42–43
  negative, 41, 42, 43
  positive, 41, 42, 43
  and religious views of nonhumans, 41–42
  as universal, 41, 135

natural history of animals. *See* realities of nonhuman
    animals
neighbors and neighborhood, v, 1, 4, 33, 78, 103,
    142, 162, 197, 219, 237, 239, 243, 245, 277,
    278, 294 (*see also* expanding circle narrative in
    ethics; local issues, neighboring nonhumans)

and critical thinking, 121–123, 151–154, 159–160
(*see also* critical thinking)
as cutting edge in Animal Studies, 120–124
and failures to engage nonhuman animals,
49, 120, 122, 143, 146–147, 153–155,
159–160
and exceptionalist tradition, 35, 120, 154, 155,
159–160
and exclusions of human groups, 153, 154
and the future of human-nonhuman relationships,
157
and human-centeredness, 154, 159–160
and ignorance about nonhuman animals, 143, 151,
153–155
interest in nonhumans reflected in, 154–155 (*see
also* Appiah, Kwame Anthony; Haraway,
Donna; Nussbaum, Martha; Rachels, James;
Regan, Tom; Rollin, Bernard; Singer, Peter)
in interspecies key, 160
and need for humility, 123, 152, 153
popular opinion driving some views of animals in,
49, 63, 151
See also American Philosophical Association;
continental philosophy; epistemology;
ethics
photographs, 131
physics, 75, 76, 81
physiology, 81
Piaget, Jean, 53, 273
Picasso, Pablo, 255
Pinker, Steven, 151
pigs, 27, 100, 243
as intelligent farm animals, 70
plants, 48, 75, 84, 148–149, 218, 222
Plato, 152, 153, 154 (*see also* cranes in Plato's *The
Statesman*)
Plutarch, 265
poets and poetry, 136–138, 142
unacknowledged legislators of the world, 136
polarized debate, 61, 81, 103, 179, 295
political science as academic field, 97, 177
politics
among primates, 97
city life as central to, 97, 101
critiques of, 98–99
defined in terms of power, 97
defined as "who gets what, when, how", 99
and the exceptionalist tradition ideology,
97–99, 260
as falling short of ideals, 97–99 (*see also*
marginalized humans)
future possibilities in, 100
science as impacted by, 97, 99
as a word, 97

See also individual humans, role of individuals;
public policy
politics of truth. *See* Foucault, Michel
Pollan, Michael, 70, 313
Pope, Alexander, 4
porcupines, 243
porpoises, 243
Posner, Richard, 59
posthumanism, 13
poverty, 45
as factor in animal protection, 202
See also economics; subaltern
Pra Barom Nakkot, 134–135, 207
Preece, Rod, 249
*Priceless: On Knowing the Price of Everything and the
Value of Nothing* (2004), 104, 106
primates (nonhuman), 37, 95, 195
and culture,
as focus of animal law, 117–118
as humans' ancestors and source of important
human traits, 37, 97
personality found in, 97
research pursued to illuminate source of human
features, 97
primatology, 25, 93, 94, 96–97
and values, 25
Machiavellian similarities between humans and
chimpanzees, 96
research pursued to illuminate human history, 97
progress, 8, 236, 239, 243, 244–245
See also city-based issues, human-centered
language of "improved," "highest and
best use," "progress"
Protagoras re humans as measure of all things, 293
psychology, 40, 93, 125
public policy, 28, 90, 102–103, 108, 109
Animal Studies enriching study of, 103–107
companion animals, 28, 29, 110–111
critiques of, 98–99, 232
and decisions of individuals, 103
definitions of, 103–104, 107
and enforcement of law, 104, 106, 107–108 (*see also*
enforcement of law)
and economics, 104, 108
and the exceptionalist tradition, 98, 101–102, 103,
106–107, 108
future of, 103, 111–112
graduate programs in, 103
and harms to humans, 98, 102
human-centeredness of, 35, 99, 102, 196 (*see
also* public policy and the exceptionalist
tradition)
and interdisciplinary approaches, 106,
107, 108